6G Wireless

6G Wireless: The Communication Paradigm Beyond 2030 offers a thorough discussion of some key emerging technologies such as Intelligent Reflecting Surface (IRS), Unmanned Aerial Vehicles (UAV), Aerial Computing, Terahertz (THz) Communications, Non-Orthogonal Multiple Access (NOMA) and Rate Splitting Multiple Access (RSMA).

The book provides a comprehensive coverage of the vision, requirements, use cases, enabling technologies, and challenges for the future 6G wireless communication systems. This includes discussions on how 6G and future IoT systems will enable extremely low latency healthcare systems, smart industry, haptic communications, programmable wireless environment (PWE), advanced VR/AR and holographic communications. IRS is expected to play a prominent role in 6G and the book thoroughly discusses the role of IRS in enabling physical layer security, UAV communications as well as D2D communications. It also explains channel modeling for IRS enabled PWE.

Another key aspect of the book is that it provides a comprehensive discussion on security challenges of emerging 6G systems and their potential solutions. Apart from this, it also explains how blockchain techniques can be used for future IoT applications such as intelligent manufacturing and asset tracking.

Written in tutorial style, the book is primarily intended for postgraduate students and researchers in the broad domain of wireless communications as well as research-active academics. The book can also be useful as a reference book for BSc/MSc project/thesis works.

6G Wireless
The Communication Paradigm Beyond 2030

Edited by
Faisal Tariq
Muhammad Khandaker
Imran Shafique Ansari

CRC Press
Taylor & Francis Group
Boca Raton London New York

CRC Press is an imprint of the
Taylor & Francis Group, an **informa** business

First edition published 2023
by CRC Press
6000 Broken Sound Parkway NW, Suite 300, Boca Raton, FL 33487-2742

and by CRC Press
4 Park Square, Milton Park, Abingdon, Oxon, OX14 4RN

CRC Press is an imprint of Taylor & Francis Group, LLC

ISBN: 978-1-032-25173-8 (hbk)
ISBN: 978-1-032-25235-3 (pbk)
ISBN: 978-1-003-28221-1 (ebk)

DOI: 10.1201/9781003282211

Typeset in Latin Modern font
by KnowledgeWorks Global Ltd.

Contents

Editors   vii

CHAPTER 1 ▪ Introduction to 6G Wireless Communications   1

FAISAL TARIQ, MUHAMMAD KHANDAKER, AND IMRAN SHAFIQUE ANSARI

CHAPTER 2 ▪ IRS-Empowered Wireless Communications   15

MING ZENG, EBRAHIM BEDEER, XINGWANG LI, QUOC-VIET PHAM,
OCTAVIA A. DOBRE, PAUL FORTIER, AND LESLIE A. RUSCH

CHAPTER 3 ▪ Role of Intelligent Reflecting Surfaces in the
Emerging 6G Technologies   39

SHAIKA MUKHTAR, UMER ASHRAF, AND GH. RASOOL BEGH

CHAPTER 4 ▪ Channel Modeling for 6G Programmable Wireless
Environment   59

PETROS KARADIMAS, MD. SAKIR HOSSAIN, AND FAISAL TARIQ

CHAPTER 5 ▪ Wireless Localization with Reconfigurable
Intelligent Surfaces   73

OMAR RINCHI, AHMED ELZANATY, AND AHMAD ALSHAROA

CHAPTER 6 ▪ The Emergence of Aerial Computing   117

QUOC-VIET PHAM, THIEN HUYNH-THE, MING ZENG, ZHAOHUI YANG,
ZHIGUO DING, AND WON-JOO HWANG

CHAPTER 7 ▪ THz-Empowered UAV Communications   149

HUMAIRAH HAMID, AAQIB RESHI, AND G R BEGH

CHAPTER 8 ▪ Performance Characterization of RSMA in THz
Networks   177

SADEQ BANI MELHEM AND HINA TABASSUM

CHAPTER 9 ▪ A Comprehensive Overview of Security and
Privacy in the 6G Era   203

SHAKILA ZAMAN, FAISAL TARIQ, MUHAMMAD KHANDAKER, AND RISALA
T KHAN

CHAPTER 10 ▪ Hybrid Massive-MIMO and Its Practical
Beamforming Implementation   259

KAI XU, JIAYU HOU, AND YUAN DING

CHAPTER 11 ▪ Blockchain Technology for 6G-Oriented IoT
Systems   283

TIANQI YU, YONGXU ZHU, AND XIANBIN WANG

CHAPTER 12 ▪ 6G and IOT Use Cases   315

ASIF ALI, SYED MUJTIBA HUSSAIN, AND G R BEGH

Index   341

Editors

Dr. Faisal Tariq earned his B.Sc.(Hons) from Rajshahi University, Bangladesh, and M.Sc. from Chalmers University of Technology, Sweden. He earned his Ph.D. from The Open University, UK. He is currently serving as a Senior Lecturer in the School of Engineering at the University of Glasgow. His main research interests include radio resource management, 5G and beyond wireless networks, and molecular communications for nano-networks. He is the recipient of the best paper award at the Wireless Personal Multimedia Conference (WPMC) in 2013. He is currently serving on the editorial board of *Elsevier Journal of Network and Computer Applications*.

Dr. Muhammad Khandaker is currently working as an Assistant Professor in the School of Engineering and Physical Sciences at Heriot-Watt University. He earned his Ph.D. in Electrical and Computer Engineering from Curtin University, Australia, in 2013. He is an Associate Editor for *IEEE Communications Letters, IEEE Wireless Communications Letters*, and *IEEE Access* and served as an Editor for *EURASIP Journal on Wireless Communications and Networking* (JWCN), the Lead Guest Editor of the special issue on *Heterogeneous Cloud Radio Access Networks* (H-CRANs) of EURASIP JWCN as well as the Managing Guest Editor of a special issue on *Self-Optimizing Cognitive Radio Technologies* of the Elsevier journal *Physical Communication*.

Dr. Imran Shafique Ansari earned his B.Sc. degree in Computer Engineering from King Fahd University of Petroleum and Minerals (KFUPM) in 2009 (with First Honors) and M.Sc. and Ph.D. degrees from King Abdullah University of Science and Technology (KAUST) in 2010 and 2015, respectively. Currently, since August 2018, he is an Assistant Professor with University of Glasgow, Glasgow, UK. Prior to this, from November 2017 to July 2018, he was an Assistant Professor with Global College of Engineering and Technology (GCET) (affiliated with University of the West of England [UWE], Bristol, UK).

From April 2015 to November 2017, he was a Postdoctoral Research Associate (PRA) with Texas A&M University at Qatar (TAMUQ). From May 2009 through Aug. 2009, he was a visiting scholar with Michigan State University (MSU), East Lansing, Michigan, USA, and from June 2010 through August 2010, he was a research intern with Carleton University, Ottawa, Ontario, Canada. Dr. Ansari has authored/co-authored 100+ journal and conference publications. His current research interests include free-space optics (FSO), underwater communications (UWC), physical layer (PHY) secrecy (PLS), and reconfigurable intelligent surfaces (RIS)/intelligent reflective surfaces (IRS).

CHAPTER 1

Introduction to 6G Wireless Communications

Faisal Tariq

James Watt School of Engineering, University of Glasgow, Glasgow, UK

Muhammad Khandaker

School of Engineering and Physical Sciences, Heriot-Watt University, Edinburgh, UK

Imran Shafique Ansari

James Watt School of Engineering, University of Glasgow, Glasgow, UK

CONTENTS

1.1	Vision for 6G Communication	2
1.2	Technology Enablers for 6G Era	6
1.3	Challenges and Opportunities	8
1.4	Organization of the Book	10
Bibliography		12

T HE DEVELOPMENT OF all previous generations of mobile communications has always been driven by some timely technologies, together with new frequency bands that facilitated the technological needs with changing demands (cf. Figure 1.1). The second-generation (2G) mobile communication made a paradigm shift from analog in the first generation (1G) to digital communications, while the third generation (3G) was built using spread-spectrum technologies, with multimedia support for the first time. Then, the fourth-generation (4G) systems became an all-IP packet-switched networks, with a host of advanced technologies such as multi-antenna and multiuser multiple-input multiple-output (MIMO), adaptive modulation, femtocells, etc. Finally,

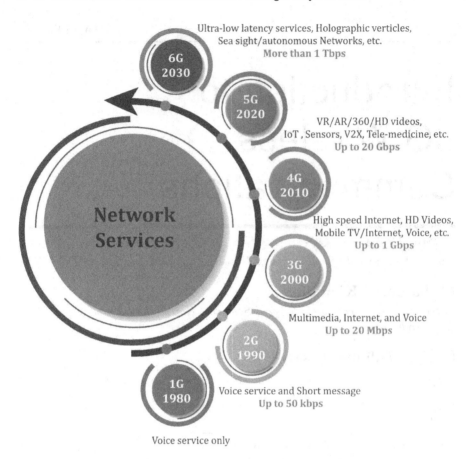

Ultra-low latency services, Holographic verticles,
Sea sight/autonomous Networks, etc.
More than 1 Tbps

VR/AR/360/HD videos,
IoT , Sensors, V2X, Tele-medicine, etc.
Up to 20 Gbps

High speed Internet, HD Videos,
Mobile TV/Internet, Voice, etc.
Up to 1 Gbps

Multimedia, Internet, and Voice
Up to 20 Mbps

Voice service and Short message
Up to 50 kbps

Voice service only

Figure 1.1 The generations of wireless communication systems.

the recently evolved fifth-generation (5G) systems are based on a number of new technologies, including massive MIMO, edge computing, software-defined networking, heterogeneous networks (HetNet), and millimeter wave frequencies. Consequently, the sixth generation (6G) is no exception that new technologies will be required to make another leap to meet the ever-increasing mobile data demands and cater unimaginable applications [7].

1.1 VISION FOR 6G COMMUNICATION

The aim of this chapter is to provide an expert view on the most trending research directions that will likely shape the technological changes needed for the 6G mobile communication systems for the next decade. Although the development of 6G is inevitably in an early stage and it

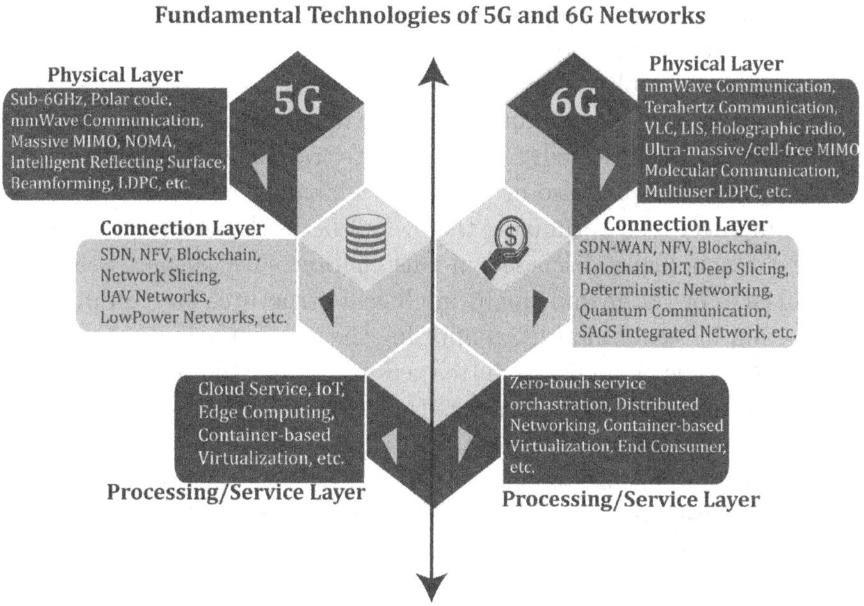

Fundamental Technologies of 5G and 6G Networks

Physical Layer
Sub-6GHz, Polar code, mmWave Communication, Massive MIMO, NOMA, Intelligent Reflecting Surface, Beamforming, LDPC, etc.

5G

Physical Layer
mmWave Communication, Terahertz Communication, VLC, LIS, Holographic radio, Ultra-massive/cell-free MIMO Molecular Communication, Multiuser LDPC, etc.

6G

Connection Layer
SDN, NFV, Blockchain, Network Slicing, UAV Networks, LowPower Networks, etc.

Connection Layer
SDN-WAN, NFV, Blockchain, Holochain, DLT, Deep Slicing, Deterministic Networking, Quantum Communication, SAGS integrated Network, etc.

Cloud Service, IoT, Edge Computing, Container-based Virtualization, etc.

Zero-touch service orchastration, Distributed Networking, Container-based Virtualization, End Consumer, etc.

Processing/Service Layer

Processing/Service Layer

Figure 1.2 Physical layer, connection layer, and service layer key technologies for 5G and 6G networks.

is expected that some ideas will only emerge in later years, this is the perfect time to study what will be needed in 6G in terms of key performance indicators (KPIs), and also to identify possible key technologies that will provide the needed step change, if sufficient advances are made in the next development cycle when the development of 5G has already come to an end.

The visionary aspects of 6G discussed in this book provide a detailed treatment of the features that they bring beyond the capabilities of 5G (cf. Figure 1.2). They provide promising directions that can help guide researchers around the world to align their research works to contribute to the 6G development. Some of the ideas discussed in this chapter are already making significant contribution in 6G research. However, there are few ideas which are not studied well and are not getting adequate attention. One major objective of this chapter is thus to provide unique insights into the key areas and bottlenecks where major efforts should be sought.

There is no doubt that artificial intelligence (AI) will play a dominant role in 6G and its presence will be pervasive. While this is hardly a surprise, particularly with partial implementation of AI in 5G, this chapter puts emphasis on where further advances are required in AI in order to tackle specific problems in 6G wireless networks. Collective AI (i.e., cooperative multi-agent learning) is a concept that is being studied actively within the AI community, and it will contribute to the radio access networks of 6G with edge caching where training can only take place locally at the edges but also needs to have global intelligence to make local caching decisions optimizing the overall network performance. Some enabling technologies discussed in this chapter such as programmable metasurfaces or (large and small) intelligent structures are gaining a lot of attention in recent months, confirming the timeliness and relevance of this chapter. The chapter also introduces radar-assisted mobile radios, a less known technology, and predicts that they will make a great impact on 6G as radars at mobiles can provide all sorts of environmental and behavioral information for physical-layer security measures. Further, a number of developing technologies that appeared in the 5G development cycle or even before 5G will potentially be included in 6G.

The vision for 6G is spearheaded by a number of technologies that push the capability far beyond what is expected for 5G. The advantages of this 6G vision lie in the technological leaps that those enabling technologies mentioned in this chapter can offer beyond 5G. Here, we focus on the dominant 6G technologies.

- First, *pervasive and collective AI* is one vision that will truly facilitate learning-based methods in improving the overall chain of the telecom system. It is true that AI is already in 5G in many aspects, e.g., by leveraging "big data" for self-organizing networks (SONs), and enabling and managing network function virtualization (NFV) and orchestration, etc. 5G is certainly revolutionary in its own right, and the term "pervasive AI" is sometimes discussed in the context of 5G. However, the introduction of AI in 5G is more about reaping the benefits of well-known machine learning techniques using data in mobile networks and becoming as pervasive as practically possible. Our 6G vision, in contrast, emphasizes the development of *new* AI algorithms to address the unique challenges of radio access that the existing AI techniques are unable to cope. The term "collective AI" highlighted in this chapter is one such concept and a step-up from the current AI techniques that will address the coexistence of multiple distributed learning agents for individual as well as *global* benefits. It should be noted that the competitive nature of mobile radios (a.k.a. "Mitola radios") suffers from poor learning efficiency and will make the existing reinforcement learning approaches perform very poorly. This vision of collective AI is therefore novel and revolutionary.

- Our second vision is the use of intelligent structures such as programmable metasurfaces, metamaterial-based antennas, fluid antennas, etc. This vision is not a part of 5G and seeks to engineer the environment to provide better conditions for wireless communications. The application is multifaceted. They can be used to improve the link quality of desired communications, block interference, enhance privacy and security, avoid adversarial attacks, and many more. Research in large intelligent surfaces (LISs), sometimes referred to as metasurfaces, has gained popularity recently. They can serve to provide a signal boost like a conventional relay or even alter the propagation environment to achieve so-called 'smart radio environments' to accommodate all sorts of propagation scenarios that are not possible in 5G. Software-defined or programmable metasurfaces can do more than adapting radio environments on a large scale. Within this vision, people further advocate that programmable metasurfaces can be used to integrate between signal processing algorithms in information science and hardware resources made of metamaterial unit cells, realizing fully adaptive hardware use for wireless transmission. This is another revolution in the proposed 6G vision. Our vision also includes the use of intelligent structures on a small scale, say in mobile handsets, where fluid-type antennas can be used to obtain diversity that is not possible for MIMO antennas.

- The third vision proposes to utilize radar technologies in mobile radios to enrich environmental awareness to the level that is not possible in 5G. This will empower 6G mobile radios to identify potential eavesdroppers or adversaries through observations from radars, and use physical-layer security approaches to adapt and protect their communications. This will also allow mobile radios to gather behavioral data of the environment, which not only helps the mobile radios themselves in many aspects but also informs the network how to proactively utilize the resources in the most efficient way. Contextual data acquired by radar technologies, combined with AI, will be another revolutionary upgrade in 6G.

- Our vision also includes more known developing technologies that are not ready to appear or appear only in a very limited scope in 5G but could flourish in the 6G cycle. For example, Li-Fi, wireless power transfer, optical angular momentum (OAM), and quantum communications are among those technologies that have been studied in recent years and their technological breakthrough might occur in 6G.

- Mitola radio, a term coined to represent the next generation of cognitive radios, will play a central role in our proposed 6G vision. The reason is that Mitola's cognitive radio idea back in 1999 emphasizes the intelligence for adaptive use of resources to meet communication needs, which aligns very well with the 6G vision empowered by AI. It is worth noting that the original 'cognitive radio' concept by Mitola is much more than spectrum sharing, although this has been the main focus in cognitive radio research in the past 20 years. Our 6G vision aims to provide a genuine realization of Mitola radio, which has exceptional awareness of the environment (by radar technologies) to make decisions using super intelligence (by collective AI), with a rich action space to adapt itself in many forms (by intelligent structures, Li-Fi, etc.).

1.2 TECHNOLOGY ENABLERS FOR 6G ERA

Our approach is to focus on the speculative elements pertaining to 6G, the revolutionary technologies, or the revolutionary aspects of the emerging technologies that may enable 6G but are less known in the research community. Here, we provide a short list of the revolutionary technologies that are being discussed as strong candidate technologies.

The notion of collective AI, in contrast to the hype of using conventional AI or machine learning methods in wireless systems, is a revolutionary element in 6G. This goes far beyond using learning-based methods exploiting big data in the mobile networks in 5G and empowers individual mobile radios (a.k.a. Mitola radio) to act intelligently based on the local data for optimizing their quality of experience (QoE) and collectively enhancing the performance of the entire network over time. To help appreciate the trends of development in this area toward 6G, it is useful to first understand the limitations of current AI techniques. The use of AI in 5G can be largely viewed as an exercise of reaping the benefits of well-known machine learning techniques using data in mobile networks, including high-mobility scenarios such as vehicle to everything (V2X) communications [2]. The fog networking architecture and its distributed nature make NFV a lot more challenging to do. A large amount of mobile network data will be stored at the edges and in distributed locations, which will handicap AI techniques such as supervised learning to be fully functional. Federated learning is a method that attempts to tackle this learning problem but more needs to be done to improve the learning efficiency and prediction accuracy. One trend of development for NFV in the 6G cycle is therefore to develop new AI algorithms that can achieve efficient training with decentralized datasets, which is suitable for the fog-RAN architecture.

Also, radar-assisted wireless communication is another revolutionary addition to 6G as advocated in this chapter. This will give 6G mobile radios the environmental awareness that has not been possible before, not even in 5G, and such enriched awareness will truly empower AI and other advanced techniques to perform at the device level. Physical-layer security and authentication approaches will finally thrive as a result.

Programmable wireless environments enabled by metasurfaces and intelligent structures are other revolutionary elements that will lit up 6G. While large intelligent surfaces (LISs) mounted outside buildings or in indoor environments are predicted by some to make their mark in 6G, this chapter further advocates that smaller metasurfaces may replace conventional wireless transmitter design if sufficient advances are made in the 6G cycle. In addition, intelligent structures such as fluid antennas in the scale of a mobile handset will become a new way to obtain extraordinary diversity and multiplexing gains.

Blockchain technology will likely play a major role in securing and authenticating future communication systems, thanks to the inherent advantages of the distributed ledger technology. It offers a number of benefits, including decentralization, transparency with adequate privacy, and alteration-proof authentication. Both decentralization and transparency ensure faster processing, which is crucial for 6G systems. As mentioned before, network resources will be virtualized, and therefore, blockchain will play a key role in simultaneous resource allocation and authentication. Also, in future, the security needs to be highly adaptive depending on the location, device capability, and application. Healthcare data, for example, need to be extremely secure and private, while emergency responses may require quick access to those data via 6G system; thus, it needs to be ensured that the right person has the right level of access to the data on the fly.

6G Mitola radios will be more than spectrum sharing in the most efficient way. They will be fully integrated with many functionalities other than communications. For example, under fog-RAN, a mobile radio can act as a local cache and an edge computing server, if this turns out to be beneficial to itself and the network in the long run. Wireless power transfer can also take place from one mobile radio to another for remote charging. Another option is for a mobile radio to act as a jammer if enhanced security is needed. All these functionalities will be part of realization of 6G Mitola radios.

The above are the revolutionary aspects of 6G as emphasized in this book. There is also glimpse of successes in other emerging areas, which are not yet making much of an impact in 5G but could become reality in 6G. For example, cell-free networking (enabled by fog-RAN), VLC/Li-Fi, wireless power transfer, energy harvesting, OAM, and quantum communications are among those technologies that have some limitations in their usage till now but might become very useful in 6G.

The last decade has already seen numerous efforts spent on wireless power transfer and energy harvesting research works, and yet these technologies do not seem to play a key role in 5G. In 6G, however, these technologies will finally shine due to a number of reasons. One of the reasons is communication distance will be much shorter, making wireless power transfer meaningful because wireless networks continue to be denser plus the use of UAVs as base stations further reduce the distances. In addition, user equipments (UEs) or any Internet-of-Things (IoT) devices in 6G will be more power hungry than ever because of the huge computation demands for AI processing. On the other hand, energy scavenging from ambient RF signals may even become a viable power source for low-power applications, as energy harvesting technologies continue to advance.

1.3 CHALLENGES AND OPPORTUNITIES

While the technologies discussed above are expected to provide the technological leaps required for the speculative 6G vision, there are many challenges researchers and engineers will need to address in order to harvest full benefits of 6G. Here, we elaborate only a few of those.

Spectrum issues: Initial discussions indicate that frequencies in the range of 1 terahertz (THz) and above will be considered for 6G, as in those bands, there are plenty of free bands and are suitable to satisfy high bit rate requirements of 6G systems. In reality, it may happen that much lower bands are considered too similar to those considered in 5G, which initially anticipated up to 300 GHz frequencies. A recent study indicates that it will not exceed 100 GHz in 5G due to a number of challenges, including lack of understanding of channel and propagation modeling, device inability to operate at such high frequencies, etc. Recent announcement by the Federal Communications Commission (FCC) reveals that the highest licensed spectrum for 5G in the US will be 48.2 GHz though the unlicensed 5G band can go up to 71 GHz. A recent article, however, surmises that 6G will consider spectrum beyond 140 GHz with particular application in very short-range communication or 'whisper radio'. However, the susceptibility of the THz band to blockage, molecular absorption, sampling and circuits for A/D D/A conversion, and communication range is among the major challenges that researchers need to address over the coming years. Another issue is with higher frequencies, the size of antenna and associated circuitry becomes miniaturized, and hence, becomes very difficult to fabricate on chip while ensuring noise and inter-component interference minimization. Also, so far the actual propagation characteristics are not really known due to the unavailability of appropriate technologies. A number of recent attempts to address these bottlenecks have reported encouraging results. A good example could be the recent design of complementary metal-oxide-semiconductor (CMOS)-based modulation circuit at 300 GHz, which demonstrates significant performance improvement.

Device challenges: While every generation of mobile communications has been defined by the UE capability, this will be more so in 6G due to the THz operating frequencies. Although some proof-of-concepts have been demonstrated, THz technologies are still in their infancy. Major challenges with THz source/detector, THz modulator/demodulator as well as THz antennas are still unresolved. Graphene-based plasmonic nano-antenna arrays can accommodate hundreds of antenna elements in a few millimeters showing lights to make ultra-massive MIMO more practical.

Industry's perspectives of 6G: Massive incorporation of robots into automation and warehouse transportation will become key elements in major industries. The emerging concept of Industry X.0 aims to enhance the Industry 4.0 by exploiting social, mobile, analytics, and cloud (SMAC). There will be a widespread proliferation of AI-enabled automated manufacturing and automated detection of faults in industrial production system [5]. The radio environment with a very complex network comprising hundreds or thousands of robots is a challenge. 6G will fully support the Industry X.0 revolution by offering massive ultra-reliable low-latency communication (URLLC) as well as massive IoT and embedded AI capability.

Users' perspectives of 6G: From users' perspective, one of the most exciting 6G technologies will be wireless virtual reality (VR), replacing wired VR. In order to deliver 360° wireless VR contents, URLLC, as well as ultra-high data rate, will be indispensable. This requirement for VR will be much more demanding than for the classical 5G URLLC requirements. One approach to address this challenge is to explore the terahertz frequency bands. However, the susceptibility of the THz band to blockage, molecular absorption, and communication range is a major challenge that researchers need to address over the coming years. A number of recent attempts to address this bottleneck have reported encouraging results [1]. Line rates of up to 10 Gbit/s on a single channel and up to 30 Gbit/s on multiple channels over a distance of 58 m have been demonstrated using THz transmission links that exploited optoelectronic signal processing techniques both at the transmitter and the receiver [3]. Ultra-massive MIMO technology is being considered to solve the distance problem in the THz band [4].

1.4 ORGANIZATION OF THE BOOK

Based on our vision of 6G shared so far, broadly speaking, this book is organized considering three major themes of excellence that we envision are the directions of relative evolution as we progress toward the next generation of cellular networks. These are evolutions around relaying technologies via utilizing reconfigurable intelligent surfaces (RIS) or intelligent reflective surfaces (IRSs) [6], advances in spectrum efficiency via utilizing the tera hertz (THz) band, and implementing security and privacy most reliably. Specifically, each of these are addressed over a sequential collection of chapters detailed as follows.

A comprehensive overview of IRS-empowered wireless communication systems will be provided in Chapters 2 and 3 beginning with their basic fundamentals, followed by understanding their possible integration into existing technologies via identifying respective challenges and key techniques for resource allocation. To top these discussions, Chapter 2 will further highlight the effects of hardware impairments on IRS-empowered wireless communication systems as well.

As we build our understanding on IRS emerging as a promising contender within electromagnetic structural domain or physical (PHY) layer aiming to enhance wireless propagation for optimum performance via maximizing received signal-to-noise ratio (SNR), an approach and/or model to physically characterize the variations in IRS-enabled communication channels will be developed in Chapter 4. Specifically, to design optimal IRS topologies and transceiver systems to achieve desired performance requirements while minimizing the complexity, such as wireless propagation channel modeling, will surely be an integral need.

Having addressed PHY layer challenges based on channel modeling in Chapter 4, it is worthy to introduce the RISs as an energy-efficient solution in Chapter 5 to overcome heavily shadowed line-of-sight (LoS) communications, especially under higher frequency bands. Hence, this chapter addresses network layer issues by proposing RIS-empowered localization algorithms, thereby minimizing errors while being applicable over various channel models. Moreover, the SNR is maximized at the base station (BS) via a specific RIS phase design.

Shifting our gear vertically toward aerial networks, it is worth mentioning unmanned aerial vehicle (UAV) is a possible key technology that will revolutionize the conventional networks. Moreover, its integration with edge computing gives birth to aerial computing as a novel concept. Hence, Chapter 6 will frame this concept within the infrastructure of 6G systems. Subsequently, this chapter will also describe the role of this concept in various vertical domains followed by its realization via standardization efforts.

Being understandably aware, owing to their flexible 3D deployment and their technological advancements with respect to cost, UAVs are envisioned to play a leading role in various aspects. Interestingly, to enhance spectral efficiency as we evolve toward 6G, THz surely stands as a challenging contender to facilitate the same. Therefore, Chapter 7 will discuss THz-empowered UAVs.

Further benefiting from THz-enabled technologies, we can surely tackle the challenge of accommodating trillions of devices within the existing highly congested and limited sub-6GHz spectrum. Equally beneficial are non-orthogonal multiple access (NOMA) schemes that can support multiple users within the same frequency and time resource blocks while guaranteeing efficient self-interference cancellation (SIC) mechanisms. To address this paradigm, Chapter 8 will provide a comprehensive analytical framework to analyze the performance of NOMA and rate-splitting multiple access (RSMA) schemes applicable in 6 GHz and THz networks.

By now, we surely realize that security is an undeniable aspect that requires equal attention if not more, thereby motivating us to enhance our efficiency in this direction. For this, Chapter 9 will clearly share this need and understanding in detail. Within the domain of security, many a time, beamforming is ought to be quite beneficial, which is also becoming a promising solution to implement massive multiple-input and multiple-output (m-MIMO). In light of this, Chapter 10 will comprehensively review three varying hybrid m-MIMO transmitter variants describing their structures, design algorithms, and implementations, thereby proposing a practical dynamic subarray beamforming structure followed by demonstrating their performance as well.

Finally, as we may relate to, wireless communication is experiencing an unprecedented exponential increase in data traffic with ever-increasing demand for secure and spectrally efficient exchange of information, especially in the presence of smart devices. IoT-based 6G systems are expected to transform usage habits and their respective applications. Initially, Chapter 11 will introduce and explain the applicability of blockchain technology toward 6G-oriented IoT systems followed by Chapter 12 that will provide a comprehensive study on the inception and coexistence of 6G and IoT to justify their conjoined potential.

BIBLIOGRAPHY

[1] M Fujishima. 300-GHz-band CMOS transceiver for ultrahigh-speed terahertz communication. In *Terahertz, RF, Millimeter, and Submillimeter-Wave Technology and Applications XII*, volume 10917, pages 68–73. SPIE, 2019.

[2] Jin Gao, Muhammad R A Khandaker, Faisal Tariq, Kai-Kit Wong, and Risala T. Khan. Deep neural network based resource allocation for v2x communications. In *2019 IEEE 90th Vehicular Technology Conference (VTC2019-Fall)*, pages 1–5, 2019.

[3] Tobias Harter, Sandeep Ummethala, Matthias Blaicher, Sascha Muehlbrandt, Stefan Wolf, Marco Weber, Md Mosaddek Hossain

Adib, Juned N Kemal, Marco Merboldt, Florian Boes, et al. Wireless THz link with optoelectronic transmitter and receiver. *Optica*, 6(8):1063–1070, 2019.

[4] Hadi Sarieddeen, Mohamed-Slim Alouini, and Tareq Y Al-Naffouri. Terahertz-band ultra-massive spatial modulation mimo. *IEEE Journal on Selected Areas in Communications*, 37(9):2040–2052, 2019.

[5] Muhammad Sohaib, Shahid Munir, M M Islam, Jungpil M Shin, Faisal Tariq, Mamun Rashid, and Jong-Myon Kim. Gearbox fault diagnosis using improved feature representation and multitask learning. *Frontiers in Energy Research*, 10:998760, 2022.

[6] Yizhuo Song, Muhammad R A Khandaker, Faisal Tariq, Kai-Kit Wong, and Apriana Toding. Truly intelligent reflecting surface-aided secure communication using deep learning. In *2021 IEEE 93rd Vehicular Technology Conference (VTC2021-Spring)*, pages 1–6, 2021.

[7] Faisal Tariq, Muhammad R A Khandaker, Kai-Kit Wong, Muhammad A Imran, Mehdi Bennis, and Merouane Debbah. A speculative study on 6G. *IEEE Wireless Communications*, 27(4):118–125, 2020.

CHAPTER 2

IRS-Empowered Wireless Communications
State-of-the-Art, Key Techniques, and Open Issues

Ming Zeng
Université Laval, Quebec, Canada

Ebrahim Bedeer
University of Saskatchewan, Saskatoon, Canada

Xingwang Li
Henan Polytechnic University, Jiaozuo, China

Quoc-Viet Pham
Trinity College Dublin, Dublin, Ireland

Octavia A. Dobre
Memorial University, St. John's, Canada

Paul Fortier
Université Laval, Quebec, Canada

Leslie A. Rusch
Université Laval, Quebec, Canada

CONTENTS

2.1 Introduction 17
2.2 IRS-Assisted Wireless Transmission 18

DOI: 10.1201/9781003282211-2

	2.2.1	Motivation	18
	2.2.2	State-of-the-Art and Key Techniques	18
2.3	Integration of IRS with Advanced Transmission Technologies		21
	2.3.1	IRS-Assisted MmWave/THz Communication	21
		2.3.1.1 Motivation	21
		2.3.1.2 State-of-the-Art and Key Techniques	21
	2.3.2	IRS-Assisted NOMA Transmission	23
		2.3.2.1 Motivation	23
		2.3.2.2 State-of-the-Art and Key Techniques	23
	2.3.3	IRS-Assisted MEC Systems	25
		2.3.3.1 Motivation	25
		2.3.3.2 State-of-the-Art and Key Techniques	26
	2.3.4	IRS-Assisted PLS Systems	27
		2.3.4.1 Motivation	27
		2.3.4.2 State-of-the-Art and Key Techniques	27
2.4	Non-Ideal Transmission of IRS-Assisted Communication Systems		29
	2.4.1	Hardware Impairments	30
		2.4.1.1 Finite-Resolution Phase Shifters	30
		2.4.1.2 RF Chain Impairments	30
	2.4.2	Imperfect CSI	31
2.5	Open Issues		31
	2.5.1	Unmanned Aerial Vehicles-Integrated IRS Systems	32
	2.5.2	Machine Learning-Empowered IRS Systems	32
	2.5.3	Sensing and Localization	32
2.6	Conclusion		33
Bibliography			33

INTELLIGENT REFLECTING SURFACE (IRS) is an emerging paradigm for future 6G wireless communication system. An overview of IRS-empowered wireless communication systems is presented in this chapter. We first present the fundamentals of IRS-assisted wireless transmission. On this basis, we explore the integration of IRS with various advanced transmission technologies, such as millimeter wave/terahertz, non-orthogonal multiple access, mobile edge computing, and physical layer security. Following this, we discuss the effects of hardware impairments and imperfect channel state information on the IRS system performance. Finally, we highlight several open issues to be addressed.

2.1 INTRODUCTION

Recently, intelligent reflecting surface (IRS) has drawn great attention as a promising physical layer transmission technology for next-generation communication systems [1–5]. An IRS is a planar surface equipped with massive low-cost passive reflecting elements; each can induce a phase and/or amplitude change to the impinging signals to achieve fine-grained reflective beamforming. By judiciously deploying an IRS in the environment, an extra communication link that goes through the IRS can be built between the transmitter (Tx) and receiver (Rx), and thus, better support diverse user requirements, such as extended coverage, increased data rate, minimized power consumption, and enhanced secure transmissions [1–5]. Not only theoretically attractive, IRS also possesses various advantages in terms of practical implementation. It is of low hardware and energy cost and can be easily deployed on environment objects, e.g., the facades of buildings. Moreover, IRS can operate in a full-duplex mode without self-interference and noise amplification.

To fully reap the benefits provided by IRS, it is necessary to investigate the integration of IRS with other transmission technologies for next-generation communication systems, such as higher frequencies (millimeter wave (mmWave) and Terahertz (THz)), non-orthogonal multiple access (NOMA), mobile edge computing (MEC), and physical layer security (PLS). The integration of IRS into mmWave/THz is natural, since IRS can establish additional line-of-sight (LoS) links to extend the coverage of high frequency, which suffers from severe signal attenuation and poor diffraction [6–8]. The application of IRS into NOMA is also promising, as IRS can be utilized to introduce desirable channel gain differences among users as well as to suppress inter-user interference; both can lead to performance improvement of NOMA systems [9–11]. IRS can also be exploited to mitigate propagation-induced impairments for MEC systems such that the benefits of MEC can be maintained under hostile communications link [12–20]. Lastly, IRS can be used to enhance the signal strength at the legitimate users while nulling the signal reception at the eavesdroppers, and thus, improve the PLS of wireless systems [21–23].

The aim of this chapter is to provide a comprehensive survey on IRS-assisted wireless transmission. We cover both conventional IRS-assisted systems and more advanced ones where IRS is integrated with other candidate technologies, such as mmWave/THz, NOMA, MEC, and PLS. For all the considered scenarios, we not only present the state-of-the-art research progress but also identify the challenges and key techniques for resource allocation. Motivated by the practical challenges of implementing IRS, we further discuss the effects of hardware impairments (HWIs) and imperfect channel state information (CSI) on the IRS system performance. Finally, several open issues are highlighted.

2.2 IRS-ASSISTED WIRELESS TRANSMISSION

2.2.1 Motivation

Conventional network optimization in wireless communication systems has been limited to transmission control at transceivers, with little attention has been paid to the wireless propagation environment. Indeed, the wireless propagation environment has long been perceived as an uncontrollable and randomly behaving entity between the transceivers. Aside from being uncontrollable, the environment usually has an adverse impact on communication efficiency, owing to the signal attenuation, fading, and interference introduced. As a result, the propagation environment itself becomes a major limiting factor that hinders further performance improvement of wireless networks. Recently, there has been an increasing demand for novel communication paradigms that can smartly tune the propagation environment either to increase the communication efficiency or to simplify the transceiver architecture. In this regard, IRS has received great attention owing to its ability to reconfigure the propagation environment via software-controlled reflection [1–5]. As shown in Figure 2.1, an IRS is a planar surface consisting of many low-cost passive reflecting elements; each can induce a phase and/or amplitude change to the incident signal to achieve fine-grained reflective beamforming. When the direct link between the transceivers fails due to unfavorable channel conditions, IRS can be deployed to build a cascaded link and resumes the communication, as illustrated in Figure 2.1(a). Even when the direct link exists, IRS can still be used to add an extra communication link between the transceivers to improve the system performance, as presented in Figure 2.1(b).

2.2.2 State-of-the-Art and Key Techniques

The simplest IRS-assisted communication system consists of three nodes, namely the Tx, Rx, and IRS, which is referred to as point-to-point communication. Depending on how many antennas the transceivers are equipped with, such a system can be further classified into three

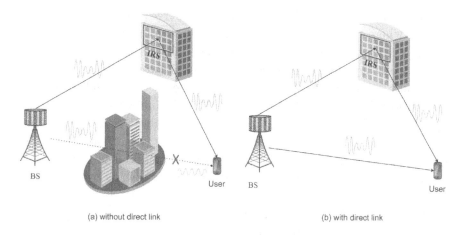

(a) without direct link   (b) with direct link

Figure 2.1 IRS-assisted wireless transmission: (a) without direct link; (b) with direct link.

categories, i.e., single-input single-output (SISO), multiple-input single-output (MISO), and multiple-input multiple-output (MIMO) systems. Existing works on MISO systems have revealed that IRS can achieve squared power gain under an asymptotically large number of reflecting elements [1]. Such a gain shows the great potential of IRS and its superiority over conventional massive MIMO systems, where only a linear gain is achieved [1]. It is further shown that by judiciously controlling the IRS reflection coefficients, the IRS-assisted MIMO channel can be substantially enhanced in terms of channel power, condition number, or rank.

In addition to point-to-point communication, IRS-assisted multi-user systems have also been studied [2, 3]. The performance analysis of multi-user systems is undoubtedly much more challenging than that of a point-to-point system, owing to the existence of inter-user interference. To evaluate the fundamental capacity limits of IRS-assisted multi-user SISO systems, [2] characterizes the capacity and rate regions subject to the constraint on IRS reconfiguration times. Presented simulation results show that the capacity and rate regions can be greatly improved using IRS. The more general MISO scenario is studied in [3], and it is shown that employing IRS can significantly improve the energy efficiency of wireless networks.

The works mentioned above focus on single-cell systems. The use of IRS in multi-cell systems is, however, investigated in [4, 5]. Compared with the single-cell counterpart, new issues emerge in multi-cell systems, such as where to deploy the IRSs and how to coordinate available resources among different base stations (BSs). The authors in [4] consider a large-scale deployment of IRSs in wireless networks and characterize the achievable spatial throughput averaged over both channel fading and random locations of the deployed BSs/IRSs. It is unveiled that deploying distributed IRSs can greatly boost the received signal power but only cause marginal extra interference in the network. Note that SISO is assumed in [4]. The authors in [5] study the weighted sum rate maximization problem for a MIMO multi-cell system. Numerical results reveal that employing IRSs can notably enhance the cell-edge performance.

In IRS-assisted systems, the IRS phase shifts need to be optimized in addition to the conventional transceiver optimization. As the IRS-assisted user channels are cascaded, variables to be optimized are often coupled, and thus, resulting in non-trivial joint resource optimization. Moreover, IRS optimization needs to satisfy the highly non-convex constant modulus constraint, since the IRS can only reflect the incident signal without amplifying it. Existing works often apply the block coordinate descent (BCD) method (also referred to as alternating optimization (AO) when there are only two types of variables) to resolve the coupling among the optimization variables [1, 3, 5]. On this basis, existing approaches developed for systems without IRS can often be borrowed for transceiver optimization. Meanwhile, the semidefinite relaxation (SDR) technique is widely used for addressing the passive beamforming at the IRS. The framework combining BCD/AO with SDR is shown to be effective in handling the joint resource allocation of various IRS-assisted systems [1, 3]. Nonetheless, it still suffers from two drawbacks: (i) the obtained solution can only be considered as a lower bound and (ii) the complexity may be too high, since the high-complexity SDR operation needs to be performed many times until convergence. To obtain a tight upper bound, a potential solution is to apply the successive convex approximation (SCA) technique to construct a convex approximation for joint optimization. This however could be non-trivial due to the coupling among the variables. Nevertheless, for certain cases, by exploiting closed-form solutions existing for transceiver optimization under given IRS phase shifts, the problem could be simplified, and it becomes relatively easy to apply the SCA technique. Additionally, the following two approaches could be adopted to replace SDR, and thus, lower the complexity: one is SCA while the other is the complex circle manifold (CCM) method. For SCA, the majorization-minimization (MM) algorithm appears to be quite promising, and the key then will be to find the appropriate surrogate function [5]. The usage of CCM, on the other hand, is motivated by the complex forms of IRS phases, and the main challenge lies in how to design a gradient descent algorithm based on the manifold space [5].

2.3 INTEGRATION OF IRS WITH ADVANCED TRANSMISSION TECHNOLOGIES

To further exploit the potential of IRS, it is of interest to investigate the integration of IRS with other advanced transmission technologies, including mmWave/THz, NOMA, MEC, and PLS.

2.3.1 IRS-Assisted MmWave/THz Communication

2.3.1.1 Motivation

MmWave/THz communication has drawn considerable attention recently owing to its ability to provide ultra-wide bandwidth [24–26]. Nonetheless, mmWave/THz communication suffers from severe signal attenuation and poor diffraction, which significantly limits its applications in mobile cellular systems. MIMO represents an effective technology to enhance the mmWave/THz signal strength owing to the high beamforming gain. However, the property of poor diffraction still makes mmWave/THz vulnerable to blocking by obstacles that break the LoS links. To address this, IRS can be deployed to create additional LoS links, and thus, extend the coverage of mmWave/THz systems [6–8, 25, 26].

2.3.1.2 State-of-the-Art and Key Techniques

Earlier works on IRS-assisted mmWave/THz MIMO focus on full digital beamforming at the BS, and have shown that deploying IRS can alleviate the blockage effect and enhance the performance of mmWave/THz systems in coverage and throughput [6]. To lower the number of radio frequency (RF) chains, one can either adopt the hybrid analog/digital beamforming structure (shown in Figure 2.2 [7, 25]) or the lens antenna array (illustrated in Figure 2.3 [8]). The former consists of two parts, namely, the baseband digital beamforming under a limited number of RF chains and the RF band analog beamforming via a network of phase shifters. Likewise, the latter also comprises two main components, namely the electromagnetic (EM) lens and the matching antenna array with elements located in the focal region of the lens. The EM lenses

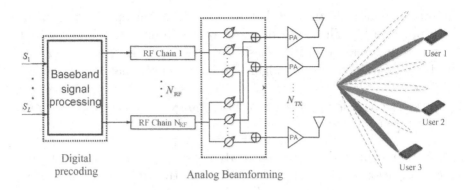

Figure 2.2   Structure of hybrid analog/digital beamforming.

provide controllable phase shifting to obtain angle-dependent energy-focusing property. It is shown that using IRS can enhance the performance for both hybrid beamforming and lens antenna array-based mmWave/THz systems [7, 8].

The BCD framework for microwave can be applied to mmWave/THz as well. However, the channel model should be updated using the appropriate mmWave/THz channel models, e.g., the geometric channel model [6, 7]. Considering the poor diffraction and penetration abilities of mmWave/THz, direct links between the users and BS under mmWave/THz are often assumed blocked [7, 8]. Furthermore, Tx-IRS and IRS-Rx channels can be approximated as a rank-one matrix/vector, as they are LoS dominated. This rank-one structure can be exploited for

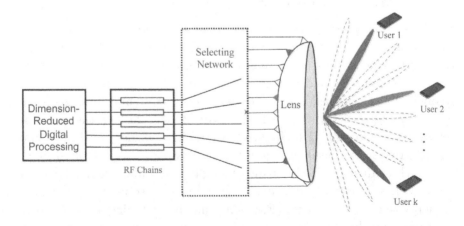

Figure 2.3   Structure of lens antenna array.

further simplifying the passive beamforming design at the IRS [6] and the active beamforming at the BS [7]. Additionally, the analog beamforming under hybrid beamforming in general can be handled using SDR as the phase shift optimization at the IRS, since both are limited by the same constant-modulus constraint. To lower the complexity, beam search based on pre-defined codebooks can also be adopted. However, note that the beam split may exist for conventional hybrid precoding schemes, especially under ultra-wideband mmWave/THz systems [26]. In this case, beam directions at different subcarriers vary, which leads to severe loss in beam gain, and thus, calls for novel hybrid beamforming solutions. In terms of lens antenna array, the key lies in how to perform an appropriate antenna/beam selection to significantly lower the RF chain cost, without sacrificing the system performance too much [8]. Interference-aware beam selection could be of interest.

2.3.2 IRS-Assisted NOMA Transmission

2.3.2.1 Motivation

NOMA is envisioned as a promising radio access technique for next-generation communication systems [27, 28]. By enabling multiple users to access the same time/frequency resources, NOMA can achieve higher spectral efficiency and energy efficiency and better support massive connectivity when compared to orthogonal multiple access (OMA) [29]. Nevertheless, to obtain a decent performance gain of NOMA over OMA, users are required to have a large channel gain disparity. Such a requirement may be violated in conventional NOMA systems, since user channels are determined by highly stochastic propagation environments. To overcome this, IRS can be utilized to introduce desirable channel gain differences among the users via constructively or destructively adding user signals (shown in Figure 2.4). Meanwhile, IRS can also be used to suppress the inter-user interference, and thus, lead to improved throughput or fairness of NOMA systems.

2.3.2.2 State-of-the-Art and Key Techniques

A body of research works has emerged very recently which investigate the design of IRS-assisted NOMA systems [9–11]. The authors in [9] investigate the sum rate maximization for an IRS-assisted multi-user SISO-NOMA system, requiring to jointly optimize the channel assignment, decoding order of NOMA users, power allocation, and reflection

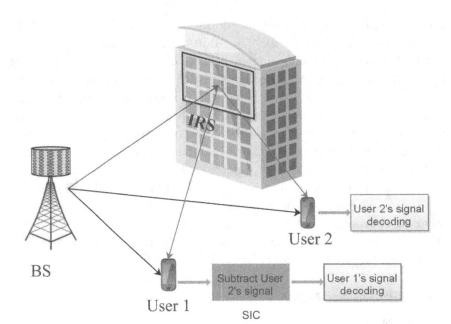

Figure 2.4 Illustration of IRS-assisted NOMA transmission.

coefficients. A three-step resource allocation algorithm is proposed and presented numerical results demonstrate the superiority of IRS-assisted NOMA over conventional NOMA without IRS and IRS-assisted OMA in terms of system throughput. To further enhance the system performance, [10] aims to maximize the minimum rate of all users for an IRS-assisted MIMO-NOMA system by jointly optimizing the transmit beamforming at the BS and phase shifts at the IRS. An efficient algorithm based on the framework of BCD with SDR is proposed to address the formulated non-convex problem. It is shown that the IRS-assisted MIMO-NOMA system can greatly boost the rate performance when compared with conventional NOMA without IRS and OMA with/without IRS. Note that the above works focus on the downlink. Uplink NOMA however is investigated by [11].

In IRS-assisted NOMA systems, resource allocation becomes more complicated due to the extra need for optimizing the decoding order [10]. Moreover, the optimization of the decoding order is coupled with that of the IRS phase shifts, since the optimal decoding order cannot be determined without knowing the IRS phase shifts while the IRS phase shifts cannot be properly configured without fixing the decoding order.

To decompose the coupling among them, a viable solution could be to iteratively update the decoding order and IRS phase shifts, by fixing the other. However, the complexity may become prohibited when the number of iterations required for convergence is large. An alternative is to exhaustively search all decoding orders and, on this basis, optimize the IRS phase shifts. Likewise, the resulting complexity of an exhaustive search may be too high, especially when the number of users is large. To lower the complexity, one can greedily set the decoding order by fixing the IRS phase matrix to certain values, such as all zeros or ones. When the IRS phase matrix is set to all zeros, it means only the direct link is considered. Clearly, this may be highly suboptimal due to the neglect of the effect of the cascaded Tx-IRS-Rx link. In contrast, when the IRS phase matrix is set to all ones, both the direct link and the cascaded Tx-IRS-Rx link are considered. Nevertheless, the resulting decoding order may still be quite different from the optimal one for the optimized IRS phase matrix. To address this, the authors in [10] propose a combined-channel strength-based user ordering scheme, where users are ordered based on their maximally achievable combined channel strengths via optimizing the IRS phase shifts. Numerical results show that the proposed user ordering scheme achieves near-optimal performance with much lower complexity. Except for user ordering, extra constraints should be imposed on users' achievable rates in IRS-assisted NOMA systems, to ensure the success of successive interference cancellation (SIC). That is, each user's achievable rate cannot exceed the minimum rates decodable at all users that need to decode its signal. Such extra constraints may make it more challenging to identify the initial feasible IRS phase shifts required for the BCD-based optimization framework.

2.3.3 IRS-Assisted MEC Systems

2.3.3.1 Motivation

The advancement of the mobile information industry is accelerating the emergence of various novel mobile applications, such as virtual reality, autonomous driving, and health care [30–32]. Such emerging applications usually have stringent computation and delay requirements that may exceed the processing capabilities of mobile devices. An effective way to enable these applications on mobile devices is MEC, which supports mobile users to offload their computation-intensive and delay-sensitive tasks to edge servers in vicinity [33–35]. Nonetheless, the promised benefits of MEC cannot be fully harvested, if the communication links utilized for

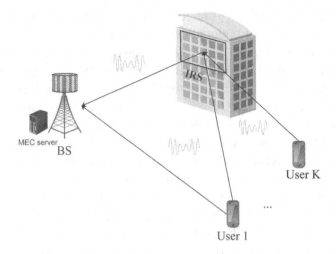

Figure 2.5 Illustration of IRS-assisted MEC systems.

computation offloading are hostile. To address this, IRS can be deployed to mitigate propagation-related impairments, and thus, facilitate offloading (as shown in Figure 2.5).

2.3.3.2 State-of-the-Art and Key Techniques

A number of works have appeared recently, targeting various scenarios of IRS-assisted MEC systems [12–20]. Based on the considered objective function, we could categorize them into four classes, namely latency minimization [12, 13], sum computational rate maximization [14–16], energy consumption minimization [17–19], and energy efficiency maximization [20]. The number of users in these existing works ranges from single user [18], two users [13] to multiple users [14–17]. Under the two-/multi-user scenarios, different multiple access schemes have been investigated or compared, including OMA [12, 14, 16, 17], NOMA [15, 19, 20], and hybrid of the two [13]. Considering the limited batteries for the users, e.g., for internet-of-things devices, wireless power transfer has also been added into IRS-assisted MEC systems to ensure that sufficient power is available for offloading the data and/or local processing [16, 17].

In IRS-assisted MEC systems, often a joint allocation of the radio and computational resources is required, which makes the analysis more complicated. To simplify the analysis, the BCD technique/AO has been widely adopted as before [12, 14, 15, 17]. On this basis, various techniques have been used to solve the corresponding subproblems, such as the Lagrange dual method [14], the penalty method [19], and the SDR method [20]. To further simplify the system model, the so-called sum computational rate maximization has been studied in [14–16], where it is assumed that the edge server is so powerful that its computational time is negligible. As a result, there is no need to determine how to appropriately allocate computing resources among the offloading users at the MEC server. Overall, these existing studies have shown that IRS-assisted MEC systems can significantly outperform conventional ones without IRS.

2.3.4 IRS-Assisted PLS Systems

2.3.4.1 Motivation

Owing to the broadcasting nature of wireless transmission media, wireless systems are vulnerable to impersonation attacks and eavesdropping [36–38]. Encryption techniques represent an effective way to ensure communication confidentiality; however, they may not be suitable for some Internet-of-things applications that cannot afford their complexity and/or have stringent delay requirements. By exploiting the randomness nature of wireless propagation channels, PLS can help secure wireless communication confidentiality without consuming much of the resources. To ensure a non-negative secrecy rate, it is, however, often required that the legitimate users experience better channel conditions than the eavesdroppers. Clearly, such a requirement does not always hold in conventional PLS systems. A simple workaround is to deploy IRS in PLS systems to reconfigure the channels for the legitimate users and eavesdroppers. In particular, IRS can be used to enhance the signal strength at the legitimate users while nulling the signal reception at the eavesdroppers, thereby enhancing secrecy transmission rates.

2.3.4.2 State-of-the-Art and Key Techniques

Research on IRS-assisted PLS systems has attracted great attention, and existing works cover various scenarios, e.g., [21–23]. As shown in Figure 2.6, both [21] and [22] consider the case where the BS is equipped with multiple antennas, while the legitimate user and eavesdropper are with a single antenna. The authors [21] aim to maximize the secrecy transmission rate under the maximum transmit power constraint, whereas [22] studies the transmit power minimization subject to secrecy rate constraint at the legitimate user. Presented simulation results show

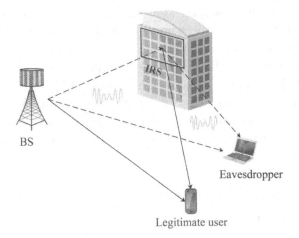

Figure 2.6   Illustration of IRS-assisted PLS systems.

that the proposed schemes with IRS outperform their counterparts without IRS in terms of secrecy rate and transmit power. In addition, the authors in [23] analyze the secrecy outage probability of an IRS-assisted PLS system, where all network nodes are equipped with a single antenna. Numerical results illustrate that deploying the IRS can lower the secrecy outage probability as well.

Introducing PLS into IRS-assisted systems often leads to the original non-convex objective function being more complicated, and thus harder to handle. As in conventional IRS systems, the BCD method can be used to decompose the coupling among the optimization variables, thereby making the problem more tractable. On this basis, there exist several ways to optimize the transmit beamforming. First, a closed-form solution may be derived for the optimal transmit beamformer, e.g., [21, 22]. Second, the SCA technique, e.g., difference-of-convex (DC) programming, could be used, considering the expression of the secrecy rate. Last, fractional programming may also be employed, by removing the $\log(\cdot)$ operation in the objective function. For the IRS phase shift optimization, semi-closed form solutions may exist in certain cases, e.g., [21]. Besides, SDR has been widely used to obtain a high-quality solution, e.g. [22]. Last, the SCA technique, e.g., the MM algorithm can also be adopted.

A summary of resource allocation for conventional and advanced IRS-assisted systems is given in Figure 2.7.

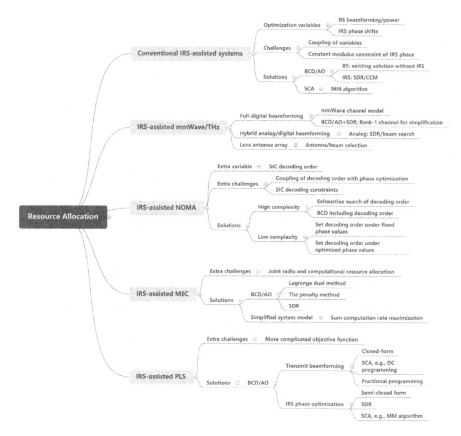

Figure 2.7 Summary of research allocation for conventional and advanced IRS-assisted systems.

2.4 NON-IDEAL TRANSMISSION OF IRS-ASSISTED COMMUNICATION SYSTEMS

IRS-assisted communication systems, like any other communication systems, suffer from non-ideal transmission conditions (mainly result from HWIs and/or imperfect CSI) that can deteriorate their performance if not properly taken into consideration. HWIs in IRS-assisted communications systems can result from the finite resolution of phase shifters at the IRS reflecting elements and/or the RF front end mismatches at the transceivers. In the rest of this section, we will discuss HWIs and imperfect CSI and their effect on the performance of IRS-assisted systems.

2.4.1 Hardware Impairments

2.4.1.1 *Finite-Resolution Phase Shifters*

IRS reflecting elements need to adjust their phase shifts in real time to compensate for the time-varying nature of the wireless channel. Such phase shift adjustments of IRS reflecting elements can be achieved via using positive intrinsic-negative (PIN) diodes, micro-electromechanical system-based switches, or field-effect transistors. A large number of studies on IRS consider that the phase shifts of the IRS reflecting elements can change continuously, which is hard to achieve in practice. For instance, the phase shift levels of the PIN diode are typically adjusted to two levels (0 and π radians) by changing the applied biasing voltage between two levels. To achieve M different phase-shift levels, $\log_2 M$ PIN diodes are required for each reflecting element. Alternatively, a single varactor diode can be used to achieve more than two phase shifts; however, it requires a high number of biasing voltages which will increase the complexity of the IRS controller. Manufacturing IRS reflecting elements to support a high number of phase shifts will increase its cost, and hence, will not be a scalable solution given that IRS typically has a very high number of reflecting elements. The effect of the finite-resolution phase shifters of the IRS reflecting elements has been investigated in a number of recent works. It is shown that for an asymptotically large number of IRS reflecting elements, a 1-bit phase shifter approaches the same squared power gain when compared to the ideal continuous phase shifters. However, as the number of IRS reflecting elements decreases, the power loss increases, and it depends on the number of the available phase-shift levels.

2.4.1.2 *RF Chain Impairments*

RF front-end impairments in IRS-assisted communication systems include in-phase and quadrature imbalance (IQI) at the transceiver, phase noise of IRS, and other transmission non-linearities. A general model to capture such RF impairments in IRS-assisted communication systems is the extended error vector magnitude (EEVM) proposed in [39]. In the EEVM model, RF chain impairments are effectively modeled as multiplicative and additive random variables terms, with a certain mean and variance. This model has been applied to both single- and multi-antenna transmitters in IRS-assisted communication systems. It is shown that aggregate RF front-end impairments at both Tx and Rx can be accurately modeled as a zero-mean complex Gaussian process with a variance that depends on the transmit power, the effective channel gain of the Tx-IRS-Rx, and the square of the error vector magnitude at both the Tx and Rx. Analysis of IRS-assisted communication systems employing the EEVM model shows that at a high signal-to-noise ratio, the performance (in terms of spectral efficiency or outage probability) is independent of the number of IRS reflecting elements, and is mainly constrained by the level of RF chain impairments at the Tx when compared to the phase distortion at IRS elements. Such analysis reveals that modest IRS elements with low-resolution phase shifters can be used as affordable deployment solutions for IRS without significantly degrading the performance.

2.4.2 Imperfect CSI

Estimating the indirect channel (between the Tx and IRS) and the reflection channel (between the IRS and Rx) is challenging, given the fact that the IRS consists of passive reflecting elements with limited signal processing capabilities. The indirect channel estimation problem can be partially solved with the knowledge of the angle of arrival given that the IRS elements are mounted on buildings, and hence, considered of fixed location. The reflection channel estimation is challenging, given the expected end users' mobility, and errors in the estimation of the reflection channel can significantly deteriorate the performance of IRS-assisted communication systems if their effect is not properly considered in the IRS system design [25]. In general, imperfect CSI estimation can be addressed through a worst-case/robust design, considering knowledge of the channel statistics rather than the instantaneous channel coefficients or allowing a controlled outage in the IRS system performance [25]. In a robust design, the transmit power will be typically increased, when compared to its counterpart assuming perfect CSI, to compensate for estimation inaccuracy. If the statistics of the CSI are known, it is possible to improve the performance of IRS systems (e.g., achievable rate) on average rather than instantaneously. However, recent studies in the literature reported that the performance of IRS systems under statistical CSI knowledge deteriorates with increasing the number of IRS elements serving a particular end user [40].

2.5 OPEN ISSUES

In this section, we highlight several open issues that are worthy of investigation.

2.5.1   Unmanned Aerial Vehicles-Integrated IRS Systems

Unmanned aerial vehicles (UAVs) have shown several benefits, as relays or flying BSs, to improve the performance of communication systems, and hence, they are currently being considered as a key enabler of next-generation wireless systems [41]. UAVs typically have a strong LoS and favorable propagation conditions to terrestrial BSs, which could lead radio waves from terrestrial BSs to interfere with UAVs in adjacent cells. IRS represents a promising candidate solution to mitigate the inter-cell interference problem of future UAV networks, owing to its ability to efficiently control the travel direction of radio waves, through joint beamforming with terrestrial BSs. A few recent works have exploited IRS to improve the performance of UAV systems, but they are limited to single-cell. More investigations into this field are required, especially for IRS-assisted multi-cell UAV systems.

2.5.2   Machine Learning-Empowered IRS Systems

Due to the coupling of optimization variables and the non-convex nature of the underlying problems, joint resource allocation in IRS systems is challenging to solve, and often sophisticated solutions with high complexity are required to obtain near-optimal performance. However, the time-varying and highly dynamic nature of wireless networks requires the proposed solutions to be of low complexity and execute easily. Such a dilemma is non-trivial to overcome using conventional optimization-based methods. A promising way to tackle this is to employ machine learning techniques [42], which have been shown as an effective tool to obtain near-optimal solutions for non-convex and sophisticated optimization problems under highly dynamic wireless environments. Machine learning also holds the potential to learn the channel indirectly from the data during training, without the need for explicit CSI. Accordingly, machine learning-empowered IRS systems are of practical interest.

2.5.3   Sensing and Localization

Next-generation wireless communication systems will operate at higher frequencies (mmWave or THz band) to support applications that require sensing of the surrounding environment and accurate localization. Such high frequencies have a limited number of propagation paths (mainly due to large penetration losses, high values of path loss, and low scattering), which may reduce the accuracy of sensing and localization. Hence, IRS as a controlled and dynamic scattering is considered a promising solution to the sensing and localization problem for next-generation wireless communication systems. One of the main research challenges is that for such high frequencies and large-size IRS, users are no longer in the far-field and conventional sensing and localization models are no longer valid. That said, sensing and localization models in the near-field that exploit the information in the wavefront curvature needs to be developed.

2.6 CONCLUSION

In this chapter, we surveyed IRS-empowered wireless networks. We first showed that judiciously deploying IRS can substantially improve the spectral efficiency, energy efficiency, and coverage of wireless networks. On this basis, we validated that IRS can be further used to enhance the performance of mmWave/THz, NOMA, MEC, and PLS systems. However, the promised gains of IRS are often obtained under ideal assumptions on channel estimation and hardware configuration. Motivated by this, we further discussed the effects of HWIs and imperfect CSI on the performance of IRS. Lastly, we identified three open issues for future research.

BIBLIOGRAPHY

[1] Q. Wu and R. Zhang. Intelligent reflecting surface enhanced wireless network via joint active and passive beamforming. *IEEE Trans. Wireless Commun.*, 18(11):5394–5409, Nov. 2019.

[2] X. Mu, Y. Liu, L. Guo, J. Lin, and N. Al-Dhahir. Capacity and optimal resource allocation for IRS-assisted multi-user communication systems. *IEEE Trans. Commun.*, 69(6):3771–3786, Jun. 2021.

[3] M. Zeng, E. Bedeer, O. A. Dobre, P. Fortier, P. Quoc-Viet, and W. Hao. Energy-efficient resource allocation for IRS-assisted multi-antenna uplink systems. *IEEE Wireless Commun. Lett.*, 10(6):1261–1265, Jun. 2021.

[4] J. Lyu and R. Zhang. Hybrid active/passive wireless network aided by intelligent reflecting surface: System modeling and performance analysis. *IEEE Trans. Wireless Commun.*, 20(11):7196–7212, Nov. 2021.

[5] C. Pan, H. Ren, K. Wang, W. Xu, M. Elkashlan, A. Nallanathan, and L. Hanzo. Multicell MIMO communications relying on intelligent reflecting surfaces. *IEEE Trans. Wireless Commun.*, 19(8):5218–5233, Aug. 2020.

[6] P. Wang, J. Fang, X. Yuan, Z. Chen, H. Li. Intelligent reflecting surface-assisted millimeter wave communications: Joint active and passive precoding design. *IEEE Trans. Veh. Technol.*, 69(12):14960–14973, Dec. 2020.

[7] J. Qiao and M. Alouini. Secure transmission for intelligent reflecting surface-assisted mmwave and terahertz systems. *IEEE Wireless Commun. Lett.*, 9(10):1743–1747, Jun. 2020.

[8] Y. Wang, H. Lu, D. Zhao, and H. Sun. Energy efficiency optimization in IRS-enhanced mmwave systems with lens antenna array. In *Proc. IEEE GLOBECOM*, pages 1–6, Dec. 2020.

[9] J. Zuo, Y. Liu, Z. Qin, and N. Al-Dhahir. Resource allocation in intelligent reflecting surface assisted NOMA systems. *IEEE Trans. Commun.*, 68(11):7170–7183, Nov. 2020.

[10] G. Yang, X. Xu, and Y.-C. Liang. Intelligent reflecting surface assisted non-orthogonal multiple access. In *Proc. IEEE WCNC*, pages 1–6, May 2020.

[11] M. Zeng, X. Li, G. Li, W. Hao, and O. A. Dobre. Sum rate maximization for IRS-assisted uplink NOMA. *IEEE Commun. Lett.*, 25(1):234–238, Jan. 2021.

[12] T. Bai, C. Pan, Y. Deng, M. Elkashlan, A. Nallanathan, and L. Hanzo. Latency minimization for intelligent reflecting surface aided mobile edge computing. *IEEE J. Sel. Areas Commun.*, 38(11):2666–2682, Nov. 2020.

[13] F. Zhou, C. You, and R. Zhang. Delay-optimal scheduling for IRS-aided mobile edge computing. *IEEE Wireless Commun. Lett.*, 10(4):740–744, Apr. 2021.

[14] Z. Chu, P. Xiao, M. Shojafar, D. Mi, J. Mao, and W. Hao. Intelligent reflecting surface assisted mobile edge computing for internet of things. *IEEE Wireless Commun. Lett.*, 10(3):619–623, Mar. 2021.

[15] G. Chen, Q. Wu, W. Chen, D. W. K. Ng, and L. Hanzo. IRS-aided wireless powered mec systems: TDMA or NOMA for computation offloading? *arXiv preprint arXiv:2108.06120*, Aug. 2021.

[16] S. Mao, N. Zhang, L. Liu, J. Wu, M. Dong, K. Ota, T. Liu, and D. Wu. Computation rate maximization for intelligent reflecting surface enhanced wireless powered mobile edge computing networks. *IEEE Trans. Veh. Technol.*, 70(10):10820–10831, Oct. 2021.

[17] T. Bai, C. Pan, H. Ren, Y. Deng, M. Elkashlan, and A. Nallanathan. Resource allocation for intelligent reflecting surface aided wireless powered mobile edge computing in OFDM systems. *IEEE Trans. Wireless Commun.*, 20(8):5389–5407, Aug. 2021.

[18] E. El Haber, M. Elhattab, C. Assi, S. Sharafeddine, and K. K. Nguyen. Latency and reliability aware edge computation offloading via an intelligent reflecting surface. *IEEE Commun. Lett.*, 25(12):3947–3951, Dec. 2021.

[19] Z. Li, M. Chen, Z. Yang, J. Zhao, Y. Wang, J. Shi, and C. Huang. Energy efficient reconfigurable intelligent surface enabled mobile edge computing networks with NOMA. *IEEE Trans. Cogn. Commun. Netw.*, 7(2):427–440, Jun. 2021.

[20] Q. Wang, F. Zhou, H. Hu, and R. Q. Hu. Energy-efficient design for irs-assisted MEC networks with NOMA. In *Proc. IEEE WCSP*, pages 1–6, Oct. 2021.

[21] H. Shen, W. Xu, S. Gong, Z. He, and C. Zhao. Secrecy rate maximization for intelligent reflecting surface assisted multi-antenna communications. *IEEE Commun. Lett.*, 23(9):1488–1492, Sep. 2019.

[22] Z. Chu, W. Hao, P. Xiao, and J. Shi. Intelligent reflecting surface aided multi-antenna secure transmission. *IEEE Wireless Commun. Lett.*, 9(1):108–112, Jan. 2020.

[23] L. Yang, J. Yang, W. Xie, M. O. Hasna, T. Tsiftsis, M. Di Renzo. Secrecy performance analysis of RIS-aided wireless communication systems. *IEEE Trans. Veh. Technol.*, 69(10):12296–12300, Oct. 2020.

[24] S. A. Busari, K. M. Saidul Huq, S. Mumtaz, L. Dai, and J. Rodriguez. Millimeter-wave massive MIMO communication for future wireless systems: A survey. *IEEE Commun. Surv. Tuts.*, 20(2):836–869, Secondquarter 2018.

[25] W. Hao, G. Sun, M. Zeng, Z. Chu, Z. Zhu, O. A. Dobre, and P. Xiao. Robust design for intelligent reflecting surface-assisted MIMO-OFDMA terahertz IoT networks. *IEEE Internet Things J.*, 8(16):13052–13064, Aug. 2021.

[26] W. Hao, F. Zhou, M. Zeng, Chu, O. A. Dobre, and N. Al-Dhahir. Ultra wide band THz IRS communications: Applications, challenges, key techniques, and research opportunities. *IEEE Netw., to appear*, 2022.

[27] S. M. R. Islam, M. Zeng, O. A. Dobre, and K. Kwak. Resource allocation for downlink NOMA systems: Key techniques and open issues. *IEEE Wireless Commun. Mag.*, 25(2):40–47, Apr. 2018.

[28] S. M. R. Islam, M. Zeng, and O. A. Dobre. NOMA in 5G systems: Exciting possibilities for enhancing spectral efficiency. *IEEE 5G Tech. Focus,* vol. 1, no. 2, May 2017. [Online]. Available: 307 http://5g.ieee.org/tech-focus.

[29] M. Zeng, A. Yadav, O. A. Dobre, G. I. Tsiropoulos, and H. V. Poor. Capacity comparison between MIMO-NOMA and MIMO-OMA with multiple users in a cluster. *IEEE J. Sel. Areas Commun.*, 35(10):2413–2424, Oct. 2017.

[30] P. Porambage, J. Okwuibe, M. Liyanage, M. Ylianttila, and T. Taleb. Survey on multi-access edge computing for internet of things realization. *IEEE Commun. Surv. Tuts.*, 20(4):2961–2991, Fourthquarter 2018.

[31] S. M. R. Islam, D. Kwak, M. H. Kabir, M. Hossain, and K. Kwak. The internet of things for health care: A comprehensive survey. *IEEE Access*, 3:678–708, Jun. 2015.

[32] S. Mumtaz, A. Alsohaily, Z. Pang, A. Rayes, K. F. Tsang, and J. Rodriguez. Massive internet of things for industrial applications: Addressing wireless iiot connectivity challenges and ecosystem fragmentation. *IEEE Ind. Electron. Mag.*, 11(1):28–33, Mar. 2017.

[33] M. Zeng, W. Hao, O. A. Dobre, Z. Ding, and H. Vincent Poor. Massive MIMO-assisted mobile edge computing: Exciting possibilities for computation offloading. *IEEE Veh. Technol. Mag.*, 15(2):31–38, Jun. 2020.

[34] M. Zeng and V. Fodor. Energy minimization for delay constrained mobile edge computing with orthogonal and non-orthogonal multiple access. *Ad Hoc Netw.*, 98:102060, 2020.

[35] M. Zeng, N. P. Nguyen, O. A. Dobre, and H. V. Poor. Delay minimization for NOMA-assisted MEC under power and energy constraints. *IEEE Wireless Commun. Lett.*, 8(6):1657–1661, Dec. 2019.

[36] M. Zeng, N. P. Nguyen, O. A. Dobre, and H. V. Poor. Securing downlink massive MIMO NOMA networks with artificial noise. *IEEE J. Sel. Topics Signal Process.*, 13(3):685–699, Jun. 2019.

[37] W. Hao, M. Zeng, G. Sun, and P. Xiao. Edge cache-assisted secure low-latency millimeter-wave transmission. *IEEE Internet of Things J.*, 7(3):1815–1825, Mar. 2020.

[38] N.-P. Nguyen, M. Zeng, O. A. Dobre, and H. V. Poor. Securing massive MIMO-NOMA networks with ZF beamforming and artificial noise. In *Proc. IEEE GLOBECOM*, pages 1–6, Puako, Hawaii, Dec. 2019.

[39] T. Schenk. *RF imperfections in high-rate wireless systems: impact and digital compensation.* Springer Science & Business Media, 2008.

[40] H. Guo, Y.-C. Liang, and S. Xiao. Model-free optimization for reconfigurable intelligent surface with statistical CSI. [Online]. Available: https://arxiv.org/abs/1912.10913, Dec. 2019.

[41] Z. Wei, Y. Cai, Z. Sun, D. W. K. Ng, J. Yuan, M. Zhou, and L. Sun. Sum-rate maximization for IRS-assisted UAV OFDMA communication systems. *IEEE Trans. Wireless Commun.*, 20(4):2530–2550, Apr. 2021.

[42] X. Gao, Y. Liu, X. Liu, and Lingyang Song. Machine learning empowered resource allocation in IRS aided MISO-NOMA networks. *IEEE Trans. Wireless Commun.*, 21(5):3478–3492, May 2022.

CHAPTER 3

Role of Intelligent Reflecting Surfaces in the Emerging 6G Technologies

Shaika Mukhtar

Advanced Communication Lab, Department of Electronics and Communication Engineering, National Institute of Technology, Srinagar, India

Umer Ashraf

Advanced Communication Lab, Department of Electronics and Communication Engineering, National Institute of Technology, Srinagar, India

Gh. Rasool Begh

Advanced Communication Lab, Department of Electronics and Communication Engineering, National Institute of Technology, Srinagar, India

CONTENTS

3.1	Introduction	40
3.2	Interplay of IRS with Different Technologies	41
3.3	IRS-Assisted Multi-Input Multi-Output (MIMO)-Non-Orthogonal Multiple access (NOMA)	41
3.4	IRS-Assisted Device-to-Device (D2D) Communication	44
3.5	IRS-Assisted Cognitive Radio Systems	46
3.6	IRS-Assisted Optical Communication Systems	48
3.7	Challenges and Future Directions	49
Bibliography		50

DOI: 10.1201/9781003282211-3

I NCESSANT RESEARCH IS carried out to improve the performance of communication systems for the 6G era. The users require higher connectivity, lower latency, and improved quality of service (QoS). In this regard, intelligent reflecting surfaces are integrated with the existing communication technologies. Such an interplay is envisioned to improve the overall system performance.

3.1 INTRODUCTION

The introduction of intelligent reflecting surfaces (IRS) in communication systems has revolutionized the users' performance. These intelligent surfaces smartly reconfigure the wireless channel leading to smart radio environments. Such initiative helps in enhancing the signal quality and coverage [1]. The intelligent surfaces consist of metasurfaces which are controlled by electronic circuits [2]. Each scattering element of the metasurface adjusts the amplitude and phase shift of the incident signals for better signal reception and coverage [3]. In this way, IRS extends the degree of freedom that can be used to fulfill different networking objectives. Moreover, these low-cost passive elements of IRS do not demand any dedicated energy supply, revealing their energy-efficient nature [4]. In order to implement IRS in 6G communication systems, the research world is analyzing most of its performance aspects. In [5], experimental proof of the potential benefits of an IRS-aided communication system is provided. The authors in [6] have presented a comprehensive description of the IRS system in terms of involved signal processing operations. In [7] and [8], two different approaches are used to provide bit error rate (BER) expression of the IRS system over diverse propagation models. In [9], the authors have presented coverage analysis by deriving the maximum coverage range of the IRS system for given network parameters. In [10], outage probability (OP) and ergodic capacity of single- and multiple-element IRS systems are derived over Fox's H fading channel along with their asymptotic analysis. The energy efficiency of IRS is studied in [11], wherein the proposed utilization of IRS for downlink multi-user communication provides higher energy efficiency. In [12], the authors have studied simultaneous transmit diversity and passive beamforming in IRS systems. All this research motivates to incorporate IRS in the emerging 6G technologies. The interplay of IRS with different communication systems is discussed in the following section.

3.2 INTERPLAY OF IRS WITH DIFFERENT TECHNOLOGIES

Owing to the spectral and energy efficiency of IRS, these intelligent surfaces have been incorporated into different communication systems as complementary devices. Such hybrid approach helps in providing better 6G services to the users. For vehicular communications, intelligent surfaces ensure low lower power consumption over full-duplex and full-band signal transmission [13]. The emerging technology of unmanned aerial vehicles (UAV) also encourages the incorporation of IRS to provide energy-efficient transmission in the internet-of-things (IoT) network. The IRS module is mounted on a UAV leading to coverage extension for far-off users [14]. With the intention of exploring millimeter wave (mmWave) and Terahertz (THz) bands, IRS plays a prominent role in mmWave and terahertz systems. The deployed IRS helps in overcoming blockage, along with securing the high-frequency signal transmission of the users [15]. IRS also shows a prominent role in simultaneous wireless information and power transfer (SWIPT) aided systems. These metasurfaces increase the energy harvesting of the energy receivers and boost the signal strength of information receivers [16]. Moreover, IRS provides the necessary advantage to optimize energy efficiency in backscatter communication. This hybrid approach opens the road to self-sustainable IoT for 6G [17]. Further, high accuracy in sensing and localization is one of the prime requirements for the implementation of 6G communication technologies. In this regard, IRS helps in controlling the scattering environment leading to better sensing and localization [18]. Such examples of the interplay of IRS with different communication technologies reveal that IRS plays a vital role in improving system performance. Despite its superiority, the passive beamforming design, channel estimation, and position of its deployment are some of the challenges which require a radical approach [19].

3.3 IRS-ASSISTED MULTI-INPUT MULTI-OUTPUT (MIMO)-NON-ORTHOGONAL MULTIPLE ACCESS (NOMA)

The NOMA is a promising technique to serve the diverse demands of the 6G era. This technique provides better spectral efficiency, link density, and user fairness. Such favorable features of NOMA have carved out its existence as a significant member of the next-generation multiple access (NGMA) family [20]. The basic notion of downlink NOMA is the transmission of the superposed signal using the same resource block from the base station to multiple NOMA users. The power domain is the add-on dimension that is explored in power NOMA, leading to higher power allocation to far users than near users. For decoding, successive interference cancellation (SIC) is carried out at the receiver side [21]. Analytically, NOMA has shown better OP and ergodic capacity than the OMA system [22]. The performances of NOMA users have also been studied over diverse environments in terms of BER, ergodic capacity, and OP [23]. Despite the encouraging utility of NOMA, there are various challenges in its implementation for 6G [24]. The prime issues are transmission distortions and interference mitigation, which need to be addressed radically. In this regard, the IRS provides a helping hand to NOMA users to boost signal reception and mitigate unwanted interference. IRS ensures improvement in spectral and energy efficiency in NOMA systems. Moreover, the performance of the NOMA system is based on the power allocation and channel strength differences of the users. In this aspect, IRS helps in providing artificial higher channel differences leading to improved rates of transmission [25]. This leads to the realization of intelligent NOMA. In [26], IRS-assisted NOMA is introduced to serve more users than spatial division multiple access (SDMA). In [27], the authors have discussed both downlink and uplink IRS-aided NOMA and OMA networks, wherein the incorporation of IRS has enhanced the performance of the cell-edge user. The IRS-assisted downlink NOMA model is discussed in [28], wherein the error performance of the NOMA users is analyzed. In terms of energy harvesting, the authors in [29] have discussed the uplink and downlink IRS-NOMA systems. In [30], the maximization of the system throughput in the IRS-NOMA network has been discussed. For multi-antenna IRS-assisted NOMA systems, an energy-efficient model for IRS-NOMA is proposed in [31] to develop a tradeoff between the sum-rate maximization and total power consumption minimization. The authors in [32] have considered a downlink IRS-assisted MISO-NOMA network to minimize power consumption. In [33], the authors have investigated the sum-rate maximization problem in IRS-NOMA systems. The effect of continuous and discrete phase shifting on the performance analysis of blocked NOMA users has been studied in [34].

The commendable feature of IRS is to manipulate channel gains, which helps in achieving better SIC-based detection at the users in NOMA transmission. So, the incorporation of the IRS has a potential effect on channel statistics. In the IRS-assisted SISO-NOMA system, the

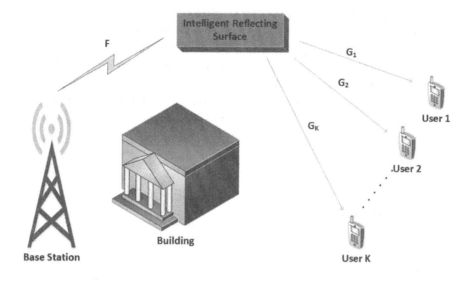

Figure 3.1   IRS-assisted MIMO-NOMA system.

change in channel gains is reflected as under

$$\mathbf{h}_i = \mathbf{g}_i^H \boldsymbol{\Theta} \mathbf{f}_i, \tag{3.1}$$

where $\mathbf{h}_i$ represents channel gain of the ith user in the IRS-assisted NOMA system. The terms $\mathbf{f}$ and $\mathbf{g}_i$ denote channel vectors from base station to IRS and IRS to the ith NOMA user, respectively. The notation $\boldsymbol{\Theta}$ is the phase shift matrix which is tuned to have higher differences in channel gains. This also helps in customizing channel gains as per users' requirements. Such control of channel gains is also explored in IRS-assisted MIMO-NOMA systems, as discussed below.

The basic model of an IRS-assisted MIMO-NOMA system is given in Figure 3.1. The base station is equipped with N transmit antennas communicating with K NOMA users. Each user is equipped with R receive antennas. There are M intelligent surfaces deployed such that line-of-sight signal transmissions exist in both BS-IRS and IRS-user links. The users are in a dead zone with respect to transmission from the base station. In such a scenario, IRS helps in ensuring improved signal reception at the NOMA users [35]. The received signal by the lth user is given by

$$\mathbf{Y}_l = (\mathbf{G}_l \boldsymbol{\Theta} \mathbf{F}) \; x \; + \; \mathbf{W}, \tag{3.2}$$

where $\mathbf{F}$ is the $M \times N$ channel matrix from BS to reflecting elements. $\mathbf{G}_l$ denotes the $M \times R$ channel matrix from reflecting elements to multi-antenna NOMA users. The notation Θ is the diagonal phase shift matrix with $\Theta = diag\{e^{j\theta_1}....e^{j\theta_m}\}$ with $\theta_m \in [0, 2\pi)$ representing phase shifts of the reflecting elements. IRS plays a prominent role in overcoming the dead zone condition by steering the signals from the BS to the users efficiently. All $K-1$ users perform SIC to retrieve their respective original signals. In the presence of IRS, NOMA users provide improved rates of transmission [36].

3.4 IRS-ASSISTED DEVICE-TO-DEVICE (D2D) COMMUNICATION

D2D communication is an additional technology for cellular networks to ensure ultra-low latency for the users. It allows users to have proximity for communicating with each other directly without the intervention of the base station. This leads to higher information transfer among the devices by efficiently using the licensed and unlicensed spectrum bands. D2D communication is also envisioned to ensure better communication in disaster-hit areas by acting as a relay to the poor signal user [37]. It promises low latency and real-time experience for machine-to-machine (M2M) communication in IoT networks [38]. Regarding its utility in the 6G era, the D2D scheme, along with artificial intelligence, can potentially improve the overall real-time experience of the users [39]. For the practical implementation of the D2D communications system, there are some challenges of synchronization, security, resource, and interference management [40]. In this regard, the IRS can be incorporated into D2D communication to overcome interference and ensure energy-efficient transmission among the users. Due to IRS's compact size, these surfaces can be attached to buildings, walls, or interior ceilings, leading to their easy deployment in D2D communication systems. The leading utility of IRS in these systems is to steer the desired signal to the specific user equipment along with mitigating interference from the other users. Such advantages of IRS are possible only because of the beamforming and phase shifts offered by the intelligent surfaces [41]. The research related to IRS-D2D communication systems is available in the literature, but it is still in its budding stage. In [42], the authors have studied the IRS-assisted uplink communication network scenario involving multiple D2D links and uplink cellular links. It has been shown that IRS helps in optimizing phase shifts leading to significant mitigation in D2D network

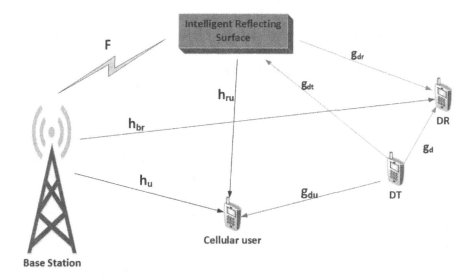

Figure 3.2 IRS-assisted D2D system.

interference. The authors in [43] have adopted a block coordinate descent algorithm and semidefinite relaxation technique to optimize the beamforming vector, power allocation, and phase shift matrix in IRS-assisted D2D communication networks. In [44], closed-form expressions for the power allocation and receive beamforming are derived in IRS-assisted D2D underlaid cellular systems. In [45], the sum rate of D2D users and cellular users is maximized by joint optimization of transmit power and passive beamforming.

The basic system model of an IRS-assisted D2D communication system is given in Figure 3.2. It consists of one BS having N transmit antennas. There is one cellular user and one D2D pair of transmitting user DT and receiving user DR. Each user is equipped with only one antenna. The system has one IRS having M passive elements. The signals received by the cellular user and receiving user of the D2D pair are given by [46]

$$\mathbf{y}_u = \left(\mathbf{h}_u{}^H + \mathbf{h}_{ru}{}^H \boldsymbol{\Theta} \mathbf{F} \right) \mathbf{s}_u + \left(g_{du} + \mathbf{h}_{ru}{}^H \boldsymbol{\Theta} \mathbf{g}_{dt} \right) \mathbf{s}_d + n, \qquad (3.3)$$

$$\mathbf{y}_{dr} = \left(g_d + \mathbf{g}_{dr}{}^H \boldsymbol{\Theta} \mathbf{g}_{dt} \right) \mathbf{s}_d + \left(\mathbf{h}_{ru}{}^H + \mathbf{g}_{dr}{}^H \boldsymbol{\Theta} \mathbf{F} \right) \mathbf{s}_u + n, \qquad (3.4)$$

where $\mathbf{h}_u$, $\mathbf{h}_{br}$, and $\mathbf{F}$ represent channel gains between the base station and cellular user, DR and IRS, respectively. Also, $\mathbf{h}_{ru}$ and $\mathbf{g}_{dr}$ denote channel gains from IRS to the cellular user and DR, respectively. Further, g_{du} and g_{dt} present channel gains from DT to the cellular user and IRS, respectively. The term g_d shows the channel gain between DT and DR of the D2D pair. The notation Θ denotes the phase shift matrix which is tuned by IRS to have required channel manipulation. In this way, IRS helps in boosting signal reception for both the cellular user and D2D pair.

For better performance of IRS-assisted D2D communication systems, interference and resource management are the main challenges because of the presence of both cellular and D2D links. In addition to these, the other challenges are optimal designing of the phase matrix along with efficient device discovery. Also, proper mode selection and power control can enhance the performance of the D2D system.

3.5 IRS-ASSISTED COGNITIVE RADIO SYSTEMS

Efficient utilization of radio spectrum capacity is a provision which helps in achieving higher throughput with low latency. The cognitive radio system is a paradigm which improves the method of spectrum sharing among users. In this system, the allotted spectrum of primary users (PUs) is shared with the secondary users (SUs), leading to an increase in spectral efficiency [47]. These cognitive radio networks are required for delivering 6G services [48]. There are various challenges in cognitive radio systems for efficient primary and SU signal transmissions. In this respect, IRS are deployed in cognitive radio systems for ensuring secure and efficient signal transmission of a SU. The signals from the base station are received by the IRS and are reflected intelligently to the SU. On the contrary, the signals of a PU are destructively superimposed on the SU leading to lesser interference [49]. For spectrum sensing, IRS aids in boosting the PU's signal at SU, which facilitates signal detection [50]. In the MIMO cognitive system, IRS also helps in the signal transmission of SU by maximizing the achievable weighted sum rate (WSR) of SUs [51]. In terms of energy efficiency and spectral efficiency, IRS enhances the performance of the users in cognitive radio systems [52]. The authors in [53] have addressed the transmission rate of SU in the presence of strong interference from the PU by jointly optimizing the SU transmit power and IRS reflect beamforming.

The basic system model of the IRS-assisted cognitive radio system is given in Figure 3.3. It consists of the both primary and secondary networks. The primary network includes one primary base station and

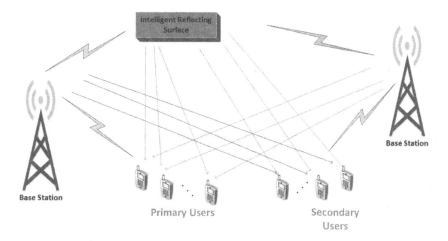

Figure 3.3 IRS-assisted cognitive radio system.

I primary receivers. The secondary network consists of one secondary base station and K SUs. The base stations are equipped with multiple antennas, while the users are single-antenna devices. The IRS is made up of M reflecting elements, which are programmable by an IRS controller. The received signals at PU i and SU k are given by [54]

$$y_i^P = s_i^P + \sum_{k \in K} \mathbf{I}_{D,i}{}^H \mathbf{w}_k d_k + \sum_{k \in K} \mathbf{I}_{R,i}{}^H \mathbf{\Theta} \mathbf{F} \mathbf{w}_k d_k + n_i^P, \quad (3.5)$$

$$y_k^S = \mathbf{g}_{D,k}{}^H \mathbf{w}_k d_k + \mathbf{g}_{R,k}{}^H \mathbf{w}_k d_k + \sum_{r \in K} \mathbf{g}_{D,k}{}^H \mathbf{w}_r d_r + \sum_{r \in K} \mathbf{g}_{R,k}{}^H \mathbf{w}_r d_r + n_k,$$

$$(3.6)$$

where s_i^P denotes the received signal from the primary transmitter. The notation d_k is the information signal for k SU with w_k as its corresponding beamformer. $\mathbf{I}_{D,i}$ represents the channel vector between i PU and the secondary BS. $\mathbf{I}_{R,i}$ denotes the channel vector between i PU and the IRS. $\mathbf{\Theta}$ represents the phase shift matrix of the IRS with $\mathbf{F}$ denotes the channel between the secondary BS and the IRS. $\mathbf{g}_{D,k}$ and $\mathbf{g}_{R,k}$ denote the channel vector between k SU and the secondary BS and the channel vector between k SU and the IRS, respectively. In (3.5), the second and third terms designate interference leakage from the secondary network. On the contrary, in (3.6), the first and second terms represent a desired signal of a SU with third and fourth terms denoting multi-user interference [54].

The incorporation of IRS in cognitive radio systems ensures improvement in system performance. The improvement in spectrum sensing is possible when interference from PUs are kept under control. Moreover, security has to be maintained in such multi-user environment where efficient resource sharing is the prime goal.

3.6   IRS-ASSISTED OPTICAL COMMUNICATION SYSTEMS

Free-space optical (FSO) communication is a potential candidate for 6G communications systems. It involves line-of-sight communication and operates over a license-free band with higher bandwidth and security than traditional radio frequency (RF) communication systems. However, the FSO channel is adversely affected by the channel impairments like atmospheric turbulence and pointing errors [55], [56]. Therefore, to leverage the advantages of RF and FSO, a cooperative communication setup known as a mixed RF-FSO system is proposed in the literature [57]. Further, with the aid of the IRS, the performance of such a setup can be enhanced by incorporating IRS on RF links. In [58], the performance of a mixed dual-hop IRS-assisted FSO-RF communication setup with hybrid automatic repeat request [H-ARQ] protocols on both FSO and RF links is studied. The effect of the number of reflecting elements, atmospheric turbulence, and H-ARQ rounds on the OP and packet error rate (PER) are presented. However, this work achieved a spectral efficiency of $1/2$ symbols per channel use. To improve the spectral efficiency, signal space diversity (SSD) based distributed IRS-assisted RF-FSO communication setup is proposed in [59]. The distributed IRS in the RF channel can be utilized in two configurations: exhaustive RIS-aided (ERA) scheme and opportunistic RIS-aided (ORA) scheme. In the case of the ERA scheme, all the IRS participate in the transmission, whereas in the ORA scheme, only the scheduled IRS participates in signal transmission. Further, it is shown that the diversity order of the SSD-based system is twice the traditional mixed RF-FSO systems. In [60], it is shown that the presence of the IRS-aided jammer severely degrades the performance of distributed IRS-based dual-hop mixed FSO RF systems. In [61], IRS is employed to study the secrecy performance of a mixed RF and underwater optical wireless communication (UOWC) link. The effect of fading parameters, number of reflecting elements, air bubbles levels, temperature gradients, and water salinity on the average secrecy capacity (ASC) and secrecy outage probability (SOP) is studied. Moreover, incorporating IRS in high-speed trains (HST) can reduce the number of free-space optical communication base station (FSO-BS) [62].

3.7 CHALLENGES AND FUTURE DIRECTIONS

IRS has opened the vision of exploiting metal surfaces for better communication. But the realization of IRS for 6G is still in its early stage. Various companies are in the race to develop IRS communication systems. The first industrial progress was made by NTT DOCOMO, which developed a 28 GHz 5G mobile communication system using meta-surface reflectarray in 2018 [63]. Another company, namely GreenerWare developed the physics-related algorithms of meta-surfaces. Regarding the university-related initiatives for IRS, various prototypes have been developed. In [64], the authors have developed the IRS-based prototype for MIMO transmission. In [65], a prototype of IRS comprising 256 two-bit elements has been developed. In addition to this, various internationally funded projects are enlisted in Table 3.1. Further, for the standardization of IRS, international collaborations are carried out by special interest groups (SIG) and emerging technology initiatives (ETI) [66]. Also, the FuTURE forum has been established to study the integration of IRS into next-generation wireless networks.

TABLE 3.1 IRS-Assisted Projects

Project	Objective
VisorSurf	Hardware setup for software-driven metasurface, 2017
ARIANDE	Artificial intelligence integrated meta-surface, 2019
META WIRELESS	Properties of meta-materials, 2020.
PathFinder	IRS-enabled wireless 2.0 wireless networks, 2021
RISE-6G	Standardization of IRS for industrial exploitation, 2021
Surfer	Surface wave-based indoor communications, 2022

BIBLIOGRAPHY

[1] Ruiqi Liu, Qingqing Wu, Marco Di Renzo, and Yifei Yuan. A path to smart radio environments: An industrial viewpoint on reconfigurable intelligent surfaces. *IEEE Wireless Communications*, 29(1):202–208, 2022.

[2] George C Alexandropoulos, Geoffroy Lerosey, Mérouane Debbah, and Mathias Fink. Reconfigurable intelligent surfaces and metamaterials: The potential of wave propagation control for 6G wireless communications. *arXiv preprint arXiv:2006.11136*, 2020.

[3] Ertugrul Basar, Marco Di Renzo, Julien De Rosny, Merouane Debbah, Mohamed-Slim Alouini, and Rui Zhang. Wireless communications through reconfigurable intelligent surfaces. *IEEE Access*, 7:116753–116773, 2019.

[4] Ertugrul Basar. Transmission through large intelligent surfaces: A new frontier in wireless communications. In *2019 European Conference on Networks and Communications (EuCNC)*, pages 112–117. IEEE, 2019.

[5] Georgios C Trichopoulos, Panagiotis Theofanopoulos, Bharath Kashyap, Aditya Shekhawat, Anuj Modi, Tawfik Osman, Sanjay Kumar, Anand Sengar, Arkajyoti Chang, and Ahmed Alkhateeb. Design and evaluation of reconfigurable intelligent surfaces in real-world environment. *IEEE Open Journal of the Communications Society*, 3:462–474, 2022.

[6] Emil Björnson, Henk Wymeersch, Bho Matthiesen, Petar Popovski, Luca Sanguinetti, and Elisabeth de Carvalho. Reconfigurable intelligent surfaces: A signal processing perspective with wireless applications. *IEEE Signal Processing Magazine*, 39(2):135–158, 2022.

[7] Ricardo Coelho Ferreira, Michelle SP Facina, Felipe AP De Figueiredo, Gustavo Fraidenraich, and Eduardo Rodrigues De Lima. Bit error probability for large intelligent surfaces under double-Nakagami fading channels. *IEEE Open Journal of the Communications Society*, 1:750–759, 2020.

[8] Monjed H Samuh and Anas M Salhab. Performance analysis of reconfigurable intelligent surfaces over Nakagami-m fading channels, 2020, arXiv.org.

[9] Hazem Ibrahim, Hina Tabassum, and Uyen T Nguyen. Exact coverage analysis of intelligent reflecting surfaces with Nakagami-m channels. *IEEE Transactions on Vehicular Technology*, 70(1):1072–1076, 2021.

[10] Trigui, Wessam Ajib, and Wei-Ping Zhu. A comprehensive study of reconfigurable intelligent surfaces in generalized fading. *arXiv preprint arXiv:2004.02922*, 2020.

[11] Chongwen Huang, Alessio Zappone, George C Alexandropoulos, Debbah, and Chau Yuen. Reconfigurable intelligent surfaces for energy efficiency in wireless communication. *IEEE Transactions on Wireless Communications*, 18(8):4157–4170, 2019.

[12] Beixiong Zheng and Rui Zhang. Simultaneous transmit diversity and passive beamforming with large-scale intelligent reflecting surface. *arXiv preprint arXiv:2202.04370*, 2022.

[13] Hongzhi Guo, Xiaoyi Zhou, Jiajia Liu, and Yanning Zhang. Vehicular intelligence in 6G: Networking, communications, and computing. *Vehicular Communications*, 33:100399, 2022.

[14] Abdulla Mahmoud, Sami Muhaidat, Paschalis C Sofotasios, Ibrahim Abualhaol, Octavia A Dobre, and Halim Yanikomeroglu. Intelligent reflecting surfaces assisted UAV communications for IoT networks: Performance analysis. *IEEE Transactions on Green Communications and Networking*, 5(3):1029–1040, 2021.

[15] Jingping Qiao and Mohamed-Slim Alouini. Secure transmission for intelligent reflecting surface-assisted mmWave and terahertz systems. *IEEE Wireless Communications Letters*, 9(10):1743–1747, 2020.

[16] Cunhua Pan, Hong Ren, Kezhi Wang, Maged Elkashlan, Arumugam Nallanathan, Jiangzhou Wang, and Lajos Hanzo. Intelligent reflecting surface aided MIMO broadcasting for simultaneous wireless information and power transfer. *IEEE Journal on Selected Areas in Communications*, 38(8):1719–1734, 2020.

[17] Sarah Basharat, Syed Ali Hassan, Aamir Mahmood, Zhiguo Ding, and Mikael Gidlund. Reconfigurable intelligent surface-assisted backscatter communication: A new frontier for enabling 6G IoT networks. *IEEE Wireless Communications*, 29(6):96–103, 2022.

[18] Carlos De Lima, Didier Belot, Rafael Berkvens, Andre Bourdoux, Davide Dardari, Maxime Guillaud, Minna Isomursu, Elena-Simona Lohan, Yang Miao, Andre Noll Barreto, et al. Convergent communication, sensing and localization in 6G systems: An overview of technologies, opportunities and challenges. *IEEE Access*, 9:26902–26925, 2021.

[19] Qingqing Wu and Rui Zhang. Towards smart and reconfigurable environment: Intelligent reflecting surface aided wireless network. *IEEE Communications Magazine*, 58(1):106–112, 2019.

[20] Yuanwei Liu, Wenqiang Yi, Zhiguo Ding, Xiao Liu, Octavia A Dobre, and Naofal Al-Dhahir. Developing NOMA to next generation multiple access (NGMA): Future vision and research opportunities. *IEEE Wireless Communications*, 29(6):120–127, 2022.

[21] SM Islam, Ming Zeng, Octavia A Dobre, and Kyung-Sup Kwak. Non-orthogonal multiple access (NOMA): How it meets 5G and beyond. *arXiv preprint arXiv:1907.10001*, 2019.

[22] Priyank Sharma, Atul Kumar, and Matadeen Bansal. Performance analysis of downlink NOMA over η–μ and κ–μ fading channels. *IET Communications*, 14(3):522–531, 2020.

[23] Shaika Mukhtar, Gh Begh, et al. Unified performance analysis of near and far user in downlink NOMA system over $\eta - \mu$ fading channel. *Journal of Communications Software and Systems*, 17(4):305–313, 2021.

[24] SM Riazul Islam, Nurilla Avazov, Octavia A Dobre, and Kyung-Sup Kwak. Power-domain non-orthogonal multiple access (NOMA) in 5G systems: Potentials and challenges. *IEEE Communications Surveys & Tutorials*, 19(2):721–742, 2016.

[25] Min Fu, Yong Zhou, Yuanming Shi, and Khaled B Letaief. Reconfigurable intelligent surface empowered downlink non-orthogonal multiple access. *IEEE Transactions on Communications*, 69(6):3802–3817, 2021.

[26] Zhiguo Ding and H Vincent Poor. A simple design of IRS-NOMA transmission. *IEEE Communications Letters*, 24(5):1119–1123, 2020.

[27] Yanyu Cheng, Kwok Hung Li, Yuanwei Liu, Kah Chan Teh, and H Vincent Poor. Downlink and uplink intelligent reflecting surface aided networks: NOMA and OMA. *IEEE Transactions on Wireless Communications*, 20(6):3988–4000, 2021.

[28] Lina Bariah, Sami Muhaidat, Paschalis C Sofotasios, Faissal El Bouanani, Octavia A Dobre, and Walaa Hamouda. Large intelligent surface assisted non-orthogonal multiple access: Performance analysis. *arXiv preprint arXiv:2007.09611*, 2020.

[29] Dongyeong Song, Wonjae Shin, and Jungwoo Lee. A maximum throughput design for wireless powered communication networks with IRS-NOMA. *IEEE Wireless Communications Letters*, 10(4):849–853, 2020.

[30] Jiakuo Zuo, Yuanwei Liu, Zhijin Qin, and Naofal Al-Dhahir. Resource allocation in intelligent reflecting surface assisted NOMA systems. *IEEE Transactions on Communications*, 68(11):7170–7183, 2020.

[31] Fang Fang, Yanqing Xu, Quoc-Viet Pham, and Zhiguo Ding. Energy-efficient design of IRS-NOMA networks. *IEEE Transactions on Vehicular Technology*, 69(11):14088–14092, 2020.

[32] Yiqing Li, Miao Jiang, Qi Zhang, and Jiayin Qin. Joint beamforming design in multi-cluster MISO NOMA intelligent reflecting surface-aided downlink communication networks. *arXiv preprint arXiv:1909.06972*, 2019.

[33] Xidong Mu, Yuanwei Liu, Li Guo, Jiaru Lin, and Naofal Al-Dhahir. Exploiting intelligent reflecting surfaces in multi-antenna aided NOMA systems. *arXiv preprint arXiv:1910.13636*, pages 0090–6778, 2019.

[34] Zeyu Sun and Yindi Jing. On the performance of multi-antenna IRS-assisted NOMA networks with continuous and discrete IRS phase shifting. *IEEE Transactions on Wireless Communications*, 21(5):3012–3023, 2021.

[35] Tianwei Hou, Yuanwei Liu, Zhengyu Song, Xin Sun, and Yue Chen. MIMO-NOMA networks relying on reconfigurable intelligent surface: A signal cancellation-based design. *IEEE Transactions on Communications*, 68(11):6932–6944, 2020.

[36] Gang Yang, Xinyue Xu, and Ying-Chang Liang. Intelligent reflecting surface assisted non-orthogonal multiple access. In *2020 IEEE wireless communications and networking conference (WCNC)*, pages 1–6. IEEE, 2020.

[37] Arash Asadi, Qing Wang, and Vincenzo Mancuso. A survey on device-to-device communication in cellular networks. *IEEE Communications Surveys & Tutorials*, 16(4):1801–1819, 2014.

[38] Oladayo Bello and Sherali Zeadally. Intelligent device-to-device communication in the internet of things. *IEEE Systems Journal*, 10(3):1172–1182, 2014.

[39] Shangwei Zhang, Jiajia Liu, Hongzhi Guo, Mingping Qi, and Nei Kato. Envisioning device-to-device communications in 6G. *IEEE Network*, 34(3):86–91, 2020.

[40] Udit Narayana Kar and Debarshi Kumar Sanyal. An overview of device-to-device communication in cellular networks. *ICT Express*, 4(4):203–208, 2018.

[41] Jindan Xu, Wei Xu, and A Lee Swindlehurst. Discrete phase shift design for practical large intelligent surface communication. In *2019 IEEE Pacific Rim Conference on Communications, Computers and Signal Processing (PACRIM)*, pages 1–5. IEEE, 2019.

[42] Yali Chen, Bo Ai, Hongliang Zhang, Yong Niu, Lingyang Song, Zhu Han, and H Vincent Poor. Reconfigurable intelligent surface assisted device-to-device communications. *IEEE Transactions on Wireless Communications*, 20(5):2792–2804, 2020.

[43] Chiya Zhang, Wenyu Chen, Chunlong He, and Xingquan Li. Throughput maximization for intelligent reflecting surface-aided device-to-device communications system. *Journal of Communications and Information Networks*, 5(4):403–410, 2020.

[44] Yashuai Cao, Tiejun Lv, Wei Ni, and Zhipeng Lin. Sum-rate maximization for multi-reconfigurable intelligent surface-assisted device-to-device communications. *IEEE Transactions on Communications*, 69(11):7283–7296, 2021.

[45] Gang Yang, Yating Liao, Ying-Chang Liang, and Olav Tirkkonen. Reconfigurable intelligent surface empowered underlaying

device-to-device communication. In *2021 IEEE Wireless Communications and Networking Conference (WCNC)*, pages 1–6. IEEE, 2021.

[46] Hongxia Zheng, Chiya Zhang, Yatao Yang, Xingquan Li, and Chunlong He. Data rate maximization in RIS-assisted D2D communication with transceiver hardware impairments. *Electronics*, 11(2):200, 2022.

[47] Faizan Qamar, Maraj Uddin Ahmed Siddiqui, MHD Nour Hindia, Rosilah Hassan, and Quang Ngoc Nguyen. Issues, challenges, and research trends in spectrum management: A comprehensive overview and new vision for designing 6G networks. *Electronics*, 9(9):1416, 2020.

[48] Muhammad Muzamil Aslam, Liping Du, Xiaoyan Zhang, Yueyun Chen, Zahoor Ahmed, and Bushra Qureshi. Sixth generation (6G) cognitive radio network (CRN) application, requirements, security issues, and key challenges. *Wireless Communications and Mobile Computing*, 2021.

[49] Xuewen Wu, Jingxiao Ma, Zhe Xing, Chenwei Gu, Xiaoping Xue, and Xin Zeng. Secure and energy efficient transmission for IRS-assisted cognitive radio networks. *IEEE Transactions on Cognitive Communications and Networking*, 8(1):170–185, 2021.

[50] Shaoe Lin, Beixiong Zheng, Fangjiong Chen, and Rui Zhang. Intelligent reflecting surface-aided spectrum sensing for cognitive radio. *IEEE Wireless Communications Letters*, 11(5):928–932, 2022.

[51] Lei Zhang, Yu Wang, Weige Tao, Ziyan Jia, Tiecheng Song, and Cunhua Pan. Intelligent reflecting surface aided MIMO cognitive radio systems. *IEEE Transactions on Vehicular Technology*, 69(10):11445–11457, 2020.

[52] Jie Yuan, Ying-Chang Liang, Jingon Joung, Gang Feng, and Erik G Larsson. Intelligent reflecting surface-assisted cognitive radio system. *IEEE Transactions on Communications*, 69(1):675–687, 2020.

[53] Xinrong Guan, Qingqing Wu, and Rui Zhang. Joint power control and passive beamforming in IRS-assisted spectrum sharing. *IEEE Communications Letters*, 24(7):1553–1557, 2020.

[54] Dongfang Xu, Xianghao Yu, and Robert Schober. Resource allocation for intelligent reflecting surface-assisted cognitive radio networks. In *2020 IEEE 21st International Workshop on Signal Processing Advances in Wireless Communications (SPAWC)*, pages 1–5. IEEE, 2020.

[55] Aman Sikri, Aashish Mathur, Manav Bhatnagar, Georges Kaddoum, Prakriti Saxena, and Jamel Nebhen. Artificial noise injection-based secrecy improvement for FSO systems. *IEEE Photonics Journal*, 13(2):1–12, 2021.

[56] Aman Sikri, Aashish Mathur, and Gyandeep Verma. Secrecy performance enhancement of artificial noise injection scheme-based FSO systems. In *2021 IEEE 94th Vehicular Technology Conference (VTC2021-Fall)*, pages 01–05. IEEE, 2021.

[57] Eunju Lee, Jaedon Park, Dongsoo Han, and Giwan Yoon. Performance analysis of the asymmetric dual-hop relay transmission with mixed RF/FSO links. *IEEE Photonics Technology Letters*, 23(21):1642–1644, 2011.

[58] Gyan Deep Verma, Aashish Mathur, Yun Ai, and Michael Cheffena. Mixed dual-hop IRS-assisted FSO-RF communication system with H-ARQ protocols. *IEEE Communications Letters*, 26(2):384–388, 2021.

[59] Aman Sikri, Aashish Mathur, and Georges Kaddoum. Signal space diversity-based distributed RIS-aided dual-hop mixed RF-FSO systems. *IEEE Communications Letters*, 26(5):1066–1070, 2022.

[60] Aman Sikri, Aashish Mathur, Gyandeep Verma, and Georges Kaddoum. Distributed RIS-based dual-hop mixed FSO-RF systems with RIS-aided jammer. In *2021 IEEE 94th Vehicular Technology Conference (VTC2021-Fall)*, pages 1–5. IEEE, 2021.

[61] T Hossain, Sarjana Shabab, ASM Badrudduza, Milton Kumar Kundu, and Imran Shafique Ansari. On the physical layer security performance over RIS-aided dual-hop RF-UOWC mixed network. *arXiv preprint arXiv:2112.06487*, 2021.

[62] Pouya Agheli, Hamzeh Beyranvand, and Mohammad Javad Emadi. High-speed trains access connectivity through RIS-assisted FSO communications. *arXiv preprint arXiv:2110.12804*, 2021.

[63] NTT DoCoMo. Metawave announce successful demonstration of 28 GHz-band 5G using world's first meta-structure technology, DoCoMo, NTT, 2018.

[64] Wankai Tang, Jun Yan Dai, Ming Zheng Chen, Kai-Kit Wong, Xiao Li, Xinsheng Zhao, Shi Jin, Qiang Cheng, and Tie Jun Cui. MIMO transmission through reconfigurable intelligent surface: System design, analysis, and implementation. *IEEE Journal on Selected Areas in Communications*, 38(11):2683–2699, 2020.

[65] Linglong Dai, Bichai Wang, Min Wang, Xue Yang, Jingbo Tan, Shuangkaisheng Bi, Shenheng Xu, Fan Yang, Zhi Chen, Marco Di Renzo, et al. Reconfigurable intelligent surface-based wireless communications: Antenna design, prototyping, and experimental results. *IEEE Access*, 8:45913–45923, 2020.

[66] Ruiqi Liu, Qingqing Wu, Marco Di Renzo, and Yifei Yuan. A path to smart radio environments: An industrial viewpoint on reconfigurable intelligent surfaces. *IEEE Wireless Communications*, 29(1):202–208, 2022.

Channel Modeling for 6G Programmable Wireless Environment

Petros Karadimas

Edinburgh Napier University, Edinburgh, UK

Md. Sakir Hossain

American International University-Bangladesh (AIUB), Dhaka, Bangladesh

Faisal Tariq

James Watt School of Engineering, University of Glasgow, Glasgow, UK

CONTENTS

4.1	Introduction	59
4.2	Reference Channel Model	61
4.3	Small-Scale Variations	64
	4.3.1 First-Order Statistical Characterization	65
	4.3.2 Second-Order Statistical Characterization	66
4.4	Path Loss and Large-Scale Variations	67
4.5	Summary	69
Bibliography		69
Notes		72

4.1 INTRODUCTION

In 6G wireless networks [1], the requirements for high data rates and low power consumption can be met by addressing the impact of wireless propagation. This can be achieved through diversity combining and multiple input multiple output (MIMO) techniques, and their performance

DOI: 10.1201/9781003282211-4

is strongly related to the propagation channel characteristics [2, 3]. The physical propagation mechanisms in wireless communication channels result in received signal variations [2 (Ch. 4), 3 (Ch. 4)]. Intelligent reflective surfaces (IRSs) have emerged as promising electromagnetic structures aiming to control wireless propagation in order to provide optimum performance such as maximum received signal-to-noise ratio (SNR) [4–6].

Channel modeling is a prerequisite to design wireless systems and components for IRS-assisted communications [6]. In [7], a wideband non-stationary channel model was proposed for IRS-assisted MIMO systems and analyzed in terms of temporal, spatial, and frequency correlation functions. A channel model was proposed in [8] for IRS-assisted un-manned aerial vehicles (UAV) communications and was analyzed using spatial correlation function. A channel model was derived in [9] for IRS-aided THz MIMO systems assuming spherical wave propagation to account for near-field communication scenarios. For one-dimensional IRS deployed in a two-dimensional space, two path loss models were derived in [10] using vector generalization of the Green's theorem [11]. Both path loss models are applicable for near-field and far-field transmissions. In [12], free-space path loss models were derived incorporating four characteristics of the transmission systems, namely distance between the IRS and transmitter/receiver, size of the IRS elements, radiation patterns of IRS elements and antennas, and near-field/far-field effects. The developed model was validated with experimental data. Isotropic radiation patterns of antennas and IRS units were assumed in [12]. The path loss models proposed in [12] were modified in [13], incorporating the directivity of transmitting antenna, receiving antenna, and IRS elements accounting for millimeter wave communications. In this chapter, a stochastic modeling approach characterizing variations in IRS-assisted communication channels is presented. Particularly, in such channels, wireless propagation channel modeling is essential to design optimum IRS topologies and transceiver systems to meet performance requirements and maintain complexity to an acceptable level [6, 14, 15].

Section 4.2 presents a reference model adaptable to IRS-assisted wireless communication channels. In Section 4.3, small-scale variations are characterized from that reference model. Distance-dependent path loss is characterized in Section 4.4. Finally, Section 4.5 summarizes the chapter.

4.2   REFERENCE CHANNEL MODEL

We consider IRS-assisted wireless communications as shown in Fig. 4.1. In such channels, wireless transmission takes place from the transmitter (Tx) to the receiver (Rx) via an IRS, i.e., from the Tx to IRS (Tx-IRS) and from the IRS to Rx (IRS-Rx) in a cascaded manner [4, 6]. Modeling can also incorporate a direct Tx-Rx channel if the link between the Tx and the Rx is unobstructed. We will start with a parameterized reference model for the Tx-IRS channel, which can be readily adapted to the IRS-Rx channel and to IRS-IRS channels in case of multi-hop transmissions via multiple IRSs [6, 16]. We will consider narrowband channel modeling as IRS-assisted wireless communications are predominantly orthogonal frequency division multiplexing (OFDM) based, leveraging the spatial characteristics of the wireless environment to create favorable propagation conditions [6, 14, 15]. Such modeling will be applicable for each OFDM subcarrier.

A number L of multipath components departs from the transmitting antenna and arrives at the nth reflecting element of the IRS via the physical mechanisms of wireless propagation channel, e.g., reflection, scattering, diffraction, etc. [2 (Ch. 4), 3 (Ch. 4)]. The IRS comprises N reflecting elements in total, i.e., $n \in \{1, 2, ..., N\}$. If $x(t)$ is the complex

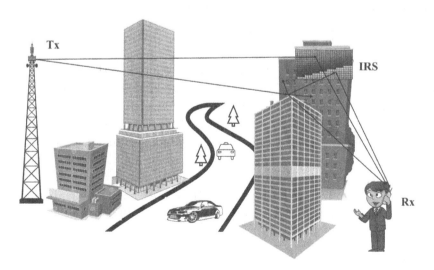

Figure 4.1   Illustration of IRS-assisted wireless communications.

baseband form of the transmitted signal, the complex baseband form of the received signal at the nth reflecting element of the IRS will be [17]

$$y_n(t) = \sum_{l=1}^{L} a_l e^{j2\pi v_l t} x(t - \tau_l) \tag{4.1}$$

where t is the time. The parameters for the lth multipath component are the following: a_l is the complex amplitude, τ_l is the time delay, and v_l is the Doppler frequency. In narrowband modeling, the delayed replicas of the transmitted signal are independent of their time delay τ_l, i.e., $x(t - \tau_l) \approx x(t)$ [18 (Ch. 6)]. Thus Eq. 4.1 becomes

$$y_n(t) = x(t). \sum_{l=1}^{L} a_l e^{j2\pi v_l t} \tag{4.2}$$

The Doppler frequency v_l results from the potential interaction of the lth multipath component with the existing mobile objects in wireless environment such as vehicular traffic, wind-blown trees/vegetation, and pedestrians [19]. If the lth multipath component interacts with a single mobile object, v_l will be derived by adding two Doppler frequency contributions, i.e., one contributed by the arrival to the mobile object (i.e., $(u_l/\lambda_o) cosa_{1,l}$) and another contributed by the departure from the mobile object (i.e., $(u_l/\lambda_o) cosa_{2,l}$). Thus [17],

$$v_l = v_{mob,l} = (u_l/\lambda_o) (cosa_{1,l} + cosa_{2,l}) \tag{4.3}$$

with λ_o being the carrier wavelength, u_l the velocity of the mobile object, $a_{1,l}$ the angle of arrival (AOA), and $a_{2,l}$ the angle of departure (AOD) with respect to the mobile object's direction of motion. We can similarly add Doppler contributions as in Eq. 4.3 when the same multipath component interacts with more than one mobile object. The notation $v_{mob,l}$ indicates Doppler frequencies induced by mobile objects such as vehicular traffic, wind-blown trees/vegetation, and pedestrians.

If the Tx and/or IRS are in motion, additional Doppler frequency contributions will be added as in a mobile-to-mobile wireless channel scenario. Thus [20],

$$v_l = v_{Tx,l} + v_{mob,l} + v_{IRS,l} \tag{4.4}$$

where $v_{Tx,l}$ and $v_{IRS,l}$ are the contributions due to Tx and IRS mobility, respectively. The situation of having an IRS in motion can arise when,

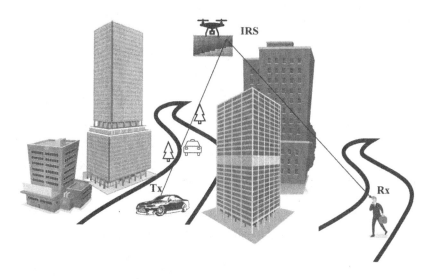

Figure 4.2   Illustration of mobile Tx and mobile IRS.

for example, the IRS is mounted on a UAV to provide over-the-air coverage [6, 9]. The Doppler frequency $v_{Tx(IRS),l}$ results from the departure (arrival) of the lth multipath component from the mobile Tx (to the mobile IRS), as illustrated in Fig. 4.2. It is defined as [18 (Ch. 1)]

$$v_{Tx(IRS),l} = \left(u_{Tx(IRS)}/\lambda_o \right) cos\alpha_{Tx(IRS),l} \qquad (4.5)$$

where $u_{Tx(IRS)}$ is the Tx(IRS) velocity and $\alpha_{Tx(IRS),l}$ is the AOD (AOA) with respect to the Tx(IRS) direction of motion.

We see from Eq. 4.2 that in narrowband modeling, the received signal is a weighted version of the transmitted signal. The complex amplitude a_l accounts for the electromagnetic field (plane wave) properties of the lth multipath component. The term $e^{j2\pi v_l t}$ characterizes phase variations due to temporal variations induced by the Tx and/or IRS mobility and/or mobility in the wireless propagation environment, i.e., vehicular traffic, wind-blown trees/vegetation, and pedestrians.

The input-output relation in Eq. 4.2 can be written in an equivalent integral form as follows [17]:[1]

$$y_n(t) = x(t) \int e^{j2\pi vt} H(v) dv \qquad (4.6)$$

where $H(v)$ is the Doppler variant channel response defined as

$$H(v) = \sum_{l=1}^{L} a_l \delta(v - v_l) \qquad (4.7)$$

with $\delta(.)$ being the Dirac delta function. Equation 4.6 can be written in a more compact form as follows:

$$y_n(t) = x(t)h(t) \qquad (4.8)$$

where $h(t)$ is the time variant channel response obtained by the inverse Fourier transform of $H(v)$ as

$$h(t) = \int H(v)e^{j2\pi vt} dv \qquad (4.9)$$

By substituting Eq. 4.7 in Eq. 4.9, we obtain the equivalent form for Eq. 4.9 based on a summation

$$h(t) = \sum_{l=1}^{L} a_l e^{j2\pi v_l t} \qquad (4.10)$$

The wireless channel responses in Eqs. 4.7 and 4.10, $H(v)$ and $h(t)$, respectively, associate time and Doppler dependencies being Fourier transform pairs according to Eq. 4.9, i.e., $h(t) \xleftrightarrow{F} H(t)$. Following exactly the same rationale, we can derive similar formulas for the responses of IRS-Rx and direct Tx-Rx channels. The received signal variations represented by Eq. 4.6 or Eq. 4.8 occur within a local area, i.e., variations in spaces with sizes of few carrier wavelengths. Those variations are characterized as small-scale variations or small-scale fading [3 (Ch. 4)]. This practically means that an IRS will comprise many adjacent local areas according to its size and carrier wavelength, where each local area will be subject to its own small-scale fading characteristics. If the transmitted signal is a single sinusoidal function as per the standard assumption in narrowband channel modeling, the complex baseband form of the received signal will be proportional to the channel response $h(t)$, as demonstrated in Eq. 4.8. Hence, in narrowband channel modeling, focus is given on modeling and characterizing $h(t)$.

4.3 SMALL-SCALE VARIATIONS

The channel response represented in Eq. 4.10 is a random time-varying complex process, i.e., a complex stochastic process due to the underlying phase variations within a local area [3 (Ch. 4)]. Thus, statistical tools should be employed to characterize the random nature of $h(t)$. The two universally accepted types of characterization are the first-order and the second-order statistical characterization. The first-order characterization arises when only one sample in Eq. 4.10 with respect to time is used to characterize channel behavior. The second-order characterization uses two samples in Eq. 4.10 with respect to time.

4.3.1 First-Order Statistical Characterization

Using Eq. 4.10, we can write the Tx-n^{th} reflecting element of the IRS channel response as a complex random variable of the form

$$h_{Tx-IRS,n} = \sum_{l=1}^{L} |a_l| e^{j\phi_n} = A_n e^{j\phi_n} \tag{4.11}$$

where $|a_l|$ is the amplitude of a_l and ϕ_l sums all the phase terms in Eq. 4.10, including that of a_l. Following exactly the same steps for deriving Eq. 4.11, we can similarly write the nth reflecting element of the IRS-Rx and direct Tx-Rx channel responses as

$$h_{IRS-Rx,n} = B_n e^{j\theta_n} \tag{4.12}$$

$$h_{Tx-Rx} = C_d e^{j\psi_d} \tag{4.13}$$

The role of the IRS is to induce phase shifts in order; the received signal components are to be added coherently at the Rx, i.e., to be aligned in phase [4, 6]. Under the existence of a direct Tx-Rx channel, the IRS will align in phase the compound Tx-IRS-Rx and the direct Tx-Rx channels [6]. Such operation maximizes the received SNR and the total channel response will be [6]:

$$h_{total} = C_d e^{j\psi_d} + \sum_{i=1}^{N} A_n e^{j\phi_n} B_n e^{j\theta_n} C_n e^{j\psi_n} \tag{4.14}$$

where $C_n \in [0,1]$ and $\psi_n \in [0, 2\pi]$ are the amplitude attenuation and the adjustable phase induced by the nth reflecting element of the IRS, respectively.

In the case of single-antenna Tx and Rx, we have $C_n = 1$ and $\psi_n = mod\,[\psi_d - (\phi_n + \theta_n)\,, 2\pi]$ for optimum performance, i.e., for maximum received SNR [6]. If there is no direct Tx-Rx channel, e.g., if it is severely obstructed, and communication takes place only through the IRS, the solution ψ_n will simply be $\psi_n = mod\left[-(\phi_n + \theta_n), 2\pi\right]$ [4, 6]. With the above definitions for C_n and ψ_n, the maximized SNR can be formulated as [4, 6]

$$\gamma = \left[C_d + \sum_{l=1}^{N} A_n B_n\right]^2 \frac{E_s}{N_o} \tag{4.15}$$

where E_s and N_o are the average transmitted energy per symbol and noise power density, respectively. Considering a sufficiently large number of IRS elements, i.e., $N >> 1$, the quantity $\left[C_d + \sum_{i=1}^{N} A_n B_n\right]$ will converge to a Gaussian-distributed random variable according to the central limit theorem (CLT) of Lyapunov [18 (Ch. 2)], i.e., $\left[C_d + \sum_{i=1}^{N} A_n B_n\right] \simeq N(\mu, \sigma)$. Thus, the received signal strength and eventually the total channel response amplitude will be a Gaussian distributed random variable, whereas the received SNR and eventually the received signal power will be a non-central chi-square random variable (see also [4]). In multi-antenna Tx and/or Rx scenarios, finding C_n and ψ_n for optimum performance, i.e., for maximum channel capacity is not so straightforward as in single-antenna settings. Various state-of-the-art methodologies were comprehensively discussed in [6].

4.3.2 Second-Order Statistical Characterization

The second-order characterization considers two samples of the channel response with respect to time. As mentioned previously, [8] and [11] can represent any of the Tx-IRS, IRS-Rx, and direct Tx-Rx channels by adjusting its parameters accordingly. Thus, the second-order characterization presented here will be valid for each such channel. Of particular importance is the temporal correlation function defined by taking the following expectation of the channel response in Eq. 4.10

$$R(t_1; t_2) = E\left[h^*(t_1) h(t_2)\right] \tag{4.16}$$

A common assumption is to consider the channel response as wide sense stationary (WSS) with respect to time. That assumption makes the ACF in Eq. 4.16 depend exclusively on the difference among the samples, i.e., $R(t_1; t_2) = R(\Delta t)$, where $\Delta t = t_2 - t_1$. Using Eq. 4.10 in Eq.4.16, we have [3 (Ch. 4), 17]

$$R(\Delta t) = \sum_{l=1}^{L} |a_l|^2 e^{j2\pi v_l \Delta t} \tag{4.17}$$

The ACF in Eq. 4.17 characterizes channel's temporal selectivity.

WSS further implies the following expectation formula when considering two samples of the channel response in Eq. 4.7, i.e., [3 (Ch. 3), 17]

$$E\left[H^*\left(v_1\right)H\left(v_2\right)\right] = P(v_2)\delta(v_2 - v_1) \qquad (4.18)$$

where the function $P(v)$ is the power spectral density (PSD) or Doppler spectrum describing how the power is distributed in the Doppler domain. The formal definition of the PSD is [3 (Ch. 4), 17]

$$P(v) = \sum_{l=1}^{L} |a_l|^2 \delta(v - v_l) \qquad (4.19)$$

The PSD in Eq. 4.19 characterizes channel's Doppler dispersion.

The formula in Eq. 4.18 simply states that the spectral components of responses $H(v_2)$ and $H(v_1)$ are uncorrelated; thus, the terminology of uncorrelated scattering (US) emerges and accordingly, the well-known terminology of WSSUS arises. This can be traced back in the classical paper of Bello where WSS channel models with respect to frequency and time were analyzed [21]. Here, we just focus on time variability as we have adopted narrowband IRS-assisted communication channel modeling. A last important property arises by a simple inspection of Eqs. 4.17 and 4.19, from which we can see that the ACF and PSD are Fourier transform pairs, i.e., $R\left(\Delta t\right) \overset{\text{F}}{\longleftrightarrow} P(v)$. This property is widely known in the literature as the Wiener–Khintchine theorem [3 (Ch. 3), 18 (Ch. 2)]. Thus, we get

$$P(v) = \int R(\Delta t)e^{-j2\pi v\Delta t}d\Delta t \qquad (4.20)$$

Equations 4.17, 4.19, and 4.20 offer a complete second-order statistical characterization for Tx-IRS, IRS-Rx, and Tx-Rx channels. The importance of modeling and compensating the detrimental effect of Doppler dispersion has recently started attracting the interest of the research community [22]. Existing techniques for PSD and Doppler dispersion modeling can be adapted from the published literature, e.g., see [19] when Doppler dispersion is only due to mobile objects in the wireless environment, and see [20] for Doppler dispersion in mobile-to-mobile scenarios.

4.4 PATH LOSS AND LARGE-SCALE VARIATIONS

Wireless communication channels exhibit deterministic variations in the received power due to the distance separating the Tx and Rx. The underlying effect is called path loss, and it results in a monotonic decrease in received power with respect to distance. In the case of wireless communications via an IRS, i.e., in Tx-IRS-Rx scenarios, the transmitted and received power, P_{T_x} and P_{R_x}, respectively, are related as [5, 12, 23].

$$P_{Rx} \propto \frac{P_{Tx}}{\left(d_{Tx-IRS}d_{IRS-Rx}\right)^2} \quad (4.21)$$

where d_{Tx-IRS} and d_{IRS-Rx} are the distances from the Tx to the center of the IRS, and from the center of the IRS to the Rx, respectively. The path loss is derived from Eq. 4.21 as

$$PL \propto \left(d_{Tx-IRS}d_{IRS-Rx}\right)^2 \quad (4.22)$$

Figure 4.3 shows the received power variation for communications via a single IRS (Eq. 4.21). The parameters given in Table 4.1 are extracted from [12] for X-band horn antennas at Tx and Rx. It is evident that the farther to the Tx (or Rx) the IRS is, the lower is the impact of the distance to the received power.

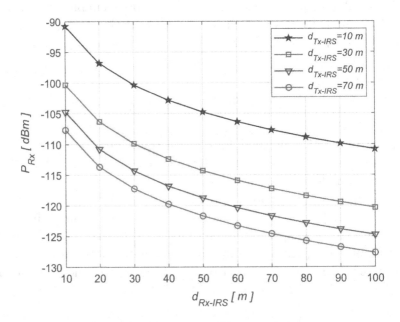

Figure 4.3 Received power variation for wireless communications via a single IRS.

TABLE 4.1   Parameters Setting

Parameters	Numerical Value
Tx and Rx antenna diameter	0.188 m
Tx and Rx antenna gain	14.5 dB
Frequency of operation	4.25 GHz
Transmitted power	0 dBm
Dimensions of IRS elements	0.012×0.012 m^2

4.5 SUMMARY

A reference narrowband stochastic model considering temporal small-scale variations was deployed for characterizing IRS-assisted wireless propagation channels. The reference model was processed to account for first- and second-order statistical characterization. Finally, path loss was modeled as a deterministic variation that IRS-assisted wireless propagation channels exhibit.

BIBLIOGRAPHY

[1] Faisal Tariq, Muhammad R. A. Khandaker, Kai-Kit Wong, Muhammad A. Imran, Mehdi Bennis, and Merouane Debbah. A speculative study on 6G. *IEEE Wireless Communications*, 27(4):118–125, 2020.

[2] Andreas F Molisch. *Wireless communications*. John Wiley & Sons, 2012.

[3] Gregory David Durgin. *Space-time wireless channels*. Prentice Hall Professional, 2003.

[4] Ertugrul Basar, Marco Di Renzo, Julien De Rosny, Merouane Debbah, Mohamed-Slim Alouini, and Rui Zhang. Wireless communications through reconfigurable intelligent surfaces. *IEEE Access*, 7:116753–116773, 2019.

[5] Shimin Gong, Xiao Lu, Dinh Thai Hoang, Dusit Niyato, Lei Shu, Dong In Kim, and Ying-Chang Liang. Toward smart wireless communications via intelligent reflecting surfaces: A contemporary survey. *IEEE Communications Surveys & Tutorials*, 22(4):2283–2314, 2020.

[6] Qingqing Wu, Shuowen Zhang, Beixiong Zheng, Changsheng You, and Rui Zhang. Intelligent reflecting surface-aided wireless communications: A tutorial. *IEEE Transactions on Communications*, 69(5):3313–3351, 2021.

[7] Hao Jiang, Chengyao Ruan, Zaichen Zhang, Jian Dang, Liang Wu, Mithun Mukherjee, and Daniel Benevides da Costa. A general wideband non-stationary stochastic channel model for intelligent reflecting surface-assisted mimo communications. *IEEE Transactions on Wireless Communications*, 20(8):5314–5328, 2021.

[8] Hao Jiang, Ruisi He, Chengyao Ruan, Jie Zhou, and Daina Chang. Three-dimensional geometry-based stochastic channel modeling for intelligent reflecting surface-assisted uav mimo communications. *IEEE Wireless Communications Letters*, 10(12):2727–2731, 2021.

[9] Konstantinos Dovelos, Stylianos D Assimonis, Hien Quoc Ngo, Boris Bellalta, and Michail Matthaiou. Intelligent reflecting surfaces at terahertz bands: Channel modeling and analysis. In *2021 IEEE International Conference on Communications Workshops (ICC Workshops)*, pages 1–6. IEEE, 2021.

[10] Fadil H Danufane, Marco Di Renzo, Julien De Rosny, and Sergei Tretyakov. On the path-loss of reconfigurable intelligent surfaces: An approach based on green's theorem applied to vector fields. *IEEE Transactions on Communications*, 69(8):5573–5592, 2021.

[11] George Green. *An essay on the application of mathematical analysis to the theories of electricity and magnetism*, volume 3. author, 1889.

[12] Wankai Tang, Ming Zheng Chen, Xiangyu Chen, Jun Yan Dai, Yu Han, Marco Di Renzo, Yong Zeng, Shi Jin, Qiang Cheng, and Tie Jun Cui. Wireless communications with reconfigurable intelligent surface: Path loss modeling and experimental measurement. *IEEE Transactions on Wireless Communications*, 20(1):421–439, 2020.

[13] Wankai Tang, Xiangyu Chen, Ming Zheng Chen, Jun Yan Dai, Yu Han, Marco Di Renzo, Shi Jin, Qiang Cheng, and Tie Jun Cui. Path loss modeling and measurements for reconfigurable intelligent surfaces in the millimeter-wave frequency band. *IEEE Transactions on Communications*, 70(9):6259–6276, 2022.

[14] Peilan Wang, Jun Fang, Huiping Duan, and Hongbin Li. Compressed channel estimation for intelligent reflecting surface-assisted millimeter wave systems. *IEEE Signal Processing Letters*, 27:905–909, 2020.

[15] Beixiong Zheng, Changsheng You, and Rui Zhang. Intelligent reflecting surface assisted multi-user OFDMA: Channel estimation and training design. *IEEE Transactions on Wireless Communications*, 19(12):8315–8329, 2020.

[16] Weidong Mei and Rui Zhang. Cooperative beam routing for multi-IRS aided communication. *IEEE Wireless Communications Letters*, 10(2):426–430, 2020.

[17] Petros Karadimas. Outdoor channels. *LTE-advanced and next generation wireless networks: Channel modelling and propagation. John Wiley & Sons*, 97–122, 2012.

[18] Matthias Pätzold *Mobile radio channels*. John Wiley & Sons, 2011.

[19] Petros Karadimas, Efstathios D Vagenas, and Stavros A Kotsopoulos. On the scatterers' mobility and second order statistics of narrowband fixed outdoor wireless channels. *IEEE Transactions on Wireless Communications*, 9(7):2119–2124, 2010.

[20] Petros Karadimas and David Matolak. Generic stochastic modeling of vehicle-to-vehicle wireless channels. *Vehicular Communications*, 1(4):153–167, 2014.

[21] Philip Bello. Characterization of randomly time-variant linear channels. *IEEE transactions on Communications Systems*, 11(4):360–393, 1963.

[22] Bho Matthiesen, Emil Björnson, Elisabeth De Carvalho, and Petar Popovski. Intelligent reflecting surface operation under predictable receiver mobility: A continuous time propagation model. *IEEE Wireless Communications Letters*, 10(2):216–220, 2020.

[23] Shih-Kai Chou, Okan Yurduseven, Hien Quoc Ngo, and Michail Matthaiou. On the aperture efficiency of intelligent reflecting surfaces. *IEEE Wireless Communications Letters*, 10(3):599–603, 2020.

NOTE

[1]It is implied that the integral limits, when not given, cover the entire range of integration of the integrated variables.

CHAPTER 5

Wireless Localization with Reconfigurable Intelligent Surfaces

Omar Rinchi

Missouri University of Science and Technology Rolla, MO, USA

Ahmed Elzanaty

University of Surrey, Guildford, United Kingdom

Ahmad Alsharoa

Missouri University of Science and Technology Rolla, MO, USA

CONTENTS

5.1	Introduction	74
	5.1.1 Wireless Localization Techniques	75
	5.1.2 Challenges in Wireless Localization	76
	5.1.3 Reconfigurable Intelligent Surface	78
	5.1.3.1 RIS Prototyping Overview	80
	5.1.3.2 RIS-Aided Localization Error Bounds	81
	5.1.3.3 RIS-Aided Localization Algorithms	85
	5.1.4 Notation	87
5.2	System Model	87
	5.2.1 Channel Model	87
	5.2.2 Steering Response for Far-Field and Near-Field	89
5.3	Problem Formulation and Solution	91
	5.3.1 RIS-Aided Localization Using Compressed Sensing (CS)	91
	5.3.1.1 Far-Field Localization Using Compressed Sensing	92
	5.3.1.2 Near-Field Localization Using Compressed Sensing	93

5.3.2   RIS-Aided Localization Using Atomic Norm
         Minimization   97
5.3.3   RIS Phase Design 101
5.4   Simulation Results 104
5.5   Conclusion and Future Work 110
Bibliography 111
Notes 115

I N THIS CHAPTER, we consider the design of localization algorithms for reconfigurable intelligent surface (RIS)-aided models under different practical channel model settings. More specifically, we utilize the compressed sensing (CS) to localize a user equipment (UE) direction and position in both far-field and near-field multipath environment, respectively; we extend our work by performing a super resolution localization using the atomic norm minimization for a user located in a single and path near-field channel. On the other hand, we propose RIS phase design that aims to minimize the localization error by maximizing the signal-to-noise ratio (SNR).

5.1   INTRODUCTION

Localization is one crucial challenge that has faced humankind for centuries. Nevertheless, localization has seen massive development in the last two decades. Different localization technologies can be used depending on the problem and the situation. For instance, the global navigation satellite system (GNSS) is the traditional current localization technology. However, GNSS suffers from some drawbacks such as line-of-sight (LOS) losses, multipath fading, and synchronization between the transmitter (Tx) and the receiver (Rx). In addition to that, GNSS signals can also be jammed, spoofed, and can't be used indoors. On the other hand, the dead reckoning (DR) localization techniques such as the inertial navigation system (INS) [1] consist of totally internal sensor-based tools that are independent of the outside wireless satellite signals, which solve the typical GNSS blockage and fading problems. However, the INS and other DR localization techniques accumulate the error with time, resulting in huge localization errors that limit their usage.

Localization using wireless systems is attracting booming interest in the last couple of years due to the recent developments in the capabilities

TABLE 5.1 Localization Techniques Comparison

	Accuracy	Robustness	Applications
GNSS	Standard GNSS can achieve sub-meter level of localization accuracy	Can't be used indoors, can be jammed, spoofed, and blocked	For typical navigation purposes
DR	Error accumulate with time resulting very high localization error	Very robust and totally independent of the outside wireless signals	For typical navigation purposes
Wireless Localization	Future wireless systems can achieve sub-cm level of accuracy	Can be used indoors, but can be jammed, spoofed, blocked and can be affected by multipath propagation and weather conditions	For typical navigation purposes, to enhance wireless navigation, extended reality, industrial internet of things (IoT), for more refer to reference [2]

of wireless networks. For instance, the next 5G+ and 6G networks aim to consider high-frequency bands (i.e., millimeter-wave (mm-wave)), wider bandwidths, multiple antennas (i.e., massive multiple-input multiple-output (MIMO)), low latency, and flexible architecture. In Table 5.1, we compare the typical localization techniques.

5.1.1 Wireless Localization Techniques

Different types of measurements can be considered for wireless localization, e.g., time-of-arrival (ToA) of the radio signal, uplink time difference-of-arrival (UTDoA), time difference-of-arrival (TDoA), phase difference-of-arrival (PDoA), angle of arrival (AoA), angle of departure (AoD), and received signal strength (RSS). Another localization method is proximity, where a rough location of the user equipment (UE) is determined by the location of some reference entities (e.g., base stations (BSs) and anchors) in the proximity of the users. For instance, cellular ID (CID) and enhanced-CID (E-CID) techniques fall within this category.

The RSS algorithm is based on estimating the received signal power and computing the distance as a function of the signal power. Taking the distances from multiple anchors can be used to develop a geometric-based solution that is based on the intersection of multiple spheres, each with a known radius and center. The RSS can also be used in fingerprint localization that requires preoffline measurements and a static system environment. On the other hand, AoA-based localization is a common approach for localization due to its high accuracy and reliability. The MUltiple SIgnal Classification (MUSIC) algorithm is a common choice to estimate the AoA. The algorithm is based on estimating the correlation matrix to estimate the ToA given that the angles are known. Although MUSIC is known as a high-accuracy algorithm, it requires multiple snapshots of the received signal to localize the user which makes it not suitable for a dynamic environment. The estimation of signal parameters by rotational invariance techniques (ESPRIT) is a similar algorithm to the MUSIC. ESPRIT requires multiple snapshots of the received signal, but the difference from that it has lower complexity with lower accuracy. Finding the sparse solution associated with localization parameters can be done using the compressed sensing (CS) algorithm. However, this algorithm suffers from quantization errors that limit its accuracy. A comparison between the commonly used AoA techniques is shown in Table 5.2.

5.1.2   Challenges in Wireless Localization

Motivated by high-accuracy localization specifications as stated by the 3rd generation partnership project (3GPP) technical reports [3], and also the future 5G+ communication services such as ultra-reliable low-latency communications (URLLC), enhanced mobile broadband (eMBB), and the massive machine type communications (mMTC), it is critical now to develop super-resolution localization algorithms that will enhance the future integrated sensing, localization, and communication (ISLAC) systems [4]. However, with high frequencies, new challenges arise such as high propagation loss, sensitivity to blockage of LOS, atmospheric attenuation, and diffraction loss [5], see Figure 5.1. Wireless localization and communication under high frequencies can lead to a high free

TABLE 5.2 Summary of Well-Known Angle of Arrival (AoA) Estimation-Based Localization Algorithms

	Advantages	Disadvantages
MUSIC	High-accuracy estimation	Requires multiple snapshots
ESPIR	High-accuracy estimation	Requires multiple snapshots
CS	Can exploit the sparsity of the signal. Single Snapshot Solution	Quantization error
Atomic Norm Minimization	High-accuracy estimation	Higher computational complexity

space attenuation, and the problem can be signified under the non-line of Sight (NLOS) paths. The current existing solutions to this problem such as beam-forming, multiple access points or repeaters, high transmitting power, or high-sensitivity receivers are all subjected to regulations, and economical and available equipment constraints make these solutions unreliable for future ISLAC systems. As a solution to these challenges, reconfigurable intelligent surfaces (RISs) have been introduced as a promising energy-efficient solution to solve high-shadowed LOS communications under high-frequency bands.

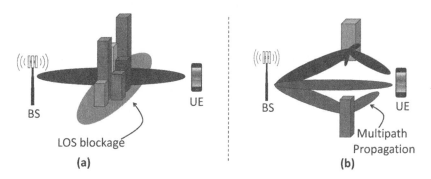

Figure 5.1 Typical challenges in wireless communications: (a) blockage of LOS; (b) multipath fading.

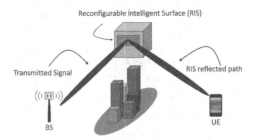

Figure 5.2 Wireless localization using reconfigurable intelligent surface.

5.1.3 Reconfigurable Intelligent Surface

The RIS is a low-cost and low-power meta-material surface that can steer the received signal toward a target direction [6], see Figure 5.2. With careful RIS optimization, the high-shadowed LOS links can be enhanced by creating strong non-direct LOS links between the transmitters and receivers through the RIS. Using RIS can enhance the achievable data rate and latency [7]. Knowing the location of the receiver is a key factor in RIS-aided networks, where the RIS can reflect the transmitted signal toward the specified receiver. Therefore, efficient localization algorithms can enhance the overall network performance.

Although most of the works introduce the new generations of wireless networks specifically the 5th-generation cellular network (5G) and beyond as enabling technologies for localization, the opposite can also be true. In other words, localization can enhance the next generation of wireless networks. This dual relationship can meet together by utilizing RIS-based communication and localization. Figure 5.3 plots the data

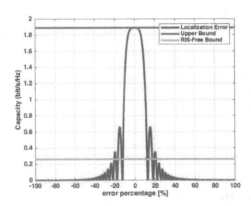

Figure 5.3 The effect of the estimation error of the localization on the channel capacity.

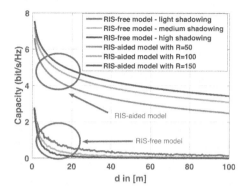

Figure 5.4 Throughput for different LOS shadowing and reflected RIS path cases.

rate throughput as a function of localization error operated at 28 GHz. The "upper bound" represents the data rate capacity using RIS with perfect localization knowledge, while the "RIS-free bound" represents the LOS link without RIS. It shows that the data capacity is a function of the localization error. For example, with a 10% localization error, the data rate can be reduced by around 50%. Therefore, localization is not only needed for navigation purposes but also will play a significant factor in communications and especially for high-frequency bands.

Figure 5.4 investigates the impact of the localization on wireless communications operated at high-frequency (mm-wave, a comparison of the achievable throughput for the case of a LOS that is blocked by obstacles without the existence of the RIS (RIS-free model) with the achievable throughput with the existence of the RIS case (RIS-aided model). The RIS-free path has been modeled as a shadowing fading channel with a log-normal random variable that describes the shadowing effect and a path loss exponent. Further, an analysis of the effect of different shadowing severity parameters on the capacity is also shown in Figure 5.4. The curves at the bottom of Figure 5.4 represent the capacity as a function of the distance for different shadowing severity cases. The curves on the top of the figure represent the capacity as a function of the distance for different numbers of RIS elements. It can be seen that the blocked paths have significantly lower performance compared to RIS-aided models.

In Figure 5.5, a comparison of the total power consumption for both RIS-free and RIS-aided models is plotted as a function of the transmit power. For the RIS-aided model, we assume that the transmitter consumes 10 W and the RIS consumes 1 W (a total of 11 W), and the

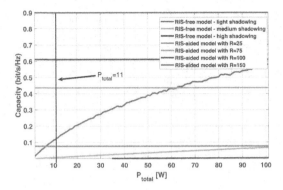

Figure 5.5 Power comparison between the RIS-model and free-RIS model for $d = 20$ m.

distance between the transmitter and receiver is equal to 20 m. The horizontal lines are the capacities for the RIS-aided model for various RIS elements, all with the same total power consumption of 11 W. While for the RIS-free, the transmit power has been varied to see when this model can achieve similar performance to the RIS-aided model. This result shows the superiority of the optimized RIS model over the RIS-free model. For example, the RIS-free model needs to use a total power of 60 W to achieve the same performance as the RIS-aided model achieved using only 11 W with 75 elements. In [6], it has been shown that with careful optimization, RIS can be self-powered or partially self-powered by integrating the RISs with renewable energy harvesting circuits.

5.1.3.1 RIS Prototyping Overview

To demonstrate the capability of localization algorithms to improve high-frequency communication, reflecting surfaces need to operate at the corresponding high frequency. In [8], the authors explain how patch array antenna can be used as reflectors. The concept of a cavity resonator is used to explain the oscillation of the incident waves. The oscillation occurs when the incident wave that has a frequency equal to the resonant frequency of the cavity resonator gets coupled into the cavity. The angle of reflection and incidence remain the same if all the properties of the elements are the same. For a different angle of reflection, the phase responses of the individual elements should change. To change the phase response of the element, the resonant frequency needs to be changed.

This could be done in many different ways. The simplest way is to change the reflection coefficient, where the PIN diode can be used as a switch to control the load impedance by adjusting the effective capacitance.

The authors in [9] propose an analytical model of transmission line circuit to represent and study RIS. In this work, the combination of a varactor's internal circuit is used to represent the circuit model of RIS. The floquet mode analysis of the unit cell is done to obtain the reflection phase for varying capacitance. In [10], the RIS design in the frequency range of 12 GHz to 18 GHz was proposed. The unit cell achieved a dynamic range of 300^0, and a 60^0 beam steering range was achieved by the unit cell. A tunable reflecting surface operating at 5.8 GHz was designed, and the bias voltage required for the varactor was controlled using a field programmable gate array (FPGA) [11]. The experimental results showed a power consumption of 1 W for the RIS. A metasurface in a frequency band from 12 to 18 GHz in the far-field was proposed via simulations only in [12]. The varactor diode is placed at the backside of the reflector to eliminate the absorption and scattering loss in the unit cell. The operating frequency of the reflective surfaces in all the works mentioned above is below 20 GHz. The challenge in using the resonating unit cell to design the RIS above 20 GHz comes from the difficulty in reducing the internal capacitance and inductance offered by each unit cell.

However, using proper structuring of the unit cells along with passive tunable elements, the reflectarray can be made to resonate in higher frequencies. The operating frequency of 5G is divided into two subfrequencies: frequency range 1 (FR1), which is less than 6 GHz, and frequency range (FR2) which has an operating range between 24 GHz and 47 GHz. Ka-band, ranging from 26.5 GHz to 40 GHz, has received the most attention for high-frequency bands for 5G. Since the Ka-band covers promising frequencies operable in the 5G network, the RIS must also be designed and operated in that frequency range.

5.1.3.2 *RIS-Aided Localization Error Bounds*

One of the key challenges for RIS-aided localization is to study the ultimate performance that any localization algorithm can reach in the presence of the RIS. The motivations that drive the authors to derive these errors are (i) to provide a benchmark to compare the accuracy of their proposed localization algorithms with the ultimate bounds; (ii) to support the current literature with the error bounds that anyone can use to judge the efficiency of his work with the RIS; and (iii) to support their theory with practical insights about the optimal or most suitable problem geometry configurations and phase design control that can be used to achieve accurate localization. Typically, these performance limits are expressed using the error bounds that can be derived using Cramér-Rao lower bound (CRLB). The CRLB can be computed by taking the inverse of the Fisher information matrix (FIM) as

$$\Lambda(\mathbf{p}^{\mathrm{U}}) \triangleq \left[\sum_{i=1}^{I} \mathbf{I}_i(\mathbf{p}^{\mathrm{U}}) \right]^{-1}, \tag{5.1}$$

where $\mathbf{I}_i(\mathbf{p}^{\mathrm{U}})$ is the FIM with the ith subcarrier for the user located at $\mathbf{p}^{\mathrm{U}}$. As a result, the position error bound of the user's location can be given as

$$\mathrm{PEB} = \sqrt{\mathrm{tr}\left(\Lambda(\mathbf{p}^{\mathrm{U}})\right)}. \tag{5.2}$$

The FIM $\mathbf{I}_i(\mathbf{p}^{\mathrm{U}})$ can be obtained using the chain rule as

$$\mathbf{I}_i(\mathbf{p}^{\mathrm{U}}) = \left(\nabla_{\mathbf{p}^{\mathrm{U}}}\Gamma\right)\mathbf{I}_i\left(\Gamma\right)\left(\nabla_{\mathbf{p}^{\mathrm{U}}}\Gamma\right)^{T}, \tag{5.3}$$

where $\mathbf{I}_n\left(\Gamma\right)$ is the FIM of Γ that can be computed using the log-likelihood function of the received signal $\log p\left(\mathbf{Y};\Gamma\right)$. In that regard, different authors in the literature have derived the CRLB for different localization scenarios and channel models. In the following sections, we provide an investigation into the existing works.

Far-Field Models: Most of the existing works consider the far-field scenario where the received signal takes a planer wavefront shape. In [13], the authors derive the error bounds of both estimating the location and the orientation for MIMO systems assuming far-field approximation and single-path model. This work suggests to use the following RIS control design:

$$\omega_i = 2\pi(i-1)\frac{d}{\lambda}\left[\sin(\theta_{\mathrm{L,M}}) - \sin(\phi_{\mathrm{B,L}})\right], \tag{5.4}$$

where ω_i is the ith element of the RIS, d is the distance between each adjacent elements in the uniform linear array (ULA), λ is the wavelength, $\theta_{\mathrm{L,M}}$ is AoD, and $\phi_{\mathrm{B,L}}$ is AoA. The authors of this work test their proposed phase design by comparing the CRLB of the proposed phase design with the random phase design, and the results show a superior performance of the proposed phase design. Finally, the authors test the effect of increasing the number of RIS elements, and they show that the error bound can be decreased by increasing the number of elements. The work in [14] considers deriving the CRLB for a far-field, single-path, positioning of a single-input single-output (SISO) system that has a series R RISs hanged on a wall. The authors use the formulated CRLB to propose using

$$\omega_{r,i} = -\pi r \left[\sin(\theta_r) - \sin(\phi_r) \right], \qquad (5.5)$$

where r is the RIS index, θ_r is the AoD, and ϕ_r is the AoA.

This work has been extended to consider multipath channels in [15, 16]. In [15], the authors derive the CRLB for far-field, multipath, and uplink single-input multiple-output (SIMO) positioning. This chapter compares the effect of removing the LOS between the transmitter and the receiver on the error bounds with the case of the existing LOS. The results show that localization through both the RIS and the LOS path can significantly reduce the errors. On the other hand, the results show that increasing the number of paths can slightly reduce the errors. The work in [16] derives the CRLB for far-field, multipath, positioning of a MIMO system. The proposed solution of this chapter suggests that increasing the number of the received snapshots will decrease the bounds of the error. However, although their model is valid for the multipath case, it is worth mentioning that the authors have conducted their simulation using only a single RIS path.

Near-Field Models: Localization and communication under short distances where the wavefront of the received signal has a considerable curvature violate the planner wavefront assumption that is used in the far-field. Such a spherical wavefront can promote the channel model to a more complicated format known as the near-field model. Large arrays and surfaces with a large number of antenna elements such as the RIS have the advantage of enhancing both localization and communications. However, surfaces with large apertures push the users into the near-field model. In addition, high frequencies that have shorter wavelengths will shorten the communication coverage to short distances only. Therefore, the near-field channel model needs to be adopted for high-frequency bands. The near-field can be defined using the Fraunhofer distance $(R = 2D^2/\lambda)$, where R is the threshold in meters that depends on the wavelength, such that any distance below this threshold is considered as a near-field, D is the antenna array length in meters, and λ is the wavelength. The wavefront in the near-field takes a spherical shape, while the far-field can be approximated in the shape of a planner wavefront. Although the near-field model is more complicated compared to the far-field model, large array surfaces such as the RIS operated in high-frequency bands consider the near-field channel model as a reasonable and practical assumption.

Several works have considered deriving the error bounds for the near-field models, for instance, in [17], the authors consider deriving the error bounds for a near-field, single path, positioning of a MIMO system. The authors also consider a 3D model and derive a closed-form solution for the RIS phase design that can work in the near-field. The results of this work suggest that the error bound can be decreased when the receiver is located near the RIS or the transmitter. The authors also compare the error bounds for different phase design methodologies and that is using the RIS as a mirror (AoD=AoA), random phase design, the proposed phase design, and the optimized phase design using maximum-likelihood. In [18], the authors derive the error bounds for a near-field, single path, positioning of a multiple-input single-output (MISO) system. This work considers utilizing the time of flight (ToF) and focuses on the synchronization problem between the transmitter and the receiver.

The work in [19] derives the error bounds for the 3D channel model. However, it is not clear what are the exact channel settings that have been used in this work. This chapter focuses on deriving the error bounds for different cases: (i) the case when the receiver is located on the perpendicular line to the surface of the RIS, (ii) the case when the receiver is not located on the perpendicular line to the surface of the RIS, and (iii) the actual hardware of the RIS produce a distortion phase. The authors successfully derive the CRLB in a closed-form solution for the first case, while an approximation is proposed for the second case. In [4], the authors derive the localization error bounds for the near-field SISO phase-dependent amplitude RIS-aided model. More specifically, the authors derive the misspecified Cramer-Rao bound (MCRB) that represents the best localization accuracy that can be achieved when using a unity amplitude RIS (ideal RIS instead of the actual phase-dependent model as derived in the author's work). The numerical results show that a severe localization performance can achieved when using the unity amplitude model rather than the actual proposed phase-dependent model. The results show that the problem can even be magnified at high signal-to-noise ratios (SNRs). Similar to [4], the work in [20] utilizes the MCRB to derive the error bounds that can be achieved for utilizing MIMO RIS-aided localization under ignored spatial non-stationarity (SNS), and spherical wave model (SWM) effects. Numerical results show that the SNS has a dominant effect in the far-field, while the SWM effect dominates under the near-field.

5.1.3.3 RIS-Aided Localization Algorithms

Designing localization algorithms for RIS-aided models is one of the challenging tasks. This is because the channel model is divided into two parts: (i) the channel between the transmitter and the RIS and (ii) the channel between the RIS and the receiver. This will lead to a complicated model that contains many variables to estimate/optimize. The problem can be further complicated, and more variables will be added when near-field and multipath models are adopted.

Far-Field Localization: User localization has been performed using dual RISs in [21], the authors consider the far-field SIMO system with a single path for every RIS. The algorithm is based on estimating the correlation matrix based on multiple snapshots of the received signal. The correlation matrix is used to estimate the ToA, and the authors assume the knowledge about the angles. The numerical results of this work show that the error can reach 3 m in the case of utilizing a single RIS while a maximum error of 0.3 m can be obtained when utilizing a dual RIS. On the other hand, the authors present a heat map that illustrates the error against the different locations of the receiver. A surprising result that can be obtained from this map is that when we place the received near the transmitter a higher error will result. This contradicts the error bounds suggested in [17] that show that placing the receiver near the transmitter will decrease the error of localization. The authors in [22] propose a SISO 3D localization for a receiver that is located in the far-field using a single path. The proposed solution is a low-complexity solution that is based on estimating the ToF. In [23], the authors propose far-field multipath localization in a MIMO system with the aid of the RIS. In the proposed model, the signal is transmitted directly from a transmitter toward the RIS which will reflect the signal toward all the users in the surrounding area which will in return reflect the signal to the receiver. In other words, the targets work as scatters between the RIS and the UE. The algorithm is based on applying a hierarchical codebook that narrows the beams of the reflected signals toward the direction of the users. It should be notated the algorithm estimates only the angles but not the distances. The estimation of the MIMO far-field, multipath environment is proposed in [24]. In this work, the authors account for a geometrical channel model that is a function of the angles in the system. The authors propose super-resolution atomic norm minimization to estimate the channel model parameters that include the angles. In [15], the authors propose a channel estimation for a far-field, multipath environment. In this work, the authors utilize the ultrawide band (UWB) communications to estimate the ToF, and AoA, and hence the user location.

Near-Field Localization: The authors in [25] propose a multiuser localization in indoor environments that is based on the RSS. In the proposed scheme, the user has a communication protocol with the RIS such that a coarse localization or a fine localization can be performed. In all the cases, the objective function is to minimize the localization loss given the RSS values and the phase shift of the RIS elements. The authors state the objective function and propose an algorithm to solve it. The authors propose a RIS phase design that is based on minimizing the localization loss that is used for estimating the user location. However, simulation parameters that have been used to generate the results imply a small localization environment with a workspace volume of 1 m^3.

Utilizing RIS as a lens has been investigated in [26]. In this configuration, the system consists of a transmitter, a single antenna receiver, and a RIS that is placed near the receiver. In such a configuration, the RIS is used as a lens rather than a reflector in such a way that it can provide a good trade-off between signal processing complexity and the utilized hardware. The authors assume a 3D model and they propose a maximum likelihood estimator to estimate the user location. The authors in [27] proposed a near-field localization for RIS-aided models. The localization algorithm is based on a two-step approach in which the ToA is first estimated and then followed by the UE localization. Additionally, in [28] the authors considered a compressed sensing (CS)-based localization for a near-field and multipath model; however, the adopted CS algorithm suffers from high-localization errors due to the quantization errors. Both [27] and [28] considered their localization algorithms for a multiple-snapshot solution only which can be impractical, especially for dynamically movable systems and low-coherence time channels.

5.1.4 Notation

We represent all the matrices by capital and bold letters $\mathbf{X}$; vectors are represented as bold and lowercase letter $\mathbf{x}$; scalars are represented as non-bold letters x or X; the transpose, conjugate, pseudo-inverse, and Hermitian transpose operators, respectively, are $(.)^T$, $(.)^*$, $(.)^\dagger$, and $(.)^H$; $(.)^\dagger$, diag$(.)$ converts a vector into a diagonal matrix, tr$(.)$ computes the trace of a matrix; rank$(.)$ computes the matrix rank; $\mathbf{I}_{N^U}$ represents an identity matrix of size N^U; $\mathbb{E}\{.\}$ is the expected value operator; $||.||_\ell$ is an operator that computes the ℓth norm; $||.||_F$ is an operator that computes the Frobenius norm; $\mathbf{x}_l$ is the lth column of $\mathbf{X}$; $\mathbf{x}_{b*}$ is the bth row of $\mathbf{X}$; $x_{b,l}$ is the bth element of $\mathbf{x}_l$; Re$(.)$ computes the real value of a complex number; $\succeq$ is the matrix semidefinite operator; $\lceil . \rceil$ is the ceiling function; and finally, $\mathcal{U} \triangleq \{-U, -U+1, \cdots, U\}$ and $\mathcal{B} \triangleq \{-B, -B+1, \cdots, B\}$.

5.2 SYSTEM MODEL

This section illustrates the localization scenario and demonstrates both the channel model and the received signal model. Further derivation of the steering response for both the far-filed and near-filed will be presented in this section.

5.2.1 Channel Model

We consider a localization system that consists of a UE, BS, and RIS which are located at $\mathbf{p}^U = [x^U, y^U]^T$, $\mathbf{p}^B = [x^B, y^B]^T$, and $\mathbf{p}^R = [x^R, y^R]^T$, respectively. We consider the MIMO system where the number of antennas in the UE, BS, and the RIS are N^U, N^B, and N^R, respectively, where all the stations are equipped with ULAs. We consider uplink localization, where the BS retrieves the location of the UE from its uplink signal that arrived through the RIS. We consider a multipath model with L^{UR} and L^{BR} paths between the (UE and RIS) and (RIS and BS), respectively. We further assume that the LOS between the BS and the UE is blocked, and the system configuration is shown in Figure 5.6. The received signal model at the BS is expressed as

$$\mathbf{Y} = \mathbf{H}\,\mathbf{X} + \mathbf{Z}, \tag{5.6}$$

where $\mathbf{Y}$ is the received signal, $\mathbf{X} \in \mathbb{C}^{N^U \times M^o}$ represents the positioning reference signal (PRS) with M^o column pilots that are orthogonal having

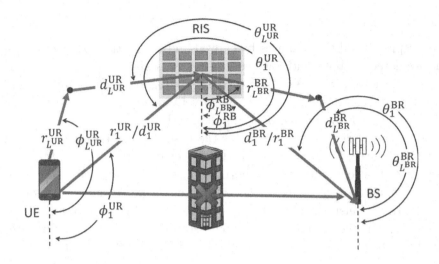

Figure 5.6 The proposed system scenario and architecture.

power P, i.e., $\mathbf{XX^H} = \frac{P}{N^U}\mathbf{I}_{N^U}$, and $\mathbf{Z} \in \mathbb{C}^{N^B \times M^o}$ represents the additive white Gaussian noise (AWGN), where $z_{i,j} \sim \mathcal{CN}(0, \sigma_z^2)$. The overall channel matrix between the UE and the BS $\mathbf{H}$, can be modeled as [24]

$$\mathbf{H} = \mathbf{H}^{BR}\mathrm{diag}(\mathbf{\Theta})\mathbf{H}^{UR}, \quad (5.7)$$

where $\mathrm{diag}(\mathbf{\Theta}) \in \mathbb{C}^{N^R \times N^R}$ is a matrix that represents the phase control of the RIS, where $\mathbf{\Theta} \triangleq \left[\zeta_1 e^{j\theta_1}, \zeta_2 e^{j\theta_2}, \cdots, \zeta_{N^R} e^{j\theta_{N^R}}\right]^T$, where $\zeta_r = 1$ as we consider ideal RIS, and $\mathbf{H}^{BR} \in \mathbb{C}^{N^B \times N^R}$ represents the channel between the RIS and the BS, while $\mathbf{H}^{RU} \in \mathbb{C}^{N^R \times N^U}$ is the channel between the UE and the RIS. More specifically,

$$\mathbf{H}^{BR} = \mathbf{A}^{BR}\,\rho^{BR}\left(\mathbf{B}^{BR}\right)^H, \quad (5.8)$$

where $\mathbf{A}^{BR} \in \mathbb{C}^{N^B \times L^{BR}}$ and $\mathbf{B}^{BR} \in \mathbb{C}^{N^R \times L^{BR}}$ represents = the steering matrices at the BS and the RIS, respectively; a detailed discussion about modeling these matrices will be presented in the next sections. $\rho^{RB} \in \mathbb{C}^{L^{BR} \times L^{BR}}$ represents the propagation gain between the RIS and BS, which can be expressed as

$$\rho_l^{BR} = \left(\frac{c}{4\pi(r_l^{BR} + d_l^{BR})f_c}\right)^{\frac{\mu}{2}} \mathcal{F}, \quad (5.9)$$

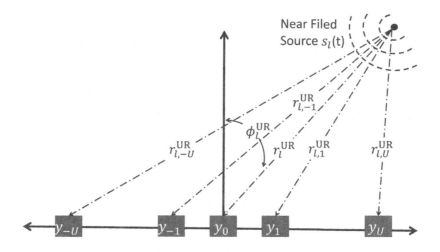

Figure 5.7 Near-field scenario.

where $\mathcal{F}$ is a random variable representing the fading and modeled as a standard complex Gaussian, μ is the path loss exponent, c is the speed of light, and r_l^{BR} and d_l^{BR} represent the distances between the (RIS and scatterers) and (scatterers and BS), respectively. The channel model between the UE and the RIS $\mathbf{H}^{\mathrm{UR}}$ is modeled in a similar way.

5.2.2   Steering Response for Far-Field and Near-Field

Consider the near-field signal impinging on the ULA in Figure 5.7. In this case, the received signal on the uth element in the array y_u is

$$y_u(t) = \sum_{l=1}^{L^{\mathrm{UR}}} s_l(t) \exp\left(j\frac{2\pi}{\lambda}\left(r_{l,u}^{\mathrm{UR}} - r_l^{\mathrm{UR}} \right) \right) + z_l(t), \tag{5.10}$$

where y_u is the received signal on the uth element, $s_l(t)$ is the complex gain of the lth source, $r_{l,u}^{\mathrm{UR}}$ is the distance between the uth antenna element and the lth source, and r_l^{UR} is the distance between the reference antenna element (i.e., $u = 0$) and the lth source. Here, we can further express $r_{l,u}^{\mathrm{UR}}$ as

$$r_{l,u}^{\mathrm{UR}} = \sqrt{r_l^{\mathrm{UR}} + u^2\delta^2 - 2u\lambda r_l^{\mathrm{UR}}\sin\left(\phi_l^{\mathrm{UR}}\right)}. \tag{5.11}$$

Due to the non-linearity of (5.11), we use second-order Taylor series approximation to achieve the Fresnel approximation [29] which will reduce the model in (5.11) to

$$r_{l,u}^{\mathrm{UR}} \approx r_l^{\mathrm{UR}} - u\delta \sin(\phi_l) + u^2\delta^2 \frac{\cos(\phi_l^{\mathrm{UR}})}{2r_l^{\mathrm{UR}}}, \tag{5.12}$$

where δ is the distance of adjacent elements in the ULA. Based on the model in (5.12), we can rewrite the received signal model in (5.10) as

$$y_u(t) \approx \sum_{l=1}^{L^{\mathrm{UR}}} s_l(t) e^{\left(u\alpha_l^{\mathrm{UR}} + u^2\beta_u^{\mathrm{UR}}\right)} + z_l, \tag{5.13}$$

where

$$\alpha_l^{\mathrm{UR}} = -\frac{2\pi\delta}{\lambda} \sin\left(\phi_u^{\mathrm{UR}}\right), \quad \beta_u^{\mathrm{UR}} = \frac{\pi\delta^2}{\lambda r_l^{\mathrm{UR}}} \cos^2\left(\phi_l^{\mathrm{UR}}\right).$$

We can further express (5.13) in the matrix format as

$$\mathbf{y} = \mathbf{B}^{\mathrm{UR}}\left(\boldsymbol{\alpha}^{\mathrm{UR}}, \boldsymbol{\beta}^{\mathrm{UR}}\right)\mathbf{s} + \mathbf{z}, \tag{5.14}$$

where

$$\boldsymbol{\alpha}^{\mathrm{UR}} = -\frac{2\pi\delta}{\lambda} \sin\left(\boldsymbol{\phi}^{\mathrm{UR}}\right), \quad \boldsymbol{\beta}^{\mathrm{UR}} = \frac{\pi\delta^2}{\lambda \mathbf{r}^{\mathrm{UR}}} \cos^2\left(\boldsymbol{\phi}^{\mathrm{UR}}\right).$$

Applying the same definitions we derived in Section 5.2, both $\mathbf{A}(\boldsymbol{\theta}^{\mathrm{BR}}, \mathbf{d}^{\mathrm{BR}}) \in \mathbb{C}^{N^{\mathrm{B}} \times L^{\mathrm{BR}}}$ and $\mathbf{B}(\boldsymbol{\phi}^{\mathrm{BR}}, \mathbf{r}^{\mathrm{BR}}) \in \mathbb{C}^{N^{\mathrm{R}} \times L^{\mathrm{BR}}}$ represent the steering matrices with AoAs, $\boldsymbol{\theta}^{\mathrm{BR}}$, and AoDs, $\boldsymbol{\phi}^{\mathrm{BR}}$. The elements of each steering matrix in (5.8) can be approximated using the Fresnel approximation of the spherical wavefront model [29] by its angle and distance as

$$a_{b,l}(\theta_l^{\mathrm{BR}}, d_l^{\mathrm{BR}}) = \exp\left(j\left[b\omega_l^{\mathrm{BR}} + b^2\gamma_l^{\mathrm{BR}}\right]\right), \tag{5.15}$$

where $\omega_l^{\mathrm{BR}} \triangleq f(\theta_l^{\mathrm{BR}})$ and $\gamma_l^{\mathrm{BR}} \triangleq g(\theta_l^{\mathrm{BR}}, r_l^{\mathrm{BR}})$ with

$$f(\theta) = -\frac{2\pi\delta}{\lambda} \sin(\theta), \quad g(\theta, d) = \frac{\pi\delta^2}{\lambda d} \cos^2(\theta), \tag{5.16}$$

where λ is the wavelength. We take the element at the center as a reference such that the distance from the center to the element of index b is δb, where $b \in \mathcal{B}$ and $B \triangleq \lceil (N^{\mathrm{B}} - 1)/2 \rceil$.

It can be noticed that the far-field is a special case of near-filed where the spherical wavefront turns into a planer shape. The far-field model can be obtained when we assume the distance d in (5.16) to be greater than the Fraunhofer distance. We can express this mathematically by assuming d to big enough (i.e., $d \rightarrow \infty$). By doing so, γ^{BR} in (5.16) will be turned to a negligible small number. The remaining part of (5.16) is ω^{BR}, which will be a function of the angle only. As a result, both $\mathbf{A}(\boldsymbol{\theta}^{\mathrm{BR}}, \mathbf{d}^{\mathrm{BR}}) \in \mathbb{C}^{N^{\mathrm{B}} \times L^{\mathrm{BR}}}$ and $\mathbf{B}(\boldsymbol{\phi}^{\mathrm{BR}}, \mathbf{r}^{\mathrm{BR}}) \in \mathbb{C}^{N^{\mathrm{R}} \times L^{\mathrm{BR}}}$ in (5.8) can be modeled in far-field as

$$\mathbf{A}^{\mathrm{BR}}(\boldsymbol{\theta}^{\mathrm{BR}}) = \left[e^{jb\omega_1^{\mathrm{BR}}}, e^{jb\omega_2^{\mathrm{BR}}}, \ldots, e^{jb\omega_{L^{\mathrm{BR}}}^{\mathrm{BR}}} \right], \tag{5.17}$$

where $\omega_1^{\mathrm{BR}} = f\left(\theta_l^{\mathrm{BR}}\right)$ is similar to the definition in (5.16) in the near-field.

We can conclude from both (5.15) and (5.17) that the received signal in the near-field will depend on two variables: (i) ω_l^{BR} that is function of the angles θ_l only and (ii) γ_l^{BR} that is a function of the angles θ_l^{BR} and the distances d_l^{BR}. On contradict, if $d_l^{\mathrm{BR}} \rightarrow \infty$ (i.e., the far-field), then γ_l^{BR} in (5.15) will approach to zero, and as a result, the steering matrix will depend on ω_l^{BR} only, which is a function of one variable only and that is the angle θ_l^{BR}.

5.3 PROBLEM FORMULATION AND SOLUTION

In this section, we aim to estimate the user location using the channel model settings that have been described in the previous section for both the far-field and the near field. The localization problem is to be solved using two different approaches, the CS and the atomic norm minimization. Each of these two approaches will have its advantages over the other, and a discussion about the motivation behind each approach will be included in this section.

5.3.1 RIS-Aided Localization Using Compressed Sensing (CS)

A typical number of paths for high-frequency bands can be in the order of 3 to 8 [30]. In general, practical environments such as urban and indoor environments exhibit multipath propagation instead of single-path. Such multipath environments can be formulated mathematically using sparse representation where a large number of possible paths, each is characterized by its localization parameters, are possible, but only few of these paths are actually utilized in the channel model. Motivated by these facts, the CS can be utilized as a sparse recovery technique where the CS aims to recover the sparsest solution that contains the actually utilized paths in the system from a set of large number of possible multipath. By doing so, the user location can be estimated from the recovered sparse signal characterized by its AoA and the distance from the other channel elements. The CS has also the advantage of estimating the user location parameters using a low-complexity single snapshot of the received signal.

5.3.1.1 Far-Field Localization Using Compressed Sensing

Problem Formulation: In this section, we aim to localize the direction of arrival (DoA) of the UE by utilizing the CS methodology. The full positioning of the UE location will be further discussed in Section 5.1.3.2. We aim to exploit the sparsity at every path to come up with a CS solution. We construct a grid measurement matrix $\overline{\mathbf{A}}\left(\boldsymbol{\theta}^{\mathrm{BR}}\right) \in \mathbb{C}^{N^{\mathrm{B}} \times N}$ as

$$\overline{\mathbf{A}}^{\mathrm{BR}}(\boldsymbol{\theta}^{\mathrm{BR}}) = \left[e^{jb\omega_1^{\mathrm{BR}}}, e^{jb\omega_2^{\mathrm{BR}}}, \ldots, e^{jb\omega_N^{\mathrm{BR}}}\right], \tag{5.18}$$

where N represents the number of all the possible angles in that grid. By doing so, we can now rewrite (5.6) as

$$\mathbf{Y} = \overline{\mathbf{A}}^{\mathrm{BR}}(\boldsymbol{\theta}^{\mathrm{BR}})\overline{\mathbf{C}^1} + \mathbf{Z}. \tag{5.19}$$

Now in (5.19) above, we define the received signal $\mathbf{Y}$ as a sum of N atoms where each atom is associated with a specified angle θ_n^{BR} in the grid. By estimating the sparse support matrix $\overline{\mathbf{C}^1}$, we can recover $\hat{\boldsymbol{\theta}}^{\mathrm{BR}}$ by locating the indices of the non-zero rows in the sparse support matrix $\overline{\mathbf{C}^1}$, where each row in $\overline{\mathbf{C}^1}$ is associated with a certain angle θ_n^{BR} as constructed in (5.19). Similarly, if we computed the conjugate of (5.6), we get

$$\mathbf{Y}^H = \overline{\mathbf{B}}^{\mathrm{UR}}(\boldsymbol{\phi}^{\mathrm{UR}})\overline{\mathbf{C}^2} + \mathbf{Z}. \tag{5.20}$$

Similar to the above discussion, we define the received signal $\mathbf{Y}^H$ using the grid measurement matrix $\overline{\mathbf{B}}\left(\boldsymbol{\phi}^{\mathrm{UR}}\right) \in \mathbb{C}^{N^U \times N}$ as as a sum of N atoms where each atom is associated with a specified angle ϕ_n^{UR}. By recovering the sparse support matrix $\overline{\mathbf{C}^2}$, we can estimate $\hat{\boldsymbol{\phi}}^{\mathrm{UR}}$ by locating the rows of non-zero elements in $\overline{\mathbf{C}^2}$.

Sparse Recovery via Compressed Sensing: The CS is a signal processing technique that aims to recover a sparse version of underdetermined linear systems. Both (5.19) and (5.20) are linear systems with either N^B or N^U equations and N unknowns, where $N^B \ll N$ and $N^U \ll N$. Such configuration can lead to an infinite number of solutions. In CS, we aim to formulate the problem in such configuration as in both (5.19) and (5.20), where we exploit the sparsity in the model (in our case, the sparsity in the possible directions) to construct a dictionary matrix that has all the possible solutions. We solve this system by recovering the sparsest solution among all the solutions, which represents the actual configuring of the model. In the case of single measurement vector (SMV), the sparsest solution can be obtained by minimizing the l_0 norm of the received signal as a function of the support matrix. However, such a problem is non-convex and hard to solve, and as a result, we relax the problem into minimizing the l_1 norm using the same formulation. For multiple measurement vectors (MMV), similar approaches can be done. In our case, we use the temporal multiple sparse Bayesian learning (T-MSBL) algorithm [31] to solve all the CS problems.

5.3.1.2 Near-Field Localization Using Compressed Sensing

In the previous section, we demonstrated AoA estimation using CS under the far-field RIS-aided model. In this section, we extend the previous work in two aspects: (i) we solve a full localization problem rather than AoA estimation; (ii) we consider a near-field model rather than the far-field due to its potential in future wireless systems.

Estimation of the Localization Parameters In the near-field, due to the spherical wavefront, CS cannot be directly applied to the received signal $\mathbf{Y}$. More precisely, the sparsifying basis, which is adopted to represent the sparse signal, requires gridding over both the angle and distance vectors, leading to higher computational complexity compared to the far-field case that requires gridding over only the angles. We propose to exploit the spatial correlation in the channel matrix $\mathbf{H}$. This allows us to decompose the two-dimensional CS problem into smaller simplified one-dimensional models, relaxing the coupling between the steering angle and distance vectors. In particular, we propose to estimate the channel and its empirical covariance matrix. Then, the measured signal in the CS model is considered as the empirically estimated covariance matrix rather than the directly received signal $\mathbf{Y}$.

More precisely, let us define $\mathbf{V}$ as the matrix containing some properly selected elements from the covariance matrix of the actual channel as[2]

$$v_{b,u} \triangleq \mathbb{E}\{h_{b,u} h_{p,n}{}^*\} = \sum_{i=1}^{L^{BR}} \sum_{k=1}^{L^{UR}} \sum_{r=1}^{N^R} \sum_{z=1}^{N^R} \sigma_i^{BR} \sigma_k^{UR} e^{2jbw_i^{BR}}$$

$$e^{-j[r\alpha_i^{BR}+r^2\beta_i^{BR}]} e^{j[z\alpha_i^{BR}+z^2\beta_i^{BR}]} e^{j[rw_k^{UR}+r^2\gamma_k^{UR}]} e^{-j[zw_k^{UR}+z^2\gamma_k^{UR}]}$$

$$e^{-2ju\alpha_k^{UR}} e^{j\theta_r} e^{-j\theta_z}, \forall u \in \mathcal{U}, b \in \mathcal{B}, p = -b, n = -u, \tag{5.21}$$

where $\boldsymbol{\sigma}^{BR}$ and $\boldsymbol{\sigma}^{UR}$ contain the variance of each path, $U \triangleq \lceil (N^U - 1)/2 \rceil$, and $\alpha_k^{UR} \triangleq f(\theta_k^{UR})$, and $\beta_k^{UR} \triangleq g(\theta_k^{UR}, d_k^{UR})$. The selected elements in the covariance matrix (with $p = -b$ and $n = -u$) are chosen such that the parts with γ_l^{BR}, which depends on the angles and distances, cancel out. This permits decoupling the angles and distances, as the remaining ω_l^{BR} depends only on the angle. We estimate the left-hand side of (5.21) as the following:

$$\mathbf{V} = \hat{\mathbf{V}} - \boldsymbol{\Gamma}, \tag{5.22}$$

where $\hat{\mathbf{V}}$ is the covariance matrix of the estimated channel and $\boldsymbol{\Gamma}$ is the covariance of the error in the channel estimation, such that $\Gamma_{b,u} = \mathbb{E}\{\mathbf{z}_{b*}\mathbf{x}_{u*}^H (\mathbf{z}_{-b*}\mathbf{x}_{-u*}^H)^H\} = 0$. The matrix $\hat{\mathbf{V}}$ can be ideally estimated using $\hat{v}_{b,u} = \mathbb{E}\{\hat{h}_{b,u}(\hat{h}_{-b,-u})^*\}$, but for a limited number of snapshots T, we have

$$\tilde{v}_{b,u} = T^{-1} \sum_{t=1}^{T} \hat{h}_{b,u}^t (\hat{h}_{-b,-u}^t)^*, \forall u \in \mathcal{U}, b \in \mathcal{B}, \tag{5.23}$$

with $\hat{\mathbf{V}} = \tilde{\mathbf{V}} + \boldsymbol{\Lambda}$ where $\boldsymbol{\Lambda}$ is the error due to the limited number of snapshots and $\hat{\mathbf{H}} = \mathbf{Y}\mathbf{X}^\dagger$ is the least square (LS) estimate of the channel. The matrix $\tilde{\mathbf{V}} \in \mathbb{C}^{B \times U}$ is written as

$$\tilde{\mathbf{V}} = \mathbf{S}^1(\boldsymbol{\theta}^{BR})\mathbf{C}^3 + \boldsymbol{\Gamma} - \boldsymbol{\Lambda}, \tag{5.24}$$

where each vector $\mathbf{s}_l^1(\boldsymbol{\theta}^{BR})$ in $\mathbf{S}^1(\boldsymbol{\theta}^{BR}) \in \mathbb{C}^{B \times L^{BR}}$ is expressed as

$$\mathbf{S}^1(\boldsymbol{\theta}^{BR}) = \left[e^{2jbw_1^{BR}}, e^{2jbw_2^{BR}}, \ldots, e^{2jbw_{L^{BR}}^{BR}} \right]. \tag{5.25}$$

In (5.25), we defined $\mathbf{S}^1(\boldsymbol{\theta}^{BR})$ to be a combination of L^{BR} atoms, each with a specific angle θ_l^{BR}. In order to estimate the angle $\hat{\theta}_l^{BR}$ using CS, we define the grid measurement matrix $\overline{\mathbf{S}}^1 \in \mathbb{C}^{B \times N}$ to be a combination of N atoms such that every single atom $\mathbf{s}_n^1(\boldsymbol{\theta}^{\mathrm{BR}})$ is associated with the angle $\frac{2\pi n - \pi(N+1)}{N-1}$, where n is the grid counter and N is the grid size. The support matrix $\overline{\mathbf{C}}^3 \in \mathbb{C}^{N \times U}$ is L^{BR}-sparse, which means that it is a matrix with N rows, but only L^{BR} of them have non-zero elements, and by finding the indices of non-zero rows, one can recover the corresponding angles $\hat{\boldsymbol{\theta}}^{\mathrm{BR}}$ up to some quantization error.

Several sparse recovery algorithms can be used to estimate angles from the MMV $\tilde{\mathbf{V}}$. For instance, angles can be recovered by minimizing the ℓ_0 quasi-norm of $\overline{\mathbf{C}}^3$ when $B \geq 2L^{\mathrm{BR}} + 1 - \mathrm{rank}(\overline{\mathbf{C}}^3)$ [32]. However, this problem is NP-hard. Alternatively, angles can be recovered using a less computationally complex optimization, i.e., ℓ_1 minimization, albeit with a higher number of measurements, e.g., $B > cL^{\mathrm{BR}}$, where $c > 1$ is an overmeasuring factor. Several algorithms have been proposed to solve the MMV problem, e.g., multiple orthogonal matching pursuit (M-OMP), joint $\ell_{2,0}$ approximation (JLZA) and T-MSBL, under some conditions on the restricted isometry constant of the measurement matrix $\overline{\mathbf{S}}^1$ [31]. The other angle $\hat{\boldsymbol{\theta}}^{\mathrm{UR}}$ can be estimated by taking the Hermitian transpose of (5.24), and that is

$$\tilde{\mathbf{V}}^H = \mathbf{S}^2(\boldsymbol{\phi}^{\mathrm{UR}})\mathbf{C}^4 + \boldsymbol{\Gamma}^H - \boldsymbol{\Lambda}^H. \tag{5.26}$$

By constructing the grid measurement matrix $\overline{\mathbf{S}}^2$ similar to $\overline{\mathbf{S}}^1$, we solve for the L^{UR} sparse support matrix $\overline{\mathbf{C}}^4 \in \mathbb{C}^{N \times B}$ using CS. To estimate the distances $\hat{\mathbf{d}}^{\mathrm{BR}}$ that correspond to the angle $\hat{\boldsymbol{\theta}}^{\mathrm{BR}}$, first, we rewrite (5.6) as

$$\mathbf{Y} = \mathbf{A}(\boldsymbol{\theta}^{\mathrm{BR}}, \mathbf{d}^{\mathrm{BR}})\mathbf{C}^5 + \mathbf{Z}. \tag{5.27}$$

Then, we apply CS on the gridded model, where the grid measurement matrix is $\overline{\mathbf{S}}^3 = \overline{\mathbf{A}}(\hat{\boldsymbol{\theta}}^{\mathrm{BR}}, \overline{\mathbf{d}}^{\mathrm{BR}})$, where $\overline{\mathbf{A}}(\hat{\boldsymbol{\theta}}^{\mathrm{BR}}, \overline{\mathbf{d}}^{\mathrm{BR}})$ is the steering matrix that is defined in (5.15), $\hat{\boldsymbol{\theta}}^{\mathrm{BR}}$ is the previously estimated angle and $\overline{\mathbf{d}}^{\mathrm{BR}}$ is the gridded distance vector with Fraunhofer distance being its known upper bound. Now $\hat{\mathbf{d}}^{\mathrm{BR}}$ can be estimated using CS. The distance $\hat{\mathbf{r}}^{\mathrm{UR}}$ can be estimated similar to $\hat{\mathbf{d}}^{\mathrm{BR}}$ by taking the Hermitian transpose of (5.27) and that is

$$\mathbf{Y}^H = \mathbf{X}^H \mathbf{A}(\boldsymbol{\phi}^{\mathrm{UR}}, \mathbf{r}^{\mathrm{UR}})\mathbf{C}^6 + \mathbf{Z}^H. \tag{5.28}$$

We define the grid measurement matrix as $\overline{\mathbf{S}}^4 = \mathbf{X}^H \overline{\mathbf{A}}(\hat{\boldsymbol{\phi}}^{\mathrm{UR}}, \overline{\mathbf{r}}^{\mathrm{UR}})$ and $\hat{\mathbf{r}}^{\mathrm{UR}}$ can be estimated similarly to all the previous parameters using CS. Similarly, the other localization parameters can be recovered.

Algorithm 1 RIS-aided near-field localization via the CS

    Input: Y

1: **Initialize:** $\Theta \leftarrow$ random phase design.

2: **for** $i = 1$ to I **do**

3:    Estimate $\tilde{\mathbf{V}}$ using (5.23).

4:    Construct the measurement matrix $\overline{\mathbf{S}}^1$ as in (5.25) with the required resolution.

5:    Estimate the $\hat{\mathbf{C}}^3$ using the sparse recovery algorithm and recover $\hat{\theta}^{\mathrm{BR}}$ from it.

6:    Compute the Hermitian transpose of $\tilde{\mathbf{V}}$ and repeat steps 4 and 5 to estimate $\hat{\phi}^{\mathrm{UR}}$.

7:    Construct the measurement matrix as in (5.27) using the required resolution..

8:    Estimate $\hat{\overline{\mathbf{C}}}^5$ using the sparse recovery algorithm and recover $\hat{\mathbf{r}}^{\mathrm{BR}}$ from it.

9:    Compute the Hermitian transpose of $\mathbf{Y}$ and repeat steps 7 and 8 to estimate $\hat{\mathbf{d}}^{\mathrm{UR}}$.

10:    Compute the mismatch using (5.30) for all the parameters.

11:    Use the estimated parameters to design the RIS according to (5.59).

12: **end for**

13: Compute the UE location $\hat{\mathbf{p}}^{\mathrm{U}}$ using the estimated parameters.

    Output $\hat{\mathbf{p}}^{\mathrm{U}}$

Off-grid estimation: CS considers that the angles can take only specific values on a grid with a certain resolution, leading to a model mismatch problem. Solving this problem requires an off-grid estimation, typically with high computational complexity [33]. The off-grid CS algorithms can provide higher estimation accuracy at the expense of higher computational complexity. Alternatively, we first consider on-grid CS, and then we represent the actual measurement matrix $\mathbf{S}$ as a sum of two parts: *(i)* the on-grid variable $\overline{\mathbf{S}}$; *(ii)* the bias mismatch error $\mathbf{E}$. For instance, to compensate for the model mismatch in (5.24), we can represent $\mathbf{S}^1(\theta^{\mathrm{BR}})$ as: $\mathbf{S}^1(\hat{\theta}^{\mathrm{BR}}, \theta) \approx \mathbf{S}^1(\hat{\theta}^{\mathrm{BR}}) + \mathbf{S}^1_{\theta^{\mathrm{BR}}}(\hat{\theta}^{\mathrm{BR}})\mathrm{diag}(\theta - \hat{\theta}^{\mathrm{BR}})$, where $\mathbf{S}^1_{\theta^{\mathrm{BR}}}(\hat{\theta}^{\mathrm{BR}})$ is the first order derivative of $\mathbf{S}^1$ around $\hat{\theta}^{\mathrm{BR}}$, i.e.,

$$\mathbf{S}^1_{\theta^{\mathrm{BR}}}\left(\hat{\theta}^{\mathrm{BR}}\right) = \left[\frac{\partial \mathbf{s}_1(\theta^{\mathrm{BR}})}{\partial \theta^{\mathrm{BR}}}\Big|_{\theta^{\mathrm{BR}}=\hat{\theta}^{\mathrm{BR}}_1}, \cdots, \frac{\partial \mathbf{s}_1(\theta^{\mathrm{BR}})}{\partial \theta^{\mathrm{BR}}}\Big|_{\theta^{\mathrm{BR}}=\hat{\theta}^{\mathrm{BR}}_{L\mathrm{BR}}}\right]. \tag{5.29}$$

$\text{diag}\left(\boldsymbol{\theta} - \hat{\boldsymbol{\theta}}^{\text{BR}}\right)$ is a diagonal matrix containing the angles that minimize the mismatch error. In order to estimate $\boldsymbol{\theta}^{\text{BR}}$, we start by solving the following optimization problem:

$$\boldsymbol{\theta}^{\star} = \arg\min_{\boldsymbol{\theta}}.\|\mathbf{S}^1(\hat{\boldsymbol{\theta}}^{\text{BR}})\hat{\mathbf{C}}^3 + \mathbf{S}^1_{\hat{\boldsymbol{\theta}}^{\text{BR}}}(\hat{\boldsymbol{\theta}}^{\text{BR}})\text{diag}(\boldsymbol{\theta} - \hat{\boldsymbol{\theta}}^{\text{BR}})\hat{\mathbf{C}}^3 - \tilde{\mathbf{V}}\|_F^2$$

$$\text{s.t. } -0.5\Delta \leq \theta_l - \hat{\theta}_l^{\text{BR}} \leq 0.5\Delta, \forall l \in \left\{1, 2, \ldots, L^{\text{BR}}\right\}, \tag{5.30}$$

where Δ is the grid resolution. Now $\boldsymbol{\theta}^{\text{BR}} \approx \tilde{\boldsymbol{\theta}}^{\text{BR}} = \hat{\boldsymbol{\theta}}^{\text{BR}} + (\boldsymbol{\theta}^{\star} - \hat{\boldsymbol{\theta}}^{\text{BR}})$. Similarly, the off-grid errors in $\boldsymbol{\phi}^{\text{UR}}$, $\mathbf{d}^{\text{BR}}$, and $\mathbf{r}^{\text{UR}}$ can be estimated. The demonstrated localization scheme is illustrated in Algorithm 1.

5.3.2 RIS-Aided Localization Using Atomic Norm Minimization

In the previous section, we solved the localization problem using CS. However, the previous methodology has two drawbacks: (i) the utilized CS methodology suffers from quantization errors as it depends on the resolution of the constructed measurement matrices; (ii) on the other hand, the adopted solution requires multiple snapshots of the received signal which can be unpractical especially in dynamic environments with short channel coherence time. In this section, we aim to solve these limitations by adopting a single-snapshot super-resolution solution that is based on atomic norm minimization. We consider a single path localization for this section, where $L^{\text{BR}} = L^{\text{UR}} = 1$; the updated single-path localization scenario is shown in Figure 5.8.

Problem Formulation: Let us define the atomic set in $\mathbf{H}^{\text{UR}}$ as

$$\mathcal{A} \triangleq \{\mathbf{a}^{\text{UR}}\left(\phi^{\text{UR}}, r^{\text{UR}}\right) | \phi^{\text{BR}} \in [-\frac{\pi}{2}, \frac{\pi}{2}), r^{\text{UR}} \in [0, R)\}. \tag{5.31}$$

The set $\mathcal{A}$ in (5.31) is called the atomic set as it is a set on the continuous domain that has all the possible atoms. Knowing the exact atoms leads to knowing the angles that this atom is made of and, hence, estimating the location. Now, define $\|\hat{\mathbf{h}}\|_{\mathcal{A},0}$ to be the atomic l_0 norm such that

$$\|\hat{\mathbf{h}}\|_{\mathcal{A},0} \triangleq \inf_{L}\left\{L : \hat{\mathbf{h}} = \sum_{l=1}^{L} \mathbf{a}^{\text{UR}}\left(\phi_l^{\text{UR}}, r_l^{\text{UR}}\right)\alpha_l, \mathbf{a}^{\text{UR}}\left(\phi_l^{\text{UR}}, r_l^{\text{UR}}\right) \in \mathcal{A}\right\},$$

$$\tag{5.32}$$

where $\hat{\mathbf{h}} = \text{vec}\left(\hat{\mathbf{H}}\right) = \text{vec}\left(\mathbf{Y}\mathbf{X}^{\dagger}\right)$ is a vectorized version of the LS estimate of the channel [34] and α_l is the complex amplitude. The infimum

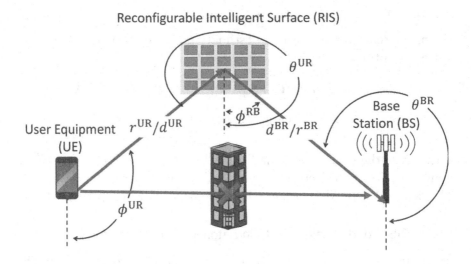

Figure 5.8 The proposed system scenario for atomic norm minimization.

function implements minimizing the number of angles L such that we get the sparsest solution. The sparsest solution is the solution that contains only the exact angle ϕ^{UR} and distance r^{UR}. This problem can be thought of as if we constructed a dictionary matrix that contains all possible combinations of ϕ^{UR} and r^{UR}. Note that only the sparsest solution is valid and all the other solutions are redundant. However, this is an NP-hard problem that cannot be solved. Instead, the following convex relaxation can be used [33]:

$$||\hat{\mathbf{h}}||_{\mathcal{A}} \triangleq \inf_{\alpha} \left\{ \sum_{l=1}^{L_{\mathrm{tot}}} |\alpha_l| : \hat{\mathbf{h}} = \sum_{l=1}^{L_{\mathrm{tot}}} \mathbf{a}^{\mathrm{UR}}\left(\phi_l^{\mathrm{UR}}, r_l^{\mathrm{UR}}\right) \alpha_l, \mathbf{a}^{\mathrm{UR}}\left(\phi_l^{\mathrm{UR}}, r_l^{\mathrm{UR}}\right) \in \mathcal{A} \right\}.$$
(5.33)

The new $||\hat{\mathbf{h}}||_{\mathcal{A}}$ in (5.33) is called the atomic norm. The atomic norm can be used to solve the localization problem with the following objective function:

$$\underset{\alpha,\mathbf{z}}{\text{minimize}} \quad ||\hat{\mathbf{h}}||_{\mathcal{A}}$$

$$\text{subject to} \quad \hat{\mathbf{h}} = \boldsymbol{\Upsilon}\alpha + \mathbf{z}, \quad \text{and} \quad ||\mathbf{z}||_2 \leq \epsilon, \qquad (5.34)$$

where $\boldsymbol{\Upsilon}$ is a matrix that contains all the possible combinations of ϕ^{UR} and r^{UR}, which yield a semi-infinite programming problem, and ϵ is the noise threshold. Solving the problem in (5.34) will result in a high-resolution estimation of the location.

The Dual Problem: Minimizing the primal problem in (5.34) is equivalent to maximizing the dual problem as strong duality hold. The first step to formulate the dual problem is to set up the Lagrangian as a weighted sum of the constraints with the objective function as

$$L(\boldsymbol{\alpha}, \mathbf{z}, \boldsymbol{\beta}, \boldsymbol{\gamma}) = ||\hat{\mathbf{h}}||_{\mathcal{A}} + \text{Re}\left[\boldsymbol{\beta}^H\left(\hat{\mathbf{h}} - \boldsymbol{\Upsilon}\boldsymbol{\alpha} - \mathbf{z}\right)\right] + \gamma\left(\mathbf{z}^H\mathbf{z} - \epsilon^2\right),$$

$$(5.35)$$

where $\boldsymbol{\beta}$ and $\boldsymbol{\gamma}$ are Lagrange multipliers. The dual function $d(\boldsymbol{\beta}, \boldsymbol{\gamma})$ is the infimum of the Lagrangian, and that is

$$d(\boldsymbol{\beta}, \boldsymbol{\gamma}) = \inf_{\boldsymbol{\alpha}, \mathbf{z}} L(\hat{\mathbf{h}}, \boldsymbol{\beta}, \boldsymbol{\gamma}) = \inf_{\boldsymbol{\alpha}, \mathbf{z}} \{\text{Re}\left[\boldsymbol{\beta}^H\hat{\mathbf{h}} - \boldsymbol{\beta}^H\mathbf{z}\right]$$
$$+ \gamma\left(\mathbf{z}^H\mathbf{z} - \epsilon^2\right) + ||y||_{\mathcal{A}} - \text{Re}\left[\boldsymbol{\beta}^H\boldsymbol{\Upsilon}\boldsymbol{\alpha}\right]\}.$$

$$(5.36)$$

To solve (5.36), we first minimize over $\mathbf{z}$ as

$$\frac{\partial d(\boldsymbol{\beta}, \boldsymbol{\gamma})}{\partial \mathbf{z}} = -\boldsymbol{\beta} + 2\gamma\mathbf{z} = 0,$$

$$(5.37)$$

which yields $\mathbf{z}^\star = \frac{\boldsymbol{\beta}}{2\gamma}$. Similarly, the dual function is maximized over the dual variable γ as where we get

$$\frac{\partial d(\boldsymbol{\beta}, \boldsymbol{\gamma})}{\partial \gamma} = \frac{||\boldsymbol{\beta}^H||_2^2}{4\gamma^2} - \epsilon^2 = 0,$$

$$(5.38)$$

which yields $\gamma^\star = \frac{||\boldsymbol{\beta}^H||_2}{2\epsilon}$. Now the dual function reduces to

$$d(\boldsymbol{\beta}) = \text{Re}\left[\boldsymbol{\beta}^H\hat{\mathbf{h}}\right] - \epsilon||\boldsymbol{\beta}||_2 + \inf_{\boldsymbol{\alpha}}\left(||\hat{\mathbf{h}}||_{\mathcal{A}} - \text{Re}\left[\boldsymbol{\beta}^H\boldsymbol{\Upsilon}\boldsymbol{\alpha}\right]\right).$$

$$(5.39)$$

In order to solve for the infimum in (5.39), consider that for every element α_i, we have $\text{Re}\left[(\boldsymbol{\beta}^H\boldsymbol{\Upsilon})_i\alpha_i\right] = \text{Re}\left[(\boldsymbol{\Upsilon}^H\boldsymbol{\beta})_i^H\alpha_i\right] = |(\boldsymbol{\Upsilon}^H\boldsymbol{\beta})_i||\alpha_i|\cos(\psi)$. Using the definition of the atomic norm in (5.33), we can get

$$|\alpha_i| - \text{Re}\left[(\boldsymbol{\Upsilon}^H\boldsymbol{\beta})_i^H\alpha_i\right] = |\alpha_i|\left[1 - |(\boldsymbol{\Upsilon}^H\boldsymbol{\beta})_i|\cos(\psi)\right]$$
$$\geq |\alpha_i|\left[1 - |(\boldsymbol{\Upsilon}^H\boldsymbol{\beta})_i|\right].$$

$$(5.40)$$

For $|(\boldsymbol{\Upsilon}^H\boldsymbol{\beta})_i| \leq 1$, the lower bound is non-negative and the infimum is zero; otherwise, the infimum is $-\infty$. As a result, we can express (5.39) as

$$d(\boldsymbol{\beta}) = \text{Re}\left[\boldsymbol{\beta}^H\hat{\mathbf{h}}\right] - \epsilon||\boldsymbol{\beta}||_2, \quad \text{s.t.} \quad ||\boldsymbol{\Upsilon}^H\boldsymbol{\beta})||_\infty \leq 1.$$

$$(5.41)$$

The dual problem in (5.34) can be reformulated as

$$\underset{\beta}{\text{maximize}} \quad \text{Re}\left[\beta^H \hat{\mathbf{h}}\right] - \epsilon||\beta||_2$$

$$\text{subject to} \quad ||\mathbf{\Upsilon}^H \beta)||_\infty \leq 1. \tag{5.42}$$

The constraints in (5.42) are again semi-infinite programming. Referring to [35], we define a trigonometric polynomial as

$$\mathbf{F}(\tau) = \sum_{l=0}^{L-1} \beta_l e^{-j\tau_l} = \mathbf{a}(\tau)^H \beta. \tag{5.43}$$

According to Theorem 4.26 in [35], if the following inequality:

$$|\mathbf{F}(\tau)| < |\mathbf{R}(\tau)| \tag{5.44}$$

is satisfied, then the following inequality is also true:

$$\begin{bmatrix} \mathbf{Q} & \beta \\ \beta^T & 1 \end{bmatrix} \succeq 0, \tag{5.45}$$

where $\mathbf{Q} = \beta\beta^H$. This can be proved using Schur's complement. According to Corollary 4.27 in [35], if a special case of $|\mathbf{R}(\tau)| = k$, then the following approximation is true:

$$\begin{bmatrix} \mathbf{Q}_k & \beta \\ \beta^T & 1 \end{bmatrix} \succeq 0, \tag{5.46}$$

where $\mathbf{Q}_k$ is a diagonal matrix of all the diagonal elements equal to k^2. Now we can relax the constraints in (5.42) as

$$||\mathbf{\Upsilon}^H \beta)||_\infty = \underset{k}{\text{minimize}} \; k \quad \text{s.t.} |\mathbf{\Upsilon}^H \beta| \leq k. \tag{5.47}$$

In our case, $k = 1$. Therefore, (5.42) can be rewritten as a semidefinite programming (SDP)

$$\underset{\beta, \mathbf{Q}_k}{\text{maximize}} \quad \text{Re}\left[\beta^H \hat{\mathbf{h}}\right] - \epsilon||\beta||_2$$

$$\text{subject to} \quad \begin{bmatrix} \mathbf{Q}_k & \beta \\ \beta^H & 1 \end{bmatrix} \succeq 0. \tag{5.48}$$

Similar to (5.46), $\mathbf{Q}_k$ is a diagonal matrix such that $\text{tr}(\mathbf{Q}_k) = 1$.

User Localization: After solving the dual problem in (5.48), localization parameters will be estimated. We can use the following null spectrum to search for ϕ^{UR} and r^{UR} as

$$P(\phi^{\mathrm{UR}}, r^{\mathrm{UR}}) = \frac{1}{|\mathbf{a}(\phi^{\mathrm{UR}} \otimes r^{\mathrm{UR}})\boldsymbol{\beta}\boldsymbol{\beta}^H \mathbf{a}(\phi^{\mathrm{UR}} \otimes r^{\mathrm{UR}})^H|}. \qquad (5.49)$$

To estimate ϕ^{UR} and r^{UR}, the objective function in (5.49) requires 2D grid search. To reduce the computational complexity, we propose utilizing an iterative solution in which we relax (5.49) into a 1D grid search for ϕ_1^{UR} given a random r_0^{UR} followed by another 1D grid search for the r_1^{UR} given ϕ_1^{UR}. We repeat the same process for k of iterations to search for the optimal $\hat{\phi}^{\mathrm{UR}} = \phi_k^{\mathrm{UR}}$ given r_{k-1}^{UR} followed by a search for the optimal $\hat{r}^{\mathrm{UR}} = r_k^{\mathrm{UR}}$ given ϕ_k^{UR}. Otherwise, (5.49) can be solved using the particle swarm optimization (PSO). For the PSO, we create a set of particles such that every particle represents a certain angle and distance, and we evaluate (5.49) at every particle location. We update the location of every individual particle based on its own location, its own optimal evaluation of (5.49), and the global optimal of (5.49) among all the particles. More specifically, let ϖ_i represents the current location of a certain particle i, let κ_i represents the best location of that certain particle i, and let g represents the best location among all the particles. Now we can update the particle i location using

$$\varpi_i(t+1) = \varpi_i(t) + \varrho_i(t+1), \qquad (5.50)$$

where $\varrho_i(t+1)$ is the updated velocity vector of particle i that can be described using

$$\varrho_i(t+1) = w\varrho_i(t) + r_1 c_1 \left(\kappa_i(t) - \varpi_i(t)\right) + r_2 c_2 \left(g(t) - \varpi_i(t)\right), \qquad (5.51)$$

where c_1 and c_2 are acceleration coefficients, r_1 and r_2 are random numbers distributed uniformly between 0 and 1, and w is the inertia coefficient.

The other channel variables θ^{BR} and d^{BR} can be similarly estimated by taking the Hermitian transpose of (5.6) and performing the same methodology again. Algorithm 1 summarizes the PSO algorithm, while Algorithm 3 summarizes the atomic norm approach.

5.3.3 RIS Phase Design

We will utilize an iterative phase design in which we first estimate the localization parameters as described in the previous sections using a

---

Algorithm 2 Particle swarm optimization (PSO) algorithm

    Input: The Lagrangian multiplier β.

1: **Initialize:** All particles positions $\varpi_i(0)$, all particles velocities $\varrho_i(0)$, acceleration coefficients c_1 and c_2, the random numbers r_1 and r_2.

2: **for** $t = 1$: maximum generation **do**

3:     **for** $i = 1$: population size **do**

4:         **if** $P(\varpi_i) < P(\kappa_i)$ **then** $\varpi_i(t) = \varpi_i(t)$ **end**

5:         Update the velocity vectors using (5.51)

6:         Update the position vectors using (5.50)

7:     **end for**

8: **end for**

    Output $P^\star(\theta^{BR}, d^{BR}) = P(\kappa_i)$

Algorithm 3 RIS-aided near-field localization via the atomic norm minimization

---

    Input: The received signal $\mathbf{Y}$, the PRS $\mathbf{X}$, the RIS location $\mathbf{p}^R$.

1: **Initialize:** $\Theta \leftarrow$ random phase design.

2: **for** $j = 1$ to J **do**

3:     Estimate $\hat{\mathbf{h}}$ using $\hat{\mathbf{h}} = \text{vec}\left(\mathbf{Y}\mathbf{X}^\dagger\right)$.

4:     Estimate β using (5.48).

5:     Estimate ϕ^{UR} and r^{UR} using (5.49).

6:     Use the estimated parameters to design the RIS according to (5.59).

7: **end for**

8: Compute the UE location using the estimated parameters.

    Output $\hat{\mathbf{p}}^U$

---

random phase design, then we use the estimated parameters to control the RIS phase matrix.

The goal is to minimize the localization error by optimizing RIS phases. Nevertheless, it is challenging to derive an analytical expression for the error. Alternatively, one can minimize the CRLB; however, the derivation of the CRLB in near-field considering reflective RIS channel with multiple scatterers is still an open problem. Therefore, we propose to maximize the SNR, as the error typically depends on it. For the SNR, it can be maximized by aiming to align the phases at the BS by minimizing the sum of the square distance of the phases from their related centroid $\bar{\epsilon}(\boldsymbol{\theta})$, i.e.,[3]

$$\boldsymbol{\theta}^{\star} = \arg\min_{\boldsymbol{\theta}\in[0,2\pi]^{N^{\mathrm{R}}}} \sum_{b,u,i,k,r} \Big[(b\hat{\omega}_i^{\mathrm{BR}} + b^2\hat{\gamma}_i^{\mathrm{BR}}) + (r\hat{\alpha}_i^{\mathrm{BR}} + r^2\hat{\beta}_i^{\mathrm{BR}})$$

$$+(r\hat{\omega}_k^{\mathrm{UR}} + r^2\hat{\gamma}_k^{\mathrm{UR}}) + (u\ \hat{\alpha}_k^{\mathrm{UR}} + u^2\hat{\beta}_k^{\mathrm{UR}}) + \theta_r - \bar{\epsilon}(\boldsymbol{\theta})\Big]^2, \qquad (5.52)$$

where $\sum_{b,u,i,k,r} \triangleq \sum_{b=-B}^{B} \sum_{u=-M}^{U} \sum_{i=1}^{L^{\mathrm{BR}}} \sum_{k=1}^{L^{\mathrm{UR}}} \sum_{r=1}^{N^{\mathrm{R}}}$, $\bar{\epsilon}(\boldsymbol{\theta})$ is defined similar to [17]. Using similar techniques to [17], after some manipulations, the proposed RIS phases can be written as

$$\boldsymbol{\theta}^{\star} = \arg\min_{\boldsymbol{\theta}\in[0,2\pi]^{N^{\mathrm{R}}}} \sum_{b,u,i,k,r} [\theta_r + C_{buikr} - \phi(\theta)]^2, \qquad (5.53)$$

where

$$C_{buikr} \triangleq j(b\ \omega_i^{\mathrm{BR}} + b^2\gamma_i^{\mathrm{BR}}) + j(r\alpha_i^{\mathrm{BR}} + r^2\beta_i^{\mathrm{BR}}) +$$
$$j(r\omega_k^{\mathrm{UR}} + r^2\gamma_k^{\mathrm{UR}}) + j(u\ \alpha_k^{\mathrm{UR}} + u^2\beta_k^{\mathrm{UR}}), \qquad (5.54)$$

and

$$\phi(\theta) = \frac{1}{N^{\mathrm{B}}N^{\mathrm{U}}N^{\mathrm{R}}L^{\mathrm{UR}}L^{\mathrm{BR}}} \sum_{b,u,i,k,r} j(b\ \omega_i^{\mathrm{BR}} + b^2\gamma_i^{\mathrm{BR}}) +$$
$$j(r\alpha_i^{\mathrm{BR}} + r^2\beta_i^{\mathrm{BR}}) + j(r\omega_k^{\mathrm{UR}} + r^2\gamma_k^{\mathrm{UR}}) + j(u\ \alpha_k^{\mathrm{UR}} + u^2\beta_k^{\mathrm{UR}}).$$
$$= \frac{1}{N^{\mathrm{R}}}\theta_z + \frac{1}{N^{\mathrm{R}}} \sum_{r=1,r\neq z}^{N^{\mathrm{R}}} \theta_r + \frac{1}{N^{\mathrm{B}}N^{\mathrm{U}}N^{\mathrm{R}}L^{\mathrm{UR}}L^{\mathrm{BR}}} \sum_{b,u,i,k,r} C_{buikr}. \qquad (5.55)$$

Now the objective function can be expressed as

$$P(\Theta) = \sum_{b,u,i,k} [\theta_z + C_{buikz} - \phi(\theta)]^2 + \sum_{b,u,i,k,r\neq z} [\theta_r + C_{buikr} - \phi(\theta)]^2 \qquad (5.56)$$

To solve this objective function, we start differentiating it with respect to θ_z

$$\frac{\partial P(\Theta)}{\partial \theta_z} = \sum_{b,u,i,k} (1 - \frac{1}{N^{\mathrm{R}}})[\theta_z + C_{buikz} - \phi(\theta)] +$$
$$\sum_{b,u,i,k,r\neq z} [\theta_r + C_{buikr} - \phi(\theta)] = 0. \qquad (5.57)$$

After some manipulation

$$\left(1 - \frac{1}{N^R}\right)\theta_z - \frac{1}{N^R}\sum_{r=1, r\neq z}^{N^R}\theta_r + \frac{1}{N^B N^U N^R L^{UR} L^{BR}} \times$$

$$\left(\sum_{b,u,i,k} C_{buikz} - \frac{1}{N^R}\sum_{b,u,i,k,r\neq z} C_{buikr}\right) = 0. \tag{5.58}$$

Now, for each value of $z \in \{1, 2, \ldots, N^R\}$, we get a linear equation with N^R unknowns. Solving the system of equations simultaneously results in

$$\theta_r^\star = \left((2U+1)(2B+1)L^{BR}L^{UR}\right)^{-1}\sum_{b,u,i,k}\Big[b\ \hat{\omega}_i^{BR}$$

$$+ b^2\hat{\gamma}_i^{BR} + r\ \hat{\alpha}_i^{BR} + r^2\hat{\beta}_i^{BR} + r\ \hat{\omega}_k^{UR} + r^2\hat{\gamma}_k^{UR}$$

$$+ u\ \hat{\alpha}_k^{UR} + u^2\hat{\beta}_k^{UR}\Big], \forall r \in \left\{1, 2, \ldots, N^R\right\}. \tag{5.59}$$

5.4 SIMULATION RESULTS

In this section, we present selected numerical results for the proposed schemes. We assume the URal noise such that $\sigma_n^2 = B_t T_k K$, where B_t is the bandwidth, $T_k = 290$ is the temperature in Kelvin, and K is the Boltzmann constant. We consider the T-MSBL algorithm to solve the CS problems, which [31] we use [36] to solve for (5.48) and (5.30) and we use MATLAB to implement the PSO. The simulation parameters are presented in Table 5.3. We set the simulation parameters as presented in Table 5.3, unless stated otherwise.

The convergence of the proposed RIS phase design for the near-field CS algorithm is shown in Figure 5.9, where the localization error vs. the number of iterations for various phase design approaches is to be compared. We compare the *(i)* proposed; *(ii)* the Eigen-beamforming (EBF) [37]; *(iii)* random; *(iv)* mirror (zero phases); *(v)* 5-level quantized phases; *(vi)* optimized with the interior point method (IPM). When comparing non-optimized phases (i.e., Mirror and Random), the suggested technique with varied optimized phases can result in an improvement of one-order-of-magnitude reduction in the positioning error. Furthermore, the suggested closed-form phase design, which does not need channel state information (CSI) knowledge, performs similarly to high computationally demanding phase design approaches, such as EBF and IPM.

TABLE 5.3   The Simulation Parameters

Description	Parameter	Value
frequency	f_c	28 GHz
Number of UE antennas	N^{U}	21
Number of BS antennas	N^{B}	51
Number of RIS elements	N^{R}	100
path loss exponent	μ	3
Number of PRS streams	M^{o}	60
BS location	$\mathbf{p}^{\mathrm{B}}$	[0 0]
RIS location	$\mathbf{p}^{\mathrm{R}}$	[5 5]
UE location	$\mathbf{p}^{\mathrm{U}}$	[10 0]
Bandwidth	B	10 MHz
Power	P	1 Watt
Noise threshold	ϵ	0.001
Number of paths between the BS and the RIS	L^{BR}	2
Number of paths between the RIS and the UE	L^{UR}	2

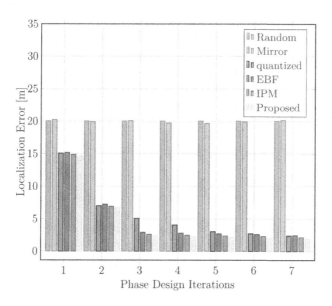

Figure 5.9   Localization error for different phase design schemes.

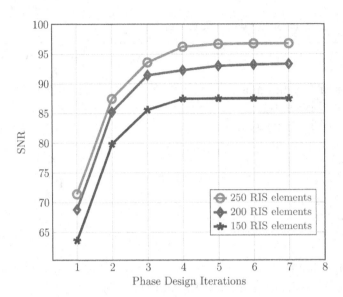

Figure 5.10   SNR vs. number of iterations.

As described in the RIS phase design subsection, the estimated user location will be used to improve the RIS phase design, while the new properly optimized RIS can lead to a better estimation of the UE location; this loop continues until convergence. In order to support the convergence claims of this approach, Figure 5.10 represents the SNR as a function of each phase design iteration. This figure supports the convergence argument as it shows an increase in the SNR with every phase design iteration. This result matches the expected theory of CS, where [38] has shown that the error $\mathbf{E}$ in the CS recovery has the following upper bound:

$$
\mathbb{E}\{||\mathbf{E}||_F^2\} \leq \left[1 - \frac{B}{N} + \frac{0.25\left(\sqrt{\frac{B}{N}}+1\right)^2}{\text{SNR}} \right] ||\hat{\mathbf{C}}^1||_F^2, \tag{5.60}
$$

where $\mathbf{C}$ is the support matrix. The above formula shows that increasing the SNR will decrease the error upper bound of the estimated support and hence will decrease the localization error.

The localization error for the near-field CS system vs. the number of RIS elements for varied numbers of BS antennas is shown in Figure 5.11. It can be shown that a larger number of elements results in improved

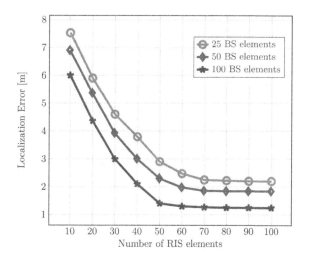

Figure 5.11 Localization error vs. numbers of RIS elements.

performance, allowing the RIS to concentrate more power on the BS. In addition, a bigger number of BS antennas means more CS measurements and better performance.

In Figure 5.12, we compute the localization error for the near-field atomic norm minimization system for different SNRs at the BS. We use $N^U = 10$, $\mu = 2$, $M^o = 20$, $B_t = 5$ MHz, and $P = 0.5$ Watt. The RIS

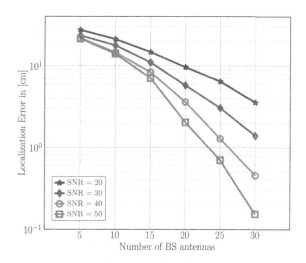

Figure 5.12 Localization error for different numbers of BS antennas.

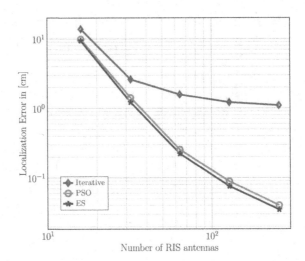

Figure 5.13 Extracting the location parameters using different approaches.

is located at $\mathbf{p}^R = [2.5, 2.5]$, the UE is located at $\mathbf{p}^U = [5, 5]$, and we set the number of paths to one (i.e., $L^{BR} = L^{UR} = 1$). We change the SNR value by changing σ_z^2. It can be seen from the figure that a lower localization error could be achieved by increasing both the SNR and the number of BS antennas. The results showed that the sub-cm level of localization error can be achieved using higher values of SNR and N^B.

In Figure 5.13, we compare the localization error for different optimization algorithms for solving (5.49), and we compare the exhaustive search (ES), PSO, and the iterative solution described in Section 5.3.2. The figure shows a similar performance between the ES and the PSO, while a slightly higher error for the iterative solution.

To judge the performance of the three optimization algorithms, we compute the computational efforts for the three cases. Table 5.4 shows hardware specifications for the workstation that is used for simulation,

TABLE 5.4 Workstation Specifications

Aspect	Specification
CPU	Intel(R) Core(TM) i5-10500 CPU @ 3.10GHz 3.10 GHz
GPU	Intel(R) UHD Graphics 360
Memory	16.0 GB DDR4-SDRAM
OS	Windows 10 Education 64-bi

TABLE 5.5 Computational Efforts

Number of RIS elements N^R	Number of BS antennas N^B	PSO	Algorithm ES Time in [s]	Iterative
64	15	3.3115	3.9251	2.5414
	35	19.4865	26.4548	9.47864
128	15	4.4866	4.4698	3.4564
	35	21.4869	27.4856	10.4564

while Table 5.5 represents the required time to solve for the UE location. The table shows that the iterative solution can significantly reduce the required simulation time, while it can produce a similar slightly larger localization error. On the other hand, the PSO can produce the same localization accuracy as the ES but with lower computational effort.

In Figure 5.14, we compare the localization performance of the proposed atomic norm minimization and CS. We fix $N^B = 35$, and we change the number of RIS elements N^R and the number of the PRS streams M^o. The results show the superior performance of the proposed atomic norm minimization in comparison to the CS. This can be justified as the CS utilizes a finite number of atoms on a discrete grid leading to a quantization error, while the proposed atomic norm uses a continuous set of atoms rather than a discrete one. The figure also shows that

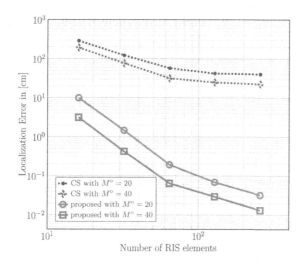

Figure 5.14 Localization error comparison between the CS and the proposed scheme.

we can achieve better performance using a RIS with a large number of elements, as with a larger N^R the RIS can make a narrow beam toward the UE location.

5.5 CONCLUSION AND FUTURE WORK

In this chapter, the uplink localization in RIS-aided models has been investigated. We considered different channel settings either in far-field, near-field, multipath, and single path. We solved the problem via sparse recovery techniques, more specifically the CS and atomic norm minimization. For the near-field model, we exploited the spatial correlation in the channel model to convert the CS problem from the complex 2D case to two easy-to-solve 1D problems. We further solved the high localization errors in near-filed by considering a super-resolution solution that is based on atomic norm minimization. In addition to that, we proposed an RIS phase design such that we maximize the SNR at the BS to minimize the localization errors. The numerical results show the validation of the proposed phase design, where we compared the proposed methodology with different phase design techniques. We also compare the localization performance of both the CS against the atomic norm minimization, and the results show that a sub-cm level of localization accuracy can be achieved using the atomic norm minimization. Up to this end, the proposed super-resolution scheme has shown great potential in terms of its performance.

RIS-aided localization is still at an early development stage, and a huge research effort is required to enable real-time reliable and accurate RIS-based localization operation. Possible research directions for RIS-aided localization might include (*i*) *real-time reliable RIS-aided localization for practical environments*, this work focused on wireless localization in far-field, near-field, single path, and multipath; however, the doors are still open for more complicated scenarios such as multi-user localization, real-time localization and dynamic channel localization. (*ii*) *Active RISs*, this work has focused on utilizing passive RIS for user localization, active RIS as an alternative does not only reflect the wireless signals toward the desired location but can also amplify the received signal to encounter the propagation losses. Utilizing the capabilities of active RIS can significantly improve the localization performance and more active RIS localization is still lacking in the literature. (*iii*) *Error bound derivation for RIS-aided localization,* this work has included some insights regarding the work that has been done in terms of deriving the error bounds. Similar to localization under complex settings, each complex scenario requires error derivations. Additionally, the model mismatch errors, which can be developed as a result of utilizing localization algorithms that solve channel models that are different than the actual model, is an interesting problem for research. Recently, literature has exhibited trendy actions to tackle this problem where special interest is developing with time.

BIBLIOGRAPHY

[1] Omar Rinchi, Sinan A. Assaid, and Hussam J. Khasawneh. Accurate Android-Based Navigation using Fuzzy Adaptive Extended Kalman Filter. In *Proc. of the 2-nd IEEE Jordan International Joint Conference on Electrical Engineering and Information Technology (JEEIT'21)*, pages 234–239, Amman, Jordan, Dec. 2021.

[2] Ahmed Elzanaty, Anna Guerra, Francesco Guidi, Davide Dardari, and Mohamed-Slim Alouini. Towards 6G Holographic Localization: Enabling Technologies and Perspectives. *arXiv preprint arXiv:2103.12415*, Feb. 2021.

[3] 3GPP Technical Specification 22.261 "Service requirements for the 5G systems", Release 18, v. 18.1.0, December, 2020.

[4] Cuneyd Ozturk, Musa Furkan Keskin, Henk Wymeersch, and Sinan Gezici. RIS-aided near-field localization under phase-dependent amplitude variations. *arXiv preprint arXiv:2204.12783*, Apr. 2022.

[5] Hussam Ibraiwish, Ahmed Elzanaty, Yazan H. Al-Badarneh, and Mohamed-Slim Alouini. EMF-aware cellular networks in RIS-assisted environments. *IEEE Communication Letters*, 1, Oct. 2021.

[6] Abdullah Almasoud, Mohamed Y. Selim, Ahmad Alsharoa, and Ahmed E. Kamal. Improvement of bi-directional communications using solar powered reconfigurable intelligent surfaces. In *Proc. of the 30-th IEEE International Conference on Computer Communications and Networks (ICCCN'21)*, pages 1–8, Athens, Greece, July 2021.

[7] Nemanja Stefan Perović, Marco Di Renzo, and Mark F. Flanagan. Channel capacity optimization using reconfigurable intelligent surfaces in indoor mmWave environments. In *Proc. of the 30-th IEEE*

International Conference on Communications (ICC'20), pages 1–7, June 2020. Virtual Conference.

[8] Visa Tapio, Ibrahim Hemadeh, Alain Mourad, Arman Shojaeifard, and Markku Juntti. Survey on reconfigurable intelligent surfaces below 10 GHz. *EURASIP Journal on Wireless Communications and Networking*, 2021(1):1–18, Sept. 2021.

[9] Filippo Costa and Michele Borgese. Electromagnetic model of reflective intelligent surfaces. *IEEE Open Journal of the Communications Society*, 2:1577–1589, June 2021.

[10] Jing Nie, Yan-Qing Tan, Chun-Lin Ji, and Ruo-Peng Liu. Analysis of Ku-band steerable metamaterials reflectarray with tunable varactor diodes. In *Proc. of the 1st Progress in Electromagnetic Research Symposium (PIERS'16)*, pages 709–713, Shanghai, China, Aug. 2016. IEEE.

[11] Xilong Pei, Haifan Yin, Li Tan, Lin Cao, Zhanpeng Li, Kai Wang, Kun Zhang, and Emil Björnson. RIS-aided wireless communications: Prototyping, adaptive beamforming, and indoor/outdoor field trials. *IEEE Transactions on Communications*, 69(12):8627–8640, Sept. 2021.

[12] David Rotshild and Amir Abramovich. Wideband reconfigurable entire Ku-band metasurface beam-steerable reflector for satellite communications. *IET Microwaves, Antennas & Propagation*, 13(3):334–339, Jan. 2019.

[13] Jiguang He, Henk Wymeersch, Long Kong, Olli Silvén, and Markku Juntti. Large intelligent surface for positioning in millimeter wave MIMO systems. In *Proc. of the 91-st IEEE Vehicular Technology Conference (VTC2020-Spring)*, pages 1–5. IEEE, May 2020. Virtual Conference.

[14] Henk Wymeersch and Benoît Denis. Beyond 5G wireless localization with reconfigurable intelligent surfaces. In *Proc. of the 30-th IEEE International Conference on Communications (ICC'20)*, pages 1–6, June 2020. Virtual Conference.

[15] Teng Ma, Yue Xiao, Xia Lei, Wenhui Xiong, and Yuan Ding. Indoor localization with reconfigurable intelligent surface. *IEEE Communication Letters*, 25(1):161–165, Sept. 2021.

[16] Yiming Liu, Erwu Liu, Rui Wang, Zhu Han, and Binyu Lu. Asymptotic achievability of the CramÃĺr-Rao lower bound of channel estimation for reconfigurable intelligent surface aided communication systems. *IEEE Wireless Communication Letters*, 10(11):2607–2611, Nov. 2021.

[17] Ahmed Elzanaty, Anna Guerra, Francesco Guidi, and Mohamed-Slim Alouini. Reconfigurable intelligent surfaces for localization: Position and orientation error bounds. *IEEE Transactions on Signal Processing*, 69:5386–5402, Aug. 2021.

[18] Alessio Fascista, Angelo Coluccia, Henk Wymeersch, and Gonzalo Seco-Granados. RIS-aided joint localization and synchronization with a single-antenna mmWave receiver. In *Proc. of the 50-th IEEE International Conference on Acoustics, Speech and Signal Processing (ICASSP'21)*, pages 4455–4459, Toronto, Ontario, Canada, June 2021.

[19] Sha Hu, Fredrik Rusek, and Ove Edfors. Beyond massive MIMO: The potential of positioning with large intelligent surfaces. *IEEE Transactions on Signal Processing*, 66(7):1761–1774, Jan. 2018.

[20] Hui Chen, Ahmed Elzanaty, Reza Ghazalian, Musa Furkan Keskin, Riku Jäntti, and Henk Wymeersch. Channel model mismatch analysis for XL-MIMO systems from a localization perspective. *arXiv preprint arXiv:2205.15417*, May 2022.

[21] Jingwen Zhang, Zhong Zheng, Zesong Fei, and Xuyan Bao. Positioning with dual reconfigurable intelligent surfaces in millimeter-wave MIMO systems. In *Proc. of the 9-th IEEE/CIC International Conference on Communications in China (ICCC'20)*, pages 800–805, Xiamen, China, July 2020.

[22] Kamran Keykhosravi, Musa Furkan Keskin, Gonzalo Seco-Granados, and Henk Wymeersch. SISO RIS-enabled joint 3D downlink localization and synchronization. In *Proc. of the 31-st IEEE International Conference on Communications (ICC'21)*, pages 1–6, Montreal, QC, Canada, Aug. 2021.

[23] Emrah Čišija, Aya Mostafa Ahmed, Aydin Sezgin, and Henk Wymeersch. Ris-aided mmWave MIMO radar system for adaptive

multi-target localization. In *Proc. of the 21-th IEEE Statistical Signal Processing Workshop (SSP'21)*, pages 196–200, Rio de Janeiro, Brazil, Aug. 2021.

[24] Jiguang He, Henk Wymeersch, and Markku Juntti. Channel estimation for RIS-aided mmWave MIMO systems via atomic norm minimization. *IEEE Transactions on Wireless Communications*, 20(9):5786–5797, Apr. 2021.

[25] Haobo Zhang, Hongliang Zhang, Boya Di, Kaigui Bian, Zhu Han, and Lingyang Song. Metalocalization: Reconfigurable intelligent surface aided multi-user wireless indoor localization. *IEEE Transactions on Wireless Communications*, 20(12):7743–7757, Dec. 2021.

[26] Zohair Abu-Shaban, Kamran Keykhosravi, Musa Furkan Keskin, George C. Alexandropoulos, Gonzalo Seco-Granados, and Henk Wymeersch. Near-field localization with a reconfigurable intelligent surface acting as lens. In *Proc. of the 31-st International Conference on Communications (ICC'21)*, pages 1–6, Montreal, QC, Canada, Aug. 2021.

[27] Davide Dardari, Nicoló Decarli, Anna Guerra, and Francesco Guidi. LOS/NLOS near-field localization with a large reconfigurable intelligent surface. *IEEE Transactions on Wireless Communications*, 21(6):4282–4294, June 2022.

[28] Omar Rinchi, Ahmed Elzanaty, and Mohamed-Slim Alouini. Compressive near-field localization for multipath RIS-aided environments. *IEEE Communication Letters*, 26(6):1268–1272, June 2022.

[29] Benjamin Friedlander. Localization of signals in the near-field of an antenna array. *IEEE Transactions on Signal Processing*, 67(15):3885–3893, Aug. 2019.

[30] Theodore S. Rappaport, Shu Sun, Rimma Mayzus, Hang Zhao, Yaniv Azar, Kevin Wang, George N. Wong, Jocelyn K. Schulz, Mathew Samimi, and Felix Gutierrez. Millimeter wave mobile communications for 5G cellular: It will work! *IEEE Access*, 1:335–349, May 2013.

[31] Zhilin Zhang and Bhaskar D. Rao. Sparse signal recovery with temporally correlated source vectors using sparse bayesian learning.

IEEE Journal of Selected Topics in Signal Processing, 5(5):912–926, June 2011.

[32] Ahmed Elzanaty, Andrea Giorgetti, and Marco Chiani. Weak RIC analysis of finite Gaussian matrices for joint sparse recovery. *IEEE Signal Processing Letters*, 24(10):1473–1477, July 2017.

[33] Wen-Gen Tang, Hong Jiang, and Shuai-Xuan Pang. Grid-free DOD and DOA estimation for MIMO radar via duality-based 2D atomic norm minimization. *IEEE Access*, 7:60827–60836, May 2019.

[34] Ahmed Alwakeel and Ahmed Mehana. Achievable rates in uplink massive MIMO systems with pilot hopping. *IEEE Transactions on Communications*, 65(10):4232–4246, June 2017.

[35] Bogdan Dumitrescu. *Positive trigonometric polynomials and signal processing applications*, volume 103. Springer, 2007.

[36] Michael Grant and Stephen Boyd. CVX: Matlab software for disciplined convex programming, version 2.1. 2014. [online: available at: www.cvxr.com/cvx]

[37] Hossein Khaleghi Bizaki. *MIMO systems: Theory and applications*. BoD–Books on Demand, 2011.

[38] Mark A Davenport, Jason N Laska, John R Treichler, and Richard G Baraniuk. The pros and cons of compressive sensing for wideband signal acquisition: Noise folding versus dynamic range. *IEEE Transactions on Signal Processing*, 60(9):4628–4642, May 2012.

NOTES

[2]In the following, we refer to $\mathbf{V}$ as the covariance matrix, albeit it represents a part of the covariance matrix.

[3]If the radio frequency chains of the RIS, ρ_l^{BR} and ρ_l^{UR}, can be estimated, then the amplitudes and phases of the paths can be considered in the optimization.

CHAPTER 6

The Emergence of Aerial Computing

Applications and Challenges

Quoc-Viet Pham

School of Computer Science and Statistics, Trinity College Dublin, Dublin, Ireland

Thien Huynh-The

Department of Computer and Communication Engineering, Ho Chi Minh City University of Technology and Education, Ho Chi Minh City, Vietnam

Ming Zeng

Department of Electrical and Computer Engineering, Laval University, Quebec, Canada

Zhaohui Yang

College of Information Science and Electronic Engineering, Zhejiang University, Hangzhou, China

Zhiguo Ding

School of Electrical and Electronics Engineering, The University of Manchester, Manchester, UK

Won-Joo Hwang

Department of Biomedical Convergence Engineering, Pusan National University, Yangsan, Republic of Korea

CONTENTS

6.1   Introduction 119
        6.1.1   Main Contributions 120

      6.1.2   Chapter Organization 121

6.2   Aerial Computing: Fundamentals and Design 121

      6.2.1   System Architecture 122

            6.2.1.1   IoT and Computing Devices 122

            6.2.1.2   Terrestrial Computing 122

            6.2.1.3   Low-Altitude Computing 124

            6.2.1.4   High-Altitude Computing 125

            6.2.1.5   Satellite Computing 125

            6.2.1.6   Vertical Domain Applications 126

      6.2.2   Desirable Features and Role of Aerial Computing in
            6G 126

6.3   Aerial Computing: Vertical Domain Applications 127

      6.3.1   Smart Cities 128

      6.3.2   Smart Vehicles 129

      6.3.3   Smart Factories 130

      6.3.4   Smart Grids 131

6.4   Aerial Computing: Projects and Standardization 133

      6.4.1   Projects 133

            6.4.1.1   6G Flagship 133

            6.4.1.2   South Korea MSIT 6G Research Program 133

            6.4.1.3   Japan 6G/B5G Promotion Strategy 134

            6.4.1.4   Industrial 6G Programs 134

            6.4.1.5   China 6G White Paper 135

            6.4.1.6   Europe 6G Program 135

            6.4.1.7   USA 6G Program 136

      6.4.2   Standardization 136

            6.4.2.1   ETSI 137

            6.4.2.2   IETF 138

            6.4.2.3   NGMN Alliance 138

            6.4.2.4   Next G Alliance 138

            6.4.2.5   3GPP 139

            6.4.2.6   IEEE 139

            6.4.2.7   ATIS 139

            6.4.2.8   ITU-T Standardization Sector 140

      6.4.3   Summary and Discussion 140

6.5   Aerial Computing: Challenges and Future Directions 141

      6.5.1   Energy Efficiency 141

      6.5.2   Resource Management 142

      6.5.3   Network Stability 142

      6.5.4   Large-Scale Network Optimization 143

      6.5.5   Security and Privacy 143

6.6   Conclusion 144

Bibliography 144

E DGE COMPUTING AND aerial access networks are two key technologies to revolutionize the conventional network and computing
paradigms. Their amalgamation introduces a novel computing concept,
namely aerial computing, which can provide more advanced computing
services at a global scale. This chapter first frames the concept of aerial
computing and presents it in a comprehensive computing infrastructure
in sixth-generation (6G) systems. Then, we analyze that aerial computing has distinctive features for providing advanced services, including
mobility, availability, scalability, flexibility, and simultaneity. After that,
we depict the role of aerial computing in important vertical domains,
including smart vehicles, smart cities, manufacturing, and smart grids.
We also summarize leading research and standardization efforts toward
the realization of aerial computing. Finally, we argue critical challenges
related to energy efficiency resource management, network scalability,
large-scale optimization, and privacy and security for the realization of
aerial computing in 6G systems.

6.1   INTRODUCTION

Many new applications and services will emerge in future sixth-
generation (6G) wireless systems. Some examples are fully autonomous
vehicles, metaverse, connected intelligence, Industry 5.0, which usually
require powerful computing capabilities to satisfy new challenging key
performance indicators (KPI) in 6G systems [1, 2]. Moreover, the availability of computing resources with favorable communication conditions
is a prerequisite for a large-scale deployment of new applications and
services at a global scale. In this context, edge computing and aerial access networks are two promising technologies, and the interplay between
these two technologies will enable many new applications, services, and
use cases in 6G wireless systems.

Edge computing, such as fog computing and multi-access edge computing (MEC), moves computing and caching resources from the central
cloud to the network edge in close proximity to end users (e.g., vehicles, sensors, Internet-of-Things (IoT) devices, and wearables). Thus,
edge computing is a key enabler of new compute-intensive applications,

such as X-reality and fully autonomous driving [3]. An easy way is to deploy MEC nodes at fixed locations (e.g., radio towers and gateways in mobile networks); however, it raises the issues of providing computing services and wireless access to unconnected areas (e.g., rural areas and mountains). On the other hand, aerial access networks can offer distinct features over conventional networks, such as line-of-sight propagation, favorable channel conditions, global coverage. In general, an aerial access network (AAN) is constituted of low-altitude platforms (LAPs, e.g., unmanned aerial vehicle [UAV]), high-altitude platforms (HAPs, e.g., airships and balloons), and Low Earth orbit (LEO) satellites [4]. Based on the significant features of edge computing and aerial access networks, we introduce the concept of aerial computing in this chapter. It is expected that aerial computing inherits amalgamated features from edge computing and aerial access networks. Thus, aerial computing can support more advanced vertical applications in 6G wireless systems and overcome the digital divide issue in a fully digital world.

6.1.1 Main Contributions

The roles of edge computing and aerial access networks for emerging applications and future networks have been provided in many studies; however, a concise discussion on aerial computing is still missing. To fill this gap, we introduce a novel computing concept and analyze its important applications and critical challenges for future research. In a nutshell, the main contributions offered by this chapter can be summarized as follows.

- **New Concept of Aerial Computing**: We introduce the concept of aerial computing, which is expected to inherit significant features from conventional edge computing paradigms and aerial access networks. In this regard, we propose a novel architecture of aerial computing toward a comprehensive computing architecture in 6G systems. Further, we analyze that aerial computing has important features for the provision of advanced wireless and computing services in the future, including ubiquity, mobility, availability, simultaneity, and scalability.

- **The Role of Aerial Computing for Vertical Domains**: After introducing the concept, we depict the role of aerial computing for vertical domain applications, such as smart cities, smart vehicles, smart factories, and smart grids. In particular, we discuss how vertical domain applications are supported by aerial computing.

- *Projects and Standardization*: To understand current research activities, we focus on reviewing leading projects and standardization efforts related to aerial computing. We review leading projects on aerial computing and 6G wireless networks, such as the 6G Flagship, Europe 6G Program, and national projects, as well as several industrial 6G programs. Standardization plays an important role in defining the technological requirements and expected timeline of development, trials, and launch of aerial computing. Therefore, we also summarize the standardization efforts and status of the standards developing organizations (SDOs).

- *Research Challenges and Directions*: We reveal that aerial computing fully complements terrestrial computing paradigms, thus enabling the provision of computing services and wireless access at a global scale. However, for the realization of aerial computing in 6G, several challenges related to energy efficiency, resource management, network stability, large-scale optimization, and security and privacy should be further investigated. These challenges are discussed in detail in this chapter.

6.1.2 Chapter Organization

The organization of this chapter is as follows. This chapter first introduces the architecture of aerial computing in a comprehensive 6G computing infrastructure and analyzes the distinct features of aerial computing. Next, the role of aerial computing for vertical domain applications, including smart cities, smart vehicles, smart factories, and smart grids, is depicted. Then, we summarize major projects and standardization efforts toward aerial computing. After that, critical challenges and potential solutions for the realization of aerial computing in 6G network systems are discussed. Finally, the conclusion is drawn.

6.2 AERIAL COMPUTING: FUNDAMENTALS AND DESIGN

In this section, we first present the system architecture of aerial computing and then analyze the fundamental features of aerial computing. Accordingly, aerial computing will be an indispensable part of 6G wireless systems and beyond.

6.2.1 System Architecture

Toward a comprehensive computing infrastructure, we present a novel computing architecture in this section. This architecture is constituted by one layer of computing and IoT devices and four computing layers: conventional terrestrial computing, low-altitude computing (LAC), high-altitude computing (HAC), and satellite computing (SC). These five layers are positioned in a hierarchical manner. On the top, vertical domain applications can be supported by aerial computing.

6.2.1.1 IoT and Computing Devices

The first layer refers to the IoT and computing devices (e.g., sensors, wearables, smartphones, and autonomous vehicles) that can generate data, perform local data processing and learning, and have compute-intensive tasks [5]. These data and tasks can be either processed locally or computed at remote computing nodes via computation offloading. For example, mobile security is critical today due to the emergence of mobile services (e.g., mobile commerce, mobile health, and mobile blockchain) [6]. For example, the work in [7] proposed that a set of IoT devices train their federated learning models locally and then share the model updates with the edge server collocated at one UAV in the air. IoT devices may not have enough on-device computing capabilities to run compute-intensive tasks. They need to offload these tasks to a target edge server, such as an MEC server located at a macro base station and a HAP/LAP platform equipped with computing capabilities. It is worth noting that IoT devices can position on the ground or the air in the case of aerial components, such as UAVs, drones, and balloons. With aerial computing, aerial devices with computing tasks have more possibilities to select the target edge server, e.g., a MEC server on the ground and a more powerful LAC/HAC server.

6.2.1.2 Terrestrial Computing

This layer refers to the deployment of conventional computing paradigms. According to [8], a terrestrial computing node can be deployed at various sites and network nodes. For example, MEC servers can be located at a network node (e.g., radio towers and optical network units) within the radio access network and in close proximity of end devices, and fog nodes can be deployed at distributed locations (e.g., IoT

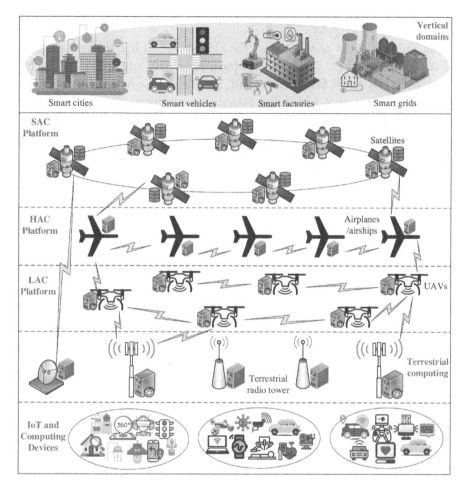

Figure 6.1 Aerial computing in a comprehensive 6G computing infrastructure.

gateways and routers). Terrestrial computing, such as fog computing and MEC, is characterized by several features, including on-premise, proximity, low latency, location awareness, and contextual information. However, the performance of terrestrial computing is limited by the immobile deployment and low coverage of edge nodes. The limitations of terrestrial computing can be overcome via a full complement to aerial computing and cloud computing, as illustrated in Figure 6.1.

6.2.1.3 Low-Altitude Computing

From Figure 6.2, as positioned at the closest tier to IoT devices with an altitude of 0–10 km, the LAC platform has clear advantages in terms of cost-effectiveness, fast deployment, and favorable network conditions over the conventional terrestrial computing platform. As a result, LAC nodes can be deployed to serve critical applications and provide temporary services to IoT and computing devices on the ground, especially when the terrestrial computing platform is not available or dysfunctional [9]. For example, the work in [10] considered that MEC-UAV has the potential for reconnaissance. In particular, the UAV, as an LAP server, can process the raw reconnaissance data (e.g., videos, images,

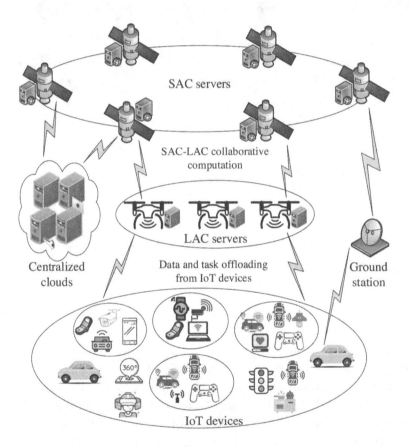

Figure 6.2 A multi-tier aerial computing network, where computation tasks can be executed by IoT/mobile devices, UAVs, satellites, and conventional centralized clouds.

and tracked locations), thus reducing the processing latency and increasing the network utility when compared with the conventional terrestrial computing deployment. Meanwhile, the platform is constrained by the limited energy capacities and low computing capabilities of LAC nodes. Recently, the work in [11] proposed to deploy an LAC server as the learning server, which helps to aggregate local model updates from wireless devices on the ground. This work also developed a deep reinforcement learning approach to maximize a long-term objective function by optimizing the LAC server placement and resource allocation, which are subject to several constraints on harvested energy of federated learning (FL) users, limited bandwidth capacity, and energy budget of the LAC server.

6.2.1.4   High-Altitude Computing

Unlike the LAC platform, which usually comprises unmanned LAC nodes, the HAC platform is constituted by both manned and unmanned components, such as airplanes and airships. Since the HAC platform operates at an altitude of 17–50 km, the coverage area becomes more significant than the LAC platform and can be 10 km. Furthermore, since HAC nodes can accommodate large solar panels, HAC nodes can be effectively powered by solar power sources and thus have a higher endurance time than that of LAC nodes [12]. As a result, the HAC platform has a higher endurance time than the LAC platform while having faster deployment and adaptability to the computing demands than the terrestrial computing platform. In addition, the HAC platform can provide wireless and computing services to end users directly instead of connecting to the SC platform via ground stations. Due to the massive investment and commercial projects of big techs, such as Google Skybender and ApusDuo, it is expected that the HAC platform will play a vital role in providing computing services in 6G systems.

6.2.1.5   Satellite Computing

Positioned at the highest computing tier, the SC platform comprises LEO satellites with computing capabilities. Typically, LEO satellites have a circular orbit at an altitude of above 80 km and less than 2000 km [13]. For example, SpaceX's LEO satellites delivered by Starlink operate at an altitude of approximately 550 km and provide wireless services with an end-to-end (E2E) latency of 20–40 ms and an expected speed of 50–150 Mbps (https://www.starlink.com/). As noted above, the SC platform does not typically provide computing services to end users directly due to the cost of embedded hardware in the satellites and the small size of IoT devices. Instead, ground stations help LEO satellites relay computing tasks between end users and computing nodes (e.g., centralized clouds) via fronthaul and backhaul links [13], as shown in Figure 6.2. However, empowering the satellites with MEC capabilities helps provide computing services globally and makes the SC platform suit large-scale applications in scenarios in which the users are located sparsely over different areas. As an example, the work in [14] studied a promising HAC-SC scenario, where computation tasks offloaded from IoT devices can be executed by both the HAC server and the SC server.

6.2.1.6 Vertical Domain Applications

With new features and advantages, aerial computing can support various vertical domain applications, including smart cities, smart vehicles, smart factories, and smart grids. For example, the Internet of Vehicles (IoV) can significantly improve the driving safety and enhance the quality of infotainment services to in-vehicle users. Typically, IoV applications are supported by the terrestrial computing platform with fixed deployment for maximum benefits of IoV service providers. In this context, vehicles can offload compute-intensive tasks to aerial computing nodes (e.g., LAC, HAC, and SC) for service continuity, and in-vehicle users can fetch cached contents from the air for enhanced infotainment services. We will articulate the role of aerial computing in four vertical domain applications. Moreover, aerial computing can support various distributed learning scenarios (i.e., federated learning) [15], where aerial components can be deployed as learning servers in the air, aerial users, aerial relays, and swarms. Therefore, aerial computing has the potential to enhance the data privacy of end users in vertical applications when the learning server does not need to collect and store the user data in a central manner.

6.2.2 Desirable Features and Role of Aerial Computing in 6G

In addition to features inherited from conventional edge computing paradigms (e.g., fog and MEC), aerial computing has several new features, as discussed below.

- *Mobility*: While terrestrial computing requires much time for planning procedures and civil works, it is quick to deploy an aerial computing node owing to the high mobility of aerial platforms, such as UAVs at the LAC platform and airships at the HAC platform.

- *Availability*: As different computing platforms are positioned hierarchically, aerial computing can offer always-available computing resources to end users in areas beyond the provision coverage of terrestrial computing, such as disaster and rural areas.

- *Scalability*: Massive IoT devices can be served thanks to the hierarchical and quick deployment of aerial platforms and the global coverage of SC. Moreover, collaborative computation among computing platforms can help execute computation tasks with a large-sized and heavy workload.

- *Flexibility*: Unlike fixed deployment in terrestrial computing, computing platforms (e.g., LAC, HAC, and SC) in aerial computing can be flexibly controlled to suit different computation tasks and network situations.

- *Simultaneity*: Computing services can be provided by multiple platforms in aerial computing, e.g., IoT devices offload computation tasks to terrestrial MEC and LAC servers simultaneously. Simultaneity is also represented by the simultaneous provision of computing services for geographical users.

The above features make aerial computing a promising concept for the design of a comprehensive computing infrastructure in 6G systems and the implementation of vertical domain applications, which are discussed in Section 6.3. It is worth emphasizing that aerial computing aims to make computing capabilities available at all times and everywhere globally, while the other computing paradigms, such as cloudlet, fog computing, and MEC [8], are to move computing resources to the network edge of certain areas and strategic locations. Therefore, aerial computing is a promising concept toward a comprehensive computing infrastructure in 6G wireless systems.

6.3   AERIAL COMPUTING: VERTICAL DOMAIN APPLICATIONS

Benefiting from distinct features, aerial computing can support and open new opportunities for many vertical domain applications. In this section, we will discuss the role of aerial computing in four domains, including smart cities, smart vehicles, smart factories, and smart grids.

6.3.1 Smart Cities

The popularity of the IoT helps cities become smarter, and the concept of smart cities is gaining a strong attraction from communities [6]. Data collected by the massive number of IoT and sensory devices can be used for various purposes, such as air quality measurement, crowd management, emotion and facial recognition, physical distancing (also known as social distancing for COVID-19 controlling), and urban planning. As a full complement to the conventional terrestrial system, the use of aerial computing platforms for data and computation processing in smart cities is a good choice. In this context, aerial computing is anticipated to support potential use cases in smart cities, as discussed below.

A potential use case of aerial computing in smart cities is illustrated in Figure 6.3, where an LAC server is dispatched to open areas for monitoring the physical distance rules and crowd management during the

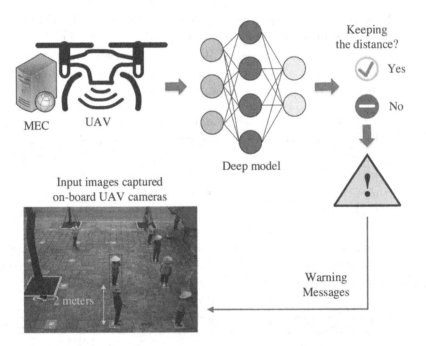

Figure 6.3 An example scenario for smart cities, where the LAC server deploys a deep learning model to process surveillance images for social distancing purposes.

COVID-19 pandemic. It has been shown that social distancing is an efficient non-pharmaceutical solution to combat the pandemic. The LAC server with onboard cameras can capture surveillance images that are then fed to a deep model to detect the existence of pedestrians in a certain area and check whether or not the physical distance is kept less than 2 m. An example of this use case is investigated in [16], where a lightweight deep model is designed to learn multi-scale representations and spatial attention maps. The designed network can proficiently capture the features of human heads and achieve a high detection accuracy. In particular, the accuracy of around 90% and the detection speed of around 75 frames per second show that the deep model in [16] is feasible for real-time social distancing detection.

6.3.2 Smart Vehicles

Many intelligent transportation systems have been investigated over the last decades to improve transportation efficiency, driving safety, real-time traffic prediction, and in-vehicle infotainment, as illustrated in Figure 6.4. With computing resources and quick deployment capabilities,

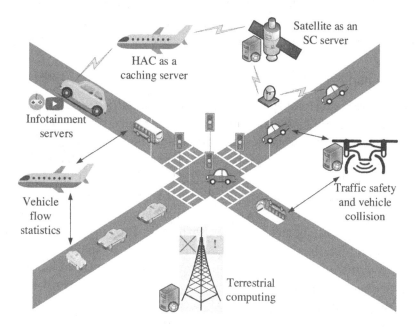

Figure 6.4 An example scenario of intelligent transportation systems supported by different computing platforms in aerial computing.

aerial components can act as computing servers in the air and provide computing services in response to the fast-growing demands of intelligent transportation applications. Moreover, in intelligent transportation systems, intelligent vehicles are typically equipped with cameras and sensors as well as localization and computing resources. Thus, intelligent vehicles generate a mass of data every day from a plethora of applications, which motivates the need for efficient aerial computing at the network edge, especially in remote and hard-to-reach areas where terrestrial computing is not available.

To illustrate the role of aerial computing in smart vehicles, we present a promising scenario considered in a recent work [17]. This work considers cooperation between an LAC server in the air and platoon vehicles on the ground. In particular, the LAC server is deployed to provide computing resources to platoon vehicles, equipped with a radio frequency (RF) transmitter to power the LAC server wirelessly and support its service duration. In such a scenario, the system optimization is constrained by network dynamics and factors, such as LAC's limited computing resources and battery capacities, complex communication channels, and time allocation. Simulation results show that the proposed iterative algorithm in [17] can achieve higher computation bits than several benchmark schemes that ignore the optimization of some design variables (e.g., time, computing resources, and power control).

6.3.3 Smart Factories

Many emerging technologies, such as artificial intelligence (AI), big data analytics, IoT, and cloud computing, have been investigated in Industry 4.0 to improve the process automation and manufacturing efficiency. Meanwhile, many have started to conceptualize Industry 5.0 and develop potential applications and use cases. Technically, the key advantage of aerial computing is the provision of available computing resources on a global scale. Moreover, the operation of smart factories hinges on efficient data processing at the network edge with critical requirements (e.g., low latency and high reliability). Thus, aerial computing can offer significant value to numerous use cases and applications in smart factories, such as supply management, manufacturing monitoring, robotic control, sensory analysis, and massive industrial IoT access.

Similar to the hierarchical architecture of aerial computing, the architecture designed for smart factories may have different layers, including the industrial IoT tier (e.g., machines, sensors, and robots), the edge

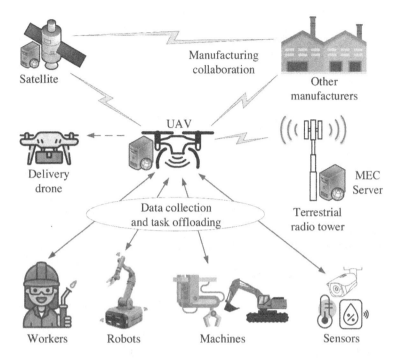

Figure 6.5 Illustration of applications of aerial computing for smart factories, where data from IoT and industrial devices are collected and processed by aerial components, such as LAC and SC platforms.

layer (e.g., LAC/HAC/SC servers and terrestrial MEC), and the cloud layer (e.g., centralized clouds). As shown in Figure 6.5, a large amount of data collected from industrial IoT devices can be processed at the LAC server via the learning ability of AI and deep learning techniques. The learning result is helpful for the factory to gain a clear perception and semantic interoperability of manufacturing operations and collected data. Moreover, a manufacturer can use enormous resources available at the other manufacturers, thus addressing limited resources and gaining added value. The aerial computing paradigm can achieve it because of a large-scale collaboration between different computing platforms.

6.3.4 Smart Grids

Smart grids have been investigated to resolve issues related to the operation and problems of traditional electrical systems. With the proliferation of massive connected and IoT devices and the popularity of emerging

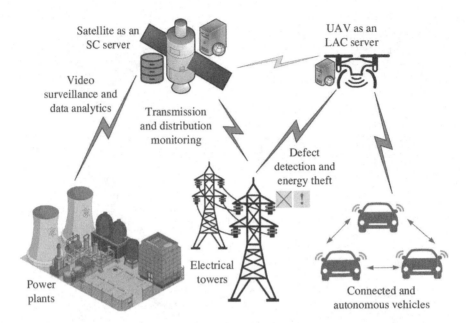

Figure 6.6 Applications of aerial computing for different scenarios in smart grids, including video surveillance and data analytics, transmission and distribution monitoring, defect detection and energy theft, and autonomous vehicles.

technologies and applications, it is expected that the use of unusual energy sources, such as fuel cells, solar cells, winds, photovoltaic, and mechanical energy, will play a more significant role in the coming years. Further, connected and autonomous vehicle is a technological evolution that may affect the electrical grid and demand powerful computing resources. Indeed, there are challenges in smart grids, such as security and privacy, data management, and monitoring and inspection. As a novel concept, aerial computing plays a vital role in providing computing resources to smart grid applications, as shown in Figure 6.6.

In aerial computing, SC servers can be utilized for video surveillance tasks of remote utilities. In addition, LAC and HAC platforms can be deployed for real-time monitoring and data preprocessing to avoid the issue of continuous data transmission to the SC servers. Aerial computing also has the potential for the implementation of compute-intensive tasks (e.g., blockchain-enabled secure energy trading) in connected and autonomous vehicle networks [6]. Moreover, since insulator defects are a serious hazard to power systems, grid operators should respond immediately to the problems, even in very hard-to-reach areas. Considering difficulties in the conventional human monitoring methods, aerial platforms (e.g., LAC and HAC) for continuous monitoring of insulator defects is a good solution. The work in [18] shows insulator defects can be efficiently detected via a two-step method. An LAC node is first dispatched to capture images of insulator strings and estimate their possible direction. The second step leverages the images and results offloaded from the LAC node to detect the insulator string and defect at the central cloud to relieve the computation burden and extend the endurance time of the LAC node.

6.4 AERIAL COMPUTING: PROJECTS AND STANDARDIZATION

Several international projects and SDOs are working on a global level. In this section, we summarize a few research projects and standardization efforts toward a comprehensive 6G computing infrastructure.

6.4.1 Projects

This part provides an overview of the international projects by major 6G groups.

6.4.1.1 6G Flagship

The 6G Flagship project focuses on the development of wireless communication techniques, implementation of fifth-generation (5G), and formulation of 6G standards. Flagship conducts large-scale pilot projects through the test network to support the industry in impelling the 5G standard to the commercialization stage [19]. In addition, it develops basic technical components required for 6G systems, including wireless access and collaborative intelligent computing, as well as new applications in these areas. Instead of communicating with people, 6G Flagship discusses the communication between objectives and devices, including LAC, HAC, and SC platforms, from the perspective of intelligence and aerial computing.

6.4.1.2 South Korea MSIT 6G Research Program

The Korea Master of Science in Information Technology (MSIT) assumes that 6G technology will realize preemptive development. The data transmission speed based on low-orbit satellites exceeds 10 km, resulting in technological evolution beyond 5G, such as remote surgery, fully auto-mated driving, and flying cars, and reducing the delay time to one-tenth of that of 5G services. The goal is to design core technology for the first commercialization of 6G services in the world and thus prepare for a leading role in the 6G global market. MSIT predicts that the initial 6G network will be deployed in 2028, and the commercialization of this technology will be realized tentatively in 2030. According to the South Korea MSIT 6G research program, the architectural requirements of future wireless networks include the limited computation capability of mobile devices, and ML starts from the initial phase of technology development [20]. This reiterates the potential of aerial computing in solving the limited computation capacity of wireless networks and the importance of integrating ML and aerial computing.

6.4.1.3 Japan 6G/B5G Promotion Strategy

The National Institute of Information and Communication Technology (NICT) released the "Beyond 5G/6G" and "Quantum Network" white papers to accelerate R&D cooperation with other research institutions [21]. The white paper summarizes social visions and use cases that are expected to be realized through the advancement of Beyond 5G/6G and quantum network technology, as well as the key technologies and R&D roadmaps needed to realize them. Moreover, in the white paper, edge computing is listed as an important application for implementing 6G, and an omid-cloud gateway is proposed, which can provide computing, communication, and power resources. A drone that stays close to the served user can be used as a gateway to provide an aerial computing service.

6.4.1.4 Industrial 6G Programs

The leading telecommunication industry has launched a number of over-the-air computing and industrial IoT projects [22]. Several industrial IoT standards have been proposed, which are still evolving. One important industrial IoT scenario is the machine-to-machine (M2M) network, which provides a service layer that can be embedded in hardware and software to connect devices. Contiki, a lightweight Unix-like operating system (OS) aimed at wireless sensor networks (WSNs), is an open-source OS for low-cost, low-power IoT devices. Owing to the low energy budget and the multitude of devices, the combination of aerial computing and over-the-air transmission is a promising solution.

6.4.1.5 China 6G White Paper

An industry organization supported by the Chinese government has released a 6G technology white paper [23], focusing on the application of next-generation wireless technology and potential key technologies, which may be ten times faster than the maximum rate of 5G wireless networks. Analysts have stated that the paper will provide clear guidance for industry participants and provide a good start for the 6G R&D (R&D) plan, which is expected to be implemented in commercial use in approximately 2030 as the competition between China and the United States is intensifying. The China 6G white paper provides the overall vision while expounding eight business application use cases, ten emerging key technologies, and views on the development of the industry. The China Academy of Information and Communications Technology (CAICT) IMT-2030 6G Promotion Group released a white paper, which pointed out that 6G applications will present three potential development directions (immersion, intelligence, and generalization), as well as eight business applications, including holographic communication, immersive extended reality (XR), digital twins, intelligent interactive communication, perception interconnection, and global interconnection. Further, in this white paper, computing resources include cloud computing (the same as aerial computing), fog computing, and edge computing, which are crucial for intelligent deployment. In addition, for aerial computing, both distributed and centralized deployments have been involved. Moreover, the trend is to implement a multi-tier aerial computing network, including LAC, HAC, and SC platforms.

6.4.1.6 Europe 6G Program

In Europe, the first batch of 6G projects worth 60 million euros was launched under the 5G-PPP [24]. Among them, the Hexa-X flagship developed the first 6G system concept, supplemented by eight research projects on 6G-specific technologies. These technologies will form the foundation of the people-centered Next Generation Internet (NGI) and achieve the Sustainable Development Goals (SDGs). The committee has passed a legislative proposal for the upcoming European Smart Networks and Services (SNS) partnership for 6G. According to Europe's vision for the 6G ecosystem, 6G will usher in a new era wherein billions of devices, people, UAVs, vehicles, and robots will produce a massive amount of digital information. In addition, 6G will manage more challenging use cases, such as holographic telepresence and immersive stringent communication requirements. The 2030s are expected to be the beginning of the era of the widespread use of personal mobile robots.

Therefore, 6G is envisioned as one of the foundations of human society. To achieve the sustainable progress of society and satisfy the requirements of the United Nations sustainable development, 6G must provide new functions while effectively addressing urgent social needs. This evolution must adhere to the main social values of Europe, such as inclusiveness, transparency, security, and privacy. Further, digital technology is also becoming an indispensable and important means of protecting national sovereignty. Thus, to meet this goal, the Europe 6G program has introduced plans for implementing UAVs and satellites to assist 6G communications, including UAV and satellite-assisted MEC, that is, aerial computing.

6.4.1.7 USA 6G Program

The National Science Foundation (NSF) has targeted 6G and launched its largest single public–private partnership, recruiting nine cloud, technology, and telecommunications giants to help academia develop technologies that will define the next generation of networks. Newly launched Resilient and Intelligent Next Generation Systems (RINGS) program partners include Apple, Ericsson, Google, IBM, Intel, Microsoft, Nokia, Qualcomm, and VMware. The US Department of Defense and the National Institute of Standards and Technology are also involved in this program. Through this plan, the NSF will allocate funding of approximately 40 million USD to promote academic research projects focused on next-generation connectivity. Moreover, participants in selected projects of the program can collaborate with the aforementioned industry and government partners as well as use the NSF's four wireless test platform facilities to test their ideas. In efforts to build a network beyond 5G, the Cloud Enhanced Open Software Defined Mobile Wireless Testbed for City-Scale Deployment (COSMOS) platform was deployed. The COSMOS platform architecture focuses on the close integration of ultra-high bandwidth, low-latency wireless communication, and edge aerial computing.

6.4.2 Standardization

Several SDOs are working or planning to work on standardization activities related to aerial computing, as summarized in Figure 6.7.

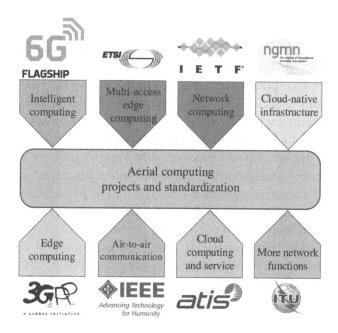

Figure 6.7 A summary of major projects and SDOs about aerial computing.

6.4.2.1 ETSI

European Telecommunications Standards Institute (ETSI) has released a white paper on MEC security [25] by several authors who participated in MEC and other related ETSI organizations. This white paper is the first initiative in this field to identify the security aspects, as edge computing renders typical industry cloud security methods inadequate for this purpose. The essential feature of the edge computing environment is a complex multi-vendor, multi-stakeholder ecosystem of hardware and software equipment. Thus, in view of the overall level of system heterogeneity, security, trust, and privacy are the key themes of the edge environment. The emergence of the edge cloud alliance and the emergence of (far) edge devices, such as in the IoT environment, solves the MEC security problem by employing an E2E approach using existing standards related to this field. However, aerial computing also faces security, trust, and privacy issues.

6.4.2.2 IETF

The Internet Engineering Task Force (IETF) is the main Internet standards organization that develops open standards through open processes. The Network Computing Research Group (COIN) of the IETF proposed research group (COINRG) will explore existing research and promote investigations into network computing and its impact on the data plane. The goal is to study the manner through which this emerging Internet architecture disruption can be exploited to improve network and application performance and user experience. The latest document [26] in COIN discusses collaborative edge cloud models and applications for XR network slicing, including AR and VR, specifically in terms of architectural framework and quality-of-service (QoS) based optimal delay-tolerant resource allocation. The collaborative edge cloud framework in COIN can be applied to a centralized aerial computing framework, which can support emerging technologies.

6.4.2.3 NGMN Alliance

The Next Generation Mobile Networks (NGMN) Alliance Committee launched a new key activity to support the adoption of cloud-native infrastructure by telecom service providers, which will solve the challenges of hybrid cloud architecture (such as edge computing and multi-tier aerial computing) and will lead to a paradigm change in the network vision of the industry. The project has been highlighted as a key development area for further optimizing and guiding the telecom industry to successfully deploy 5G after 2020. The NGMN white paper explains how 5G is expected to increasingly support new business models and many new use cases, and it provides suggestions on the manner through which all the advantages of virtualization, cloud computing, and edge computing can be utilized.

6.4.2.4 Next G Alliance

G Alliance is a bold new initiative aimed at increasing the leading position of North American mobile technology in the next ten years. Emphasis has been placed on the commercialization of technology, and the work will cover the entire life cycle of R&D, manufacturing, standardization, and market preparation. To date, Next G Alliance has united 45 leading ICT companies to jointly promote the evolution of 5G, plan for the future of 6G technology, and put North America at the forefront of wireless technology leadership in the upcoming decades. Members include infrastructure developers, semiconductor and equipment suppliers, operators, hyperscalers, and other organizations, including organizations engaged in the field of research. In the context of aerial computing, two studies on radio access network (RAN) based MEC [27], which support the integration of communication and computing technologies, have been reported.

6.4.2.5 3GPP

The 3rd Generation Partnership Project (3GPP) Releases 15 and 17 provide native support for edge computing in 3GPP networks, covering application layer architecture, core network enhancement, security, media processing, and management. The application architecture that supports edge applications has been designed based on the following architectural principles: application client portability, edge application server portability, service differentiation, and flexible deployment. The features of aerial computing (e.g., mobility, availability, scalability, flexibility, and simultaneity) should be considered in future 3GPP Releases to realize a comprehensive computing infrastructure in 6G networks.

6.4.2.6 IEEE

The Institute of Electrical and Electronics Engineers (IEEE) P1920.1 standard enhances the situational awareness of aircrafts for communication in self-organizing air networks and aerial computing systems. This standard defines the air-to-air communication of self-organizing air networks. Further, communication and network standards are independent of network type (wireless, cellular, or other) and can be used for manned and unmanned, small and large, civil, and industrial aircraft designs. The IEEE Standards Association (IEEE-SA) provides an international, open, and cooperative platform for the wireless community to participate in and support relevant, innovative, and new use cases and standards, thereby accelerating time-to-market consensus technologies, including mobile edge, fog computing, and virtualization.

6.4.2.7 ATIS

The Alliance for Telecommunications Industry Solutions (ATIS) is a cloud computing standard. As cloud computing becomes increasingly profitable, telecommunications providers are investing considerable resources to advance cloud computing solutions, which can be applied to aerial computing platforms. ATIS established a cloud service forum to focus on the rapid development of global market-driven standards in the ICT industry. In addition, ATIS provides a cloud framework including aerial computing for telepresence services.

6.4.2.8 ITU-T Standardization Sector

International Telecommunication Union-Telecommunication (ITU-T) Standardization Department S-17 provides a roadmap that is expected to expand with the work of other SDOs. Currently, ATIS, ETSI, IEEE, IETF, ISO (International Organization for Standardization)/IEC (International Electrotechnical Commission), ITU, OASIS (Organization for the Advancement of Structured Information Standards), 3GPP, 3GPP2, and other security standards are included in this roadmap. As data become available, it is expected to expand further to other organizations. In Y.3173, as the network evolves to IMT-2020, KPIs, such as transmission delay, transmission rate, and connection scale, are also crucial for aerial computing systems. Further, application scenarios are becoming more abundant, and network performance and flexibility are constantly being improved. In addition, an increasing number of network functions have been deployed on aerial computing platforms.

6.4.3 Summary and Discussion

The aforementioned projects provide a vision for implementing aerial computing, including the LAC, HAC, and SC platforms, as shown in Table 6.1. Many efforts have been made to integrate wireless communication and computing resources from both the land and sky. This trend can be explained by the following three reasons. First, with the development of LAP, HAPs, and satellites, many computing resources in the sky can be used to assist compute-intensive applications and services. Second, the air-to-ground link is dominated by LoS, which can provide high-quality communication for users. Third, the coverage rate can be greatly improved owing to the high mobility of aerial components (e.g., LAC and HAC servers in aerial computing). Thus, standardization is accordingly settled to provide a collaboration platform among different countries with issues related to security and green communication. It has been observed that the various SDOs share common knowledge to

TABLE 6.1   Contribution of Ongoing Projects and SDOs on Aerial Computing

	Projects							Standardization							
	6G Flagship	South Korea MSIT 6G Research Program	Japan 6G/B5G Promotion Strategy	Industrial 6G Programs	China 6G White Paper	Europe 6G Program	USA 6G Program	ETSI	IETF	NGMN Alliance	Next G Alliance	3GPP	IEEE	ATIS	ITU-T Standardization Sector
LAC	✓	✓	✓		✓	✓				✓		✓	✓	✓	✓
HAC	✓		✓					✓	✓	✓	✓	✓			
Satellite Computing	✓		✓					✓	✓					✓	

build a comprehensive computing network, which can combine computing resources from the ground, air, and sea.

6.5   AERIAL COMPUTING: CHALLENGES AND FUTURE DIRECTIONS

We argue that there are critical challenges to be considered in future research, which will be discussed in this section.

6.5.1   Energy Efficiency

Energy efficiency is a critical issue for supporting green mobile networks and computing applications. In contrast to edge nodes in the conventional edge computing paradigms (e.g., fog computing and MEC), which are typically connected with stable power sources, aerial computing nodes are powered by renewable energy sources and onboard batteries. Besides energy consumed for computation and communication, additional energy needs to be consumed for landing, hovering, and moving. As a result, designing energy-efficient solutions is a prerequisite for the realization of aerial computing in the future. In addition, energy refilling techniques, such as energy harvesting and wireless power transfer, can be adopted so that aerial nodes at the LAC and HAC tiers can harness external energy for efficient computing operations.

6.5.2 Resource Management

Aerial computing accommodates task offloading and execution from massive IoT devices via different computing platforms (e.g., terrestrial computing and LAC/HAC/SC servers). However, aerial nodes usually have low storage and battery capacities, limited computing capabilities, and complicated communication models. Thus, efficient resource management is required for aerial computing systems. One can foresee that the resource management problem in aerial is further exacerbated by additional optimization aspects and design constraints, such as joint computation and wireless resources, satellite constellation, LAC trajectory, and IoT device scheduling. The use of AI techniques (e.g., machine learning and deep learning) to optimize resource management is promising in heterogeneous and time-varying aerial computing systems. Moreover, optimizing on-device learning models needs to be considered in future studies to ease the computation burden of aerial computing nodes.

6.5.3 Network Stability

Mobility is a new feature of aerial computing, which makes it different from the conventional computing paradigms. However, it also introduces a critical issue in terms of stability to aerial computing systems. Actually, the network stability and topology stability of aerial computing are affected by various factors, such as the number of deployed aerial nodes and flying devices, energy harvesting capabilities, and aerial channel models. Moreover, the operational complexity and node unpredictability can render aerial computer systems more unstable than terrestrial computing. The reason for this is that since an LAC platform can be deployed for computing services at temporary events, SC nodes of different computing service providers may have different constellation designs [13]. As a result, it is necessary to achieve network stability of aerial computing systems. As a robust solution, AI techniques are helpful to learn from historical data and adapt to possible changes in the network.

6.5.4 Large-Scale Network Optimization

As a full complement to terrestrial computing, aerial computing can provide computing and wireless services to massive devices on a global scale. It is enabled by the hierarchical deployment of different computing platforms, from fog and MEC on the ground to LAC/HAC computing in the air and SC in space. Further, many emerging applications require the involvement of geographical users due to the deployment of many satellites and the global availability of computing resources. Thus, distributed and large-scale optimization techniques are urgently needed to allow scalable aerial computing systems and intelligent computing applications. In this regard, distributed optimization approaches are promising since they can work well when the number of involved devices increases and reduces the dependence on a central entity for the implementation of centralized algorithms. Moreover, distributed AI techniques, such as federated learning and multi-agent deep reinforcement learning, are research trends in aerial computing as they allow IoT devices with different characteristics to learn a collaborative model.

6.5.5 Security and Privacy

For the realization of aerial computing in 6G, how to solve privacy and security issues is crucial. There are various security attacks in aerial computing systems, ranging from physical layer security to node intrusion and data security at the computing server. For example, adversaries may attack transmission channels between aerial components and between end users and computing servers due to the broadcast nature of wireless channels [28]. Further, adversaries can deploy data breach techniques to access sensitive information computation tasks of end users. In terms of data privacy, the issue may be caused by the data collection and storage at a central node in aerial computing systems. It is also the case of central learning when end users should share computation tasks and raw data with the learning server. In this context, the use of perturbation methods (e.g., differential privacy) and emerging technologies (e.g., blockchain and federated learning) is particularly helpful to protect data privacy in aerial computing. For example, blockchain is used to record trading information (e.g., resource price and resource allocation) in a UAV-aided MEC system to improve network security [29].

6.6 CONCLUSION

In this chapter, we have first introduced the concept of aerial computing and presented a novel comprehensive computing architecture in 6G systems. Then, we have illustrated the manner in which aerial computing can support vertical domain applications. Finally, we have concluded the chapter by discussing critical challenges for the realization of aerial computing. We hope this chapter will gain more attention from the communication and computing communities and drive more research in aerial computing.

BIBLIOGRAPHY

[1] Bo Yang, Xuelin Cao, Kai Xiong, Chau Yuen, Yong Liang Guan, Supeng Leng, Lijun Qian, and Zhu Han. Edge intelligence for autonomous driving in 6G wireless system: Design challenges and solutions. *IEEE Wireless Communications*, 28(2):40–47, 2021.

[2] Chamitha de Alwis, Anshuman Kalla, Quoc-Viet Pham, Pardeep Kumar, Kapal Dev, Won-Joo Hwang, and Madhusanka Liyanage. Survey on 6G frontiers: Trends, applications, requirements, technologies and future research. *IEEE Open Journal of the Communications Society*, 2:836–886, 2021.

[3] Fang Fang, Kaidi Wang, Zhiguo Ding, and Victor CM Leung. Energy-efficient resource allocation for NOMA-MEC networks with imperfect CSI. *IEEE Transactions on Communications*, 69(5):3436–3449, 2021.

[4] Nhu-Ngoc Dao, Quoc-Viet Pham, Ngo Hoang Tu, Tran Thien Thanh, Vo Nguyen Quoc Bao, Demeke Shumeye Lakew, and Sungrae Cho. Survey on aerial radio access networks: Toward a comprehensive 6G access infrastructure. *IEEE Communications Surveys & Tutorials Computing*, 23(2):1193–1225, 2021.

[5] Dinh C. Nguyen, Ming Ding, Pubudu N. Pathirana, Aruna Seneviratne, Jun Li, Dusit Niyato, Octavia Dobre, and H. Vincent Poor. 6G internet of things: A comprehensive survey. *IEEE Internet of Things Journal*, 9(1):359–383, 2022.

[6] Thippa Reddy Gadekallu, Quoc-Viet Pham, Dinh C Nguyen, Praveen Kumar Reddy Maddikunta, N Deepa, B Prabadevi, Pubudu N Pathirana, Jun Zhao, and Won-Joo Hwang. Blockchain for edge of things: Applications, opportunities, and challenges. *IEEE Internet of Things Journal*, 9(2):964–988, 2022.

[7] Quoc-Viet Pham, Mai Le, Thien Huynh-The, Zhu Han, and Won-Joo Hwang. Energy-efficient federated learning over UAV-enabled wireless powered communications. *IEEE Transactions on Vehicular Technology*, 71(5):4977–4990, 2022.

[8] Quoc-Viet Pham, Fang Fang, Vu Nguyen Ha, Md Jalil Piran, Mai Le, Long Bao Le, Won-Joo Hwang, and Zhiguo Ding. A survey of multi-access edge computing in 5G and beyond: Fundamentals, technology integration, and state-of-the-art. *IEEE Access*, 8:116974–117017, 2020.

[9] Quoc-Viet Pham, Ming Zeng, Rukhsana Ruby, Thien Huynh-The, and Won-Joo Hwang. UAV communications for sustainable federated learning. *IEEE Transactions on Vehicular Technology*, 70(4):3944–3948, 2021.

[10] Zhen Qin, Hai Wang, Zhenhua Wei, Yuben Qu, Fei Xiong, Haipeng Dai, and Tao Wu. Task selection and scheduling in UAV-enabled MEC for reconnaissance with time-varying priorities. *IEEE Internet of Things Journal*, 8(24):17290–17307, 2021.

[11] Quang Vinh Do, Quoc-Viet Pham, and Won-Joo Hwang. Deep reinforcement learning for energy-efficient federated learning in UAV-enabled wireless powered networks. *IEEE Communications Letters*, 26(1):99–103, 2022.

[12] Gunes Karabulut Kurt, Mohammad G Khoshkholgh, Safwan Alfattani, Ahmed Ibrahim, Tasneem SJ Darwish, Md Sahabul Alam, Halim Yanikomeroglu, and Abbas Yongacoglu. A vision and framework for the high altitude platform station (HAPS) networks of the future. *IEEE Communications Surveys & Tutorials*, 23(2):729–779, 2021.

[13] Shicong Liu, Zhen Gao, Yongpeng Wu, Derrick Wing Kwan Ng, Xiqi Gao, Kai-Kit Wong, Symeon Chatzinotas, and Bjorn Ottersten. LEO satellite constellations for 5G and beyond: How will

they reshape vertical domains? *IEEE Communications Magazine*, 59(7):30–36, 2021.

[14] Changfeng Ding, Jun-Bo Wang, Hua Zhang, Min Lin, and Geoffrey Ye Li. Joint optimization of transmission and computation resources for satellite and high altitude platform assisted edge computing. *IEEE Transactions on Wireless Communications*, 21(2):1362–1377, 2022.

[15] Quoc-Viet Pham, Ming Zeng, Thien Huynh-The, Zhu Han, and Won-Joo Hwang. Aerial access networks for federated learning: Applications and challenges. *IEEE Network*, 36(3):159–166, 2022.

[16] Zhenfeng Shao, Gui Cheng, Jiayi Ma, Zhongyuan Wang, Jiaming Wang, and Deren Li. Real-time and accurate UAV pedestrian detection for social distancing monitoring in COVID-19 pandemic. *IEEE Transactions on Multimedia*, 24:2069–2083, 2021.

[17] Yang Liu, Jianshan Zhou, Daxin Tian, Zhengguo Sheng, Xuting Duan, Guixian Qu, and Victor CM Leung. Joint communication and computation resource scheduling of a UAV-assisted mobile edge computing system for platooning vehicles. *IEEE Transactions on Intelligent Transportation Systems*, 23(7):8435–8450, 2022.

[18] Chunhe Song, Wenxiang Xu, Guangjie Han, Peng Zeng, Zhongfeng Wang, and Shimao Yu. A cloud edge collaborative intelligence method of insulator string defect detection for power IIoT. *IEEE Internet of Things Journal*, 8(9):7510–7520, 2021.

[19] Marcos Katz, Pekka Pirinen, and Harri Posti. Towards 6G: Getting ready for the next decade. In *2019 16th International Symposium on Wireless Communication Systems (ISWCS)*, pages 714–718. IEEE, 2019.

[20] Juan Pedro Tomás. "South Korea to launch 6G pilot project in 2026: Report." RCR Wireless, August 10, 2020. Available at: https://www.rcrwireless.com/20200810/business/south-korea-launch-6g-pilot-project-2026-report [Accessed on: 01 January 2023]

[21] Hitoshi Asaeda et al. Quantum Network White Paper. Technical Report, National Institute of Information and Communication Technology (NIICT), Japan, August 2021. Available at: https://www2.nict.go.jp/idi/common/pdf/NICT_QN_WhitePaper EN_v1_0.pdf [Accessed on: 01 January 2023]

[22] Yifang Gao. Using artificial intelligence approach to design the product creative on 6G industrial internet of things. *International Journal of System Assurance Engineering and Management*, 12(4):696–704, 2021.

[23] Xiaohu You, Cheng-Xiang Wang, Jie Huang, Xiqi Gao, Zaichen Zhang, Mao Wang, Yongming Huang, Chuan Zhang, Yanxiang Jiang, Jiaheng Wang, et al. Towards 6G wireless communication networks: Vision, enabling technologies, and new paradigm shifts. *Science China Information Sciences*, 64(1):1–74, 2021.

[24] Peter Stuckmann, Smart Networks and Services Stepping up 6G R&I in Europe. EUCNC 6G Wireless Summit, 10 June 2022. Available at: https://www.eucnc.eu/wp-content/uploads/2022/06/Peter-Stuckmann-Closing-2022.pdf

[25] Dario Sabella, Alex Reznik, Kamal Ranjan Nayak, Diego Lopez, Fei Li, Ulrich Kleber, Alex Leadbeater, Kishen Maloor, Sheeba Backia Mary Baskaran, Luca Cominardi, Cristina Costa, Fabrizio Granelli, Vangelis Gazis, Francois Ennesser, and Xi Gu. MEC security: Standard support status and future development. *ETSI White Paper No. 46*, May 2021.

[26] Rohit Abhishek, A collaborative Edge-Cloud framework for XR applications. Internet Draft, Internet Engineering Task Force, April 2021. Available at: https://datatracker.ietf.org/doc/draft-abhishek-coin-xr-edge-cloud/

[27] Wypiór D, Klinkowski M, Michalski I. Open RAN—Radio Access Network Evolution, Benefits and Market Trends. Applied Sciences. 2022; 12(1):408. https://doi.org/10.3390/app12010408

[28] Pasika S Ranaweera, Anca D Jurcut, and Madhusanka Liyanage. Survey on multi-access edge computing security and privacy. *IEEE Communications Surveys & Tutorials Computing*, 23(2):1078–1124, 2021.

[29] Haitao Xu, Wentao Huang, Yunhui Zhou, Dongmei Yang, Ming Li, and Zhu Han. Edge computing resource allocation for unmanned aerial vehicle assisted mobile network with blockchain applications. *IEEE Transactions on Wireless Communications*, 20(5):3107–3121, 2021.

CHAPTER 7

THz-Empowered UAV Communications

Humairah Hamid

Advanced Communication Lab, Electronics and Communication Engineering Department, NIT Srinagar, Srinagar, Jammu and Kashmir, India

Aaqib Reshi

Advanced Communication Lab, Electronics and Communication Engineering Department, NIT Srinagar, Srinagar, Jammu and Kashmir, India

G R Begh

Advanced Communication Lab, Electronics and Communication Engineering Department, NIT Srinagar, Srinagar, Jammu and Kashmir, India

CONTENTS

7.1	Terahertz Communication	....	150
	7.1.1	Need for THz Communication	151
	7.1.2	Terahertz Communication Architectural Scenario ..	152
		7.1.2.1 Terahertz Wireless Transceivers	152
		7.1.2.2 Terahertz Channel Model	153
	7.1.3	Opportunities of THz Wireless Communications ...	154
		7.1.3.1 Terahertz-Optical Wireless Communication Links	154
		7.1.3.2 Terahertz-Free Space Applications	155
		7.1.3.3 Terahertz for Enhanced Data Centers Performance	156
		7.1.3.4 Terahertz 3D Hybrid Beamforming Technology	156
	7.1.4	Challenges and Open Research Issues	157
		7.1.4.1 Transceiver Design and THz Signal Generators	157
		7.1.4.2 Propagation Modeling and Path Losses ..	158
7.2	Fundamentals of UAV Communication		158

DOI: 10.1201/9781003282211-7

7.2.1   Classification of UAVs 159

7.2.2   Use Cases-Wireless Communication with UAVs 161

     7.2.2.1   UAVs as Flying Base Stations in 5G and 6G 161

     7.2.2.2   Cellular-Connected Drones as Aerial User Equipments 166

     7.2.2.3   Flying Ad-Hoc Networks with UAVs 166

7.2.3   Network Architecture for UAVs 167

7.3   THz-empowered UAV communication 168

     7.3.1   Applications of THZ-Empowered UAVs Use Cases . 169

         7.3.1.1   Interactive Aerial Telepresence 169

         7.3.1.2   Reliable and Secure Functioning of UAVs 169

         7.3.1.3   Pollution Monitoring and Tracking 169

         7.3.1.4   Time-Critical MEC-Enabled Operation .. 170

         7.3.1.5   Advanced Swarm of UAVs 171

     7.3.2   Design Challenges 172

         7.3.2.1   Wobbling of UAVs 172

         7.3.2.2   Limited Design Space 172

         7.3.2.3   Dense Deployment of UAVs 172

         7.3.2.4   Multi-Functionality Design 173

         7.3.2.5   Uncertain Medium Condition 173

     7.3.3   Trade-Offs 173

Bibliography 174

U NMANNED AERIAL VEHICLES (UAVs) are going to play leading roles in various aspects like autonomy and network intelligence, owing to their flexible 3D deployment and their technological advancements in terms of cost. Further, for satisfying the increasing demands in 6G, Terahertz (THz) band stands among the best facilitators. In this chapter, we will briefly discuss THz wireless communication in association with UAVs.

7.1 TERAHERTZ COMMUNICATION

In order to fulfill bandwidth demands, carrier frequencies for wireless communications have been growing in recent years. Wide radio bands, such as millimeter-wave (mmWave) frequencies, have been created, according to the technical community, to support the growing demand for mobile data and pave the way for 5G networks. Other researchers are interested in optical wireless communication, which allows for faster data transmission, improved physical security, and the avoidance of electromagnetic interference. Regardless, the utilization of the Terahertz (THz) frequency range (0.1–10 THz) has signaled the beginning of a new era in electromagnetic wireless technology. The next generation of wireless networks should be able to realize their full potential thanks to THz technology, which bridges the gap between radio and optical frequency ranges.

7.1.1 Need for THz Communication

Wireless data traffic has increased at an unprecedented rate during the last several years. Between 2016 and 2021, mobile data traffic is anticipated to rise sevenfold, while video traffic will triple. Researchers are looking at optimal radio frequency (RF) zones to meet consumers' growing expectations due to the rapid expansion of wireless traffic. As a result, the THz frequency band (0.1–10 THz) began to attract a lot of study interest. Big data wireless cloud, ultra-fast wireless download, and seamless data transmission are all improvements that will transform the telecommunication environment and reshape how people connect and access information. The THz frequency spectrum has a wide bandwidth that might theoretically reach multiple THz, resulting in a potential capacity of Terabits per second. As a result, the bandwidth provided is one order of magnitude more than that of mmWave systems. When compared to millimeter transmissions, THz signals provide for greater connection, directionality, and fewer eavesdropping risks [1]. THz band analysis reveals that, in comparison to optical frequencies, these frequencies provide a number of advantages. THz waves are being studied as uplink communication possibilities. They allow for non-line-of-sight (NLoS) propagation [2] and serve as good stand-ins for difficult weather conditions like as fog, dust, and turbulence [3].

Furthermore, ambient noise from optical sources has no effect on the THz frequency spectrum, and it is not related to any health restrictions or safety constraints. Table 7.1 compares the THz frequency band to various existing technologies in order to demonstrate the band's ultimate promise for the next generation of wireless networks. [4].

TABLE 7.1 Comparison between Wireless Communication Technologies

Technology	mmWave	THz Band	Infrared
Frequency range	30 GHz–300 GHz	100 GHz–10 THz	10 THz–430 THz
Range	Short range	Short/medium range	Short/Long range
Power consumption	Medium	Medium	Relatively low
Network topology	Point to multi-point	Point to multi-point	Point to point
Noise source	Thermal noise	Thermal noise	Sun/ambient light
Weather conditions	Robust	Robust	Sensitive
Security	Medium	High	High

7.1.2 Terahertz Communication Architectural Scenario

Advancements in both the devices and channel modeling are necessary to establish a functional THz communication system. This will lead to remarkable success in the THz wireless field and a significant increase in THz applications.

7.1.2.1 Terahertz Wireless Transceivers

THz waves are a portion of the electromagnetic spectrum that has been understudied and underappreciated. These waves can be produced using both electronics and photonics since they fall between the mmWave and optical frequency regions. Recent advances in nano-fabrication technology have facilitated the creation of THz-frequency semiconductor devices in the realm of electronic devices. Electronics made of gallium arsenide (GaAs) and indium phosphide (InP), as well as numerous silicon-based technologies, are examples. A new silicon architecture that enables scalability and signal synthesis as well as THz wave shaping on a single microchip was recently suggested [5]. Such contributions are critical since a large part of the issue in THz creation, particularly in microchips, is providing a wide range of wavelengths within the THz band. The feeding method of optical fibers to THz emission circuits is a critical aspect in achieving higher data rates in photonics devices. Photonics-based approaches, in particular, offer a one-of-a-kind opportunity to attract early adopters or consumers by demonstrating the technology utilizing easily accessible lasers, modulators, and photodiodes. These devices can also deal with multi-carrier THz channels thanks to their carrier switching capabilities, which help with the radio/optical interface in hybrid networks. Aside from that, the antenna requires extra care. Because THz frequencies allow for modest antenna diameters, terminal devices may accommodate a wide variety of antenna components. As a result, multiple-input-multiple-output (MIMO) techniques accomplish diversity gain as well as antenna directivity gain. In fact, the notion of 1024×1024 ultra-massive MIMO was proposed as a way to extend communication distance in THz systems. Furthermore, most typical materials used at lower frequencies, such as microwave and mmWave, are unable to respond well to higher frequencies and frequently display significant losses at THz. As a result, graphene is being employed as a THz band tiny transceiver replacement [6]. For practical applications, electromagnetically reconfigurable materials, which are the foundation of all active THz components, are also required. Because of its atomic thickness, tunability, and high kinetic inductance, graphene has earned scientific attention in this context, allowing for the reformation of THz electromagnetic waves using thin graphene layers. As a result, graphene-based THz components have demonstrated excellent performance in terms of producing, modifying, and sensing THz waves [7]. Graphene plasmonics, in particular, open the way for optoelectronic applications in the THz frequency range. Subwavelength guiding structures, nano antennas, super lenses, hyper lenses, and light concentrators are examples of innovative technologies with unrivalled functional capabilities.

7.1.2.2 Terahertz Channel Model

It is vital to develop a special model that completely explains the behavior of any wireless channel before deployment in order to be capable of recognizing its idiosyncrasies. The channel suffers from strong atmospheric absorption at the THz frequency region, which is caused by water vapor molecules. Free-space route loss, in addition to air attenuation, is considered physically inescapable. As a result, the THz band channel is thought to be very frequency selective. As a result, attenuation limits the transmission distance, and the optimal carrier frequency is established based on the application. The current research on channel modeling in the THz range is restricted and constrained to indoor conditions since we still lack models that mimic outdoor situations. The bulk of models in the literature falls into one of two categories: path loss or ray tracing. The first statistical model for THz channels, covering the frequency range of 275–325 GHz, was given in [8]. The given model uses extensive ray-tracing simulations to estimate the channel statistical parameters. The downside of this paradigm is that it is difficult to comprehend information regarding channel characteristics such as the correlation function and power-delay profile. To address these issues, the authors [9], [10] developed a geometrical statistical model for device-to-device (D2D) scatter channels in the sub-THz range. In a sub-THz D2D environment, these models replicate scattering and reflection patterns. Furthermore, unlike typical channel measurements, scenario-specific metrics are documented. In [11], the authors presented a stochastic model for kiosk applications in the THz band (between 220 and 340 GHz). The channel characteristics of three separate kiosk application scenarios were determined using a 3D ray-tracing simulator. The suggested channel model will allow THz-based close-proximity communication systems to be developed. The authors [12] present another recent work on the statistical channel characterization of a THz scenario. Despite the fact that the study covers a frequency range of 240–300 GHz, it is one of the first to produce single-sweep THz measurement results. The measured data enables for the extraction of finer temporal information, which aids in the design of reliable transceiver systems that address antenna misalignment issues.

7.1.3 Opportunities of THz Wireless Communications

In this section, we'll look at a few potential prospects that show the future of THz wireless and the push to utilize a spectral band that was originally described as a "gap".

7.1.3.1 *Terahertz-Optical Wireless Communication Links*

Recently, a hybrid RF/free space optical (FSO) technology has attracted people's curiosity. In this situation, a THz/optical connection is predicted to be a potential option for future wireless communication. Parallel and serial are two hybrid ways for such a configuration. Two communication links are provided in the parallel hybrid scenario and can be employed in either a one-way or two-way mode. The backhaul link is an example of a one-way hybrid scenario where the weather dictates whether FSO or THz technology is employed. The FSO technology is utilized to provide reliable beam tracking when the weather is clear, and there is little wind. THz, on the other hand, arrives in foggy and/or windy weather conditions, allowing a steady connection to be formed. THz can provide trustworthy performance in the uplink of visible light communication (VLC) cells without the necessity for position, acquisition, and tracking (PAT) protocols associated with infrared uplink service use in terms of two-way communication. Multi-hop communication links are employed in the serial communication scenario, with each link tailored to function using the technology that maximizes system performance.

7.1.3.2 Terahertz-Free Space Applications

Unmanned autonomous vehicles have grown more widely available in recent years, resulting in a wide range of civilian and business applications. Examples include weather management, forest fire detection, traffic detection, freight transportation, emergency search and rescue, and communication broadcasting. UAVs must have a stable communication link available at all times in order to perform such applications. Moisture has no effect at heights above 16 km; hence, THz attenuation is negligible. As a result, THz may be a viable option for establishing secure communications in a range of UAV applications. The THz frequency band is a superior option for mitigating the high-mobility environment of UAVs than the FSO frequency band because it not only provides high-capacity UAV-UAV wireless backhaul but also provides a better alternative in mitigating the high-mobility environment. In fact, as the carrier frequency increases, communication lines that suffer from the Doppler effect are diminished as a result of mobility. As a result, high-speed communication lines between two dynamic sites can be established using the THz frequency spectrum. Furthermore, UAVs require short-distance secure communications to acquire orders or deliver data before dispersing to pursue either their remote-controlled or self-governing missions. As a result, THz communications are recognized as a trustworthy channel for transferring safety-critical data between UAVs and between the UAV and ground control stations (GCSs). In fact, THz systems' huge bandwidth allows for a variety of defense mechanisms against standoff attacks such as jamming, as well as the capacity to entirely conceal information flow. In addition, THz connectivity could be used between UAVs and planes to offer internet for flights rather than relying on satellite service.

In this sense, the UAV will function as a sky-based switchboard, acting as a conduit between the ground station and the aeroplane.

7.1.3.3 Terahertz for Enhanced Data Centers Performance

Consumer demand for cloud apps has sparked competition among data centers in order to provide consumers with better performance. This is accomplished by hosting a large number of servers and providing adequate bandwidth for a diverse set of applications. However, relying entirely on fixed wired networks to handle traffic surges caused by static links and finite network interfaces is particularly difficult. Wireless networking, which is a complement to Ethernet, does, however, provide the adaptability and efficiency required to contribute to feasible solutions to the problem. Wireless communications, on the other hand, may suffer from restricted lengths and intolerance to blockage if all wires are replaced, resulting in data center capacity being depleted. THz lines are also suggested as a backup option in data centers. Performance increases in data centers can be achieved as a result of this deployment, as can huge savings in cable prices without sacrificing throughput [13]. The THz channel has been modeled using atmospheric data, with the authors recommending a bandwidth of 120 GHz for data center applications. In this application, THz networks outperform both mmWave and infrared technologies since the former has restricted bandwidth and the latter suffers from huge complexity in coherent detection as well as square law detector limitations.

7.1.3.4 Terahertz 3D Hybrid Beamforming Technology

THz wire-free appears to have a bright future with 3D MIMO technology. The real-world channel does, in fact, have 3D properties, making 2D MIMO approaches suboptimal. To combat the THz channel's inevitability of route loss, 3D beamforming appears to be a viable option for constructing directing beams, extending communication range, and reducing interference. As a consequence, each resource and user equipment have an active interaction with the vertical beam pattern. By allowing the vertical main lobe to be precisely set at any point at the receiver, 3D beamforming can boost signal strength. The change in vertical dimension has the ability to leverage increased diversity or spatial separation by using beam coordination or MIMO techniques. This will eventually result in either improved signal quality or an increase in the number of concurrent users supported.

7.1.4 Challenges and Open Research Issues

Because 6G is a new frontier in wireless communications, several obstacles are expected and must be overcome. One of the primary challenges is the effect of air and water absorption on signal transmission. Until these issues are resolved, the objective of realizing cost-effective, efficient, and pragmatic THz band communication for 6G communication systems will remain a mirage. These issues demand both specific policy changes and the advancement of existing solutions throughout numerous communication layers, from the physical layer to protocol stacks, apps, and eventually the user interface. It is important to note that fulfilling all qualities is not achievable; rather, trade-offs between these factors should be made in order to elevate all features in a balanced manner. This section focuses on a number of critical issues in THz band communication.

7.1.4.1 Transceiver Design and THz Signal Generators

One of the most critical topics with using the THz spectrum is figuring out how to make a new transceiver module that can operate at THz band frequencies. More attempts to fix this issue are planned in this regard. For example, the DARPA T-MUSIC programe is working on SiGe HBT (heterojunction bipolar transistors), CMOS (complementary metal–oxide–semiconductor)/SOI (silicon on insulator), and BiCMOS (Bipolar CMOS) circuit integration in order to attain power amplifier threshold frequencies of 500–750 GHz. To overcome the relative minute wavelengths and physical size of RF transistors matching to element spacing in THz arrays, the semiconductor industry will need to develop unique designs for incredibly dense antenna arrays. Furthermore, to overcome the substantial propagation attenuation that is usual at THz band frequencies, a great degree of ingenuity is necessary. Other factors to consider during the manufacture of THz-enabled transceivers include high-power, high-sensitivity, and low-noise figure. Furthermore, from the standpoint of transmission power, THz transceivers are performance constrained (and distance). THz signal generators and detectors are commonly made of silicon germanium (SiGe), gallium nitride (GaN), GaAs, and InP. Similarly, GaN, GaAs, and InP-based transceivers have a limited transmission distance. It has become necessary to develop a new THz band transceiver architecture. Imperfect hardware features, such as nonlinear amplifiers, phase noise, and limited modulation index, can degrade the quality of transmitted signals significantly. It's also probable that challenges with antenna design and waveforms, as well as energy-efficient signal processing, will occur. As a result, novel transceiver technologies have become critical in the construction of cutting-edge THz sources (hardware), particularly in the medium- to high-end of the THz bands (300 GHz). For innovative transceiver designs for THz-enabled devices, both traditional CMOS technology and recently discovered nanomaterials, such as graphene, can be employed [14]. Analytical testing, on the other hand, should be accompanied by real-world experimentation in order to re-evaluate the performance of any proposed design.

7.1.4.2 *Propagation Modeling and Path Losses*

Another method for aiding researchers in their exploration of the THz spectrum is to categorize THz sub-domains based on their absorption and reflection coefficients. This technique explicitly serves as a guide for deciding whether or not any of the THz spectrum is acceptable for communications and other purposes. Another issue that needs to be looked into is how to deal with circumstances that support several applications. The resulting harmonic overlaps can be minimized in this scenario by proper frequency planning. Furthermore, the sensitivity of weak signals degrades with time, making detection one of the key limits limiting this band's flexibility. The THz electromagnetic spectrum can be divided into two categories: atmospheric below 500 GHz and atmospheric above 500 GHz. As we get into the THz range, there is a large increase in free space path loss, molecule air absorption, and total signal loss. The high antenna available in this position can compensate for the high free space route loss due to the high-signal directivity components. Aside from the negative impact of free space loss, higher frequency transmission increases sophistication and parallelism in RF hardware, as well as a reduction in beam width, which causes signal collecting and beam tracking issues in mobile applications [15]. Other elements to consider when categorizing the radio spectrum include penetration through various materials and reflections from surfaces, in addition to technological restrictions.

7.2 FUNDAMENTALS OF UAV COMMUNICATION

The next-generation wireless network leading to 6G is likely to provide intelligent, secure, and reliable connectivity. From autonomous automobiles to UAVs, 6G is expected to deliver a full-fledged framework for connected things and automation systems with severe and diverse needs in terms of reliability, energy efficiency, latency, and data throughput. Drones, or UAVs, play a crucial part in a variety of use cases and situations that extend beyond 5G and 6G. Package delivery, media production, real-time surveillance, and remote constructions are all examples of UAV applications. UAVs, in particular, can provide reliable and economical wireless communication solutions for a wide range of real-world applications if correctly deployed and operated. By mounting communication transceivers on UAVs, they can be used as aerial communication platforms (e.g., flying base stations (BSs) or mobile relays) to impart communication services to ground targets in high traffic demand and overloaded situations, a practice known as UAV-assisted communications. These aerial (BSs) can provide secure, economical, and on-demand wireless connectivity to specific locations. UAVs can also be utilized as aerial nodes for varied applications, which range from surveillance to freight delivery, popularly called cellular-connected UAVs. UAVs can also be utilized as wireless relays to improve ground wireless device connectivity and coverage, as well as in surveillance scenarios.

7.2.1 Classification of UAVs

Depending on the application and aim, a suitable UAV must be used that can meet a variety of demands imposed by the desired quality-of-service (QoS), the environment, and federal directives. Accordingly, drones are classified as follows [16]:

- Based on flying mechanism:

    1. Multi-rotor drones (well-known as rotary-wing drones): They can take off and land vertically, and they can hover over a set spot to provide continuous cellular coverage in specific locations. Because of their great maneuverability, they are well suited to support cellular communications, since they can precisely distribute BSs at the necessary places or fly in a predetermined trajectory while carrying BSs. Multi-rotor drones have limited mobility and consume a lot of power since they have to constantly combat gravity.

    2. Fixed-wing drones: They can glide through the air, making them far more energy efficient and capable of carrying large payloads. Fixed-wing drones can also benefit from gliding to go faster. Fixed-wing drones have the disadvantages of (i) requiring a runway to take off and land because vertical take-off

and landing are not possible and (ii) being unable to hover over a fixed spot. Drones with fixed wings are likewise more expensive than multi-rotor drones.

3. Hybrid fixed/rotary wing drones: These drones can provide a middle ground between fixed and rotary drones listed above. For example, Parrot Swing is a type of drone which is able to take off in a vertical direction, glide across the air, and shift to hovering with the help of four rotors.

- Based on drone altitude
  The maximum height a drone can achieve, regardless of country-specified limitations, is referred to as altitude. Since UAV flying BS has to adjust the height to fulfill varied QoS requirements, the maximum flying altitude of a specific drone becomes an important metric for UAV-aided cellular communications. Flying platforms can be divided into the following categories based on the altitude:

  1. Low-altitude platforms (LAPs): LAPs can fly at heights ranging from few meters to several kilometers. They can move quickly, are adaptable to the environment, cost-effective, and are commonly used to serve cellular communications. Furthermore, LAPs typically impart short-range line-of-sight (LOS) connectivity, which can improve communication performance dramatically. Data from ground sensors can be collected using LAPs. LAPs may also be easily recharged or replaced if necessary.

  2. High-altitude platforms (HAPs): HAPs are normally quasi-stationary and have altitudes above 17 km. HAPs have a larger coverage area than LAPs and can stay in the air for much longer. They have a higher endurance and are suited for operations that last a long time (e.g., a few months). In addition, HAP systems are commonly used to provide large-area wireless coverage for huge geographic regions. They tend to be more expensive than LAPs with a much longer time for deployment.

- Based on size
  Table 7.2 demonstrates how drones are classified based on their size.

TABLE 7.2   Classification of UAVs Based on Size

Type of Drones	Weight
Micro drones	$\leqslant 100$ g
Very small drones	100 g to 2 kg
Small drones	2 kg to 25 kg
Medium drones	25 kg to 150 kg
Large drones	> 150 kg

7.2.2   Use Cases-Wireless Communication with UAVs

In order to get clarity on the employment of UAVs as flying BSs, we review a variety of potential applications for various wireless-centric UAV deployments. The applications are inspired by diverse situations, including upcoming use cases like public safety scenarios or hotspot coverage, also the revolutionary ones like the usage of drones as caching devices or enablers of IoT. Moreover, the UAV-based wireless communication also includes cellular-connected UAVs, which will be discussed later.

7.2.2.1   UAVs as Flying Base Stations in 5G and 6G

In this section, we will discuss some vital use cases of UAV-assisted aerial BSs in future-generation wireless networks which are as follows:

1. **UAVs for Capacity and Coverage Enhancement of Future Wireless Communication Networks:** The increasing demand for high-speed wireless communication has led to the rapid growth of powerful mobile equipments that include smartphones, tablets, and lately, drone-UEs and Internet-of-Things-style devices. As a result, existing wireless cellular networks' capacity and coverage have been severely drained, prompting the development of new wireless technologies to address the problem. Millimeter wave (mmWave) communications, ultra-dense small cell networks and D2D communications are among the technologies that make up the core of future-generation cellular systems. In spite of numerous advantages, these technologies have drawbacks of their own. D2D communication, for example, will likely necessitate better resource allocation and frequency planning in cellular networks. At the same time, in terms of interference, backhaul and altogether network modeling, ultra-dense small cell communication networks meet numerous challenges.

We see UAV-borne aerial BSs as a natural complement to this heterogeneous 5G scenario, allowing us to overcome some of the existing technologies' limitations. The deployment of LAP drones can prove to be an economical solution for imparting connectivity to regions having insufficient cellular network infrastructure. They can satisfy the high-speed and low-latency communication requirements. Aerial BSs can furnish on-demand wireless connectivity, high data rates, and traffic offloading in hotspot area as well as during short-term events like football games or presidential inaugurations. Verizon and AT&T have already set forth various ideas for using UAVs in providing temporary internet coverage in college football national championship and Super Bowl [17].

2. **UAVs as Aerial Base Stations for Public Safety Situations:** Natural disasters like hurricanes, floods, snow storms, tornados, etc., sometimes lead to disastrous consequences, which can damage the communication network. More specifically, the cellular BSs and terrestrial communication infrastructure can get compromised leading to malfunctioned BS or overloaded BS. In such cases, communications among the rescue team and victims become critical. In order to provide successful communications during public safety operations, a robust, rapid, and suitable emergency communication system is required which will not only improve connectivity but will also save lives. Satellite communication, 4G LTE, WiFi, and dedicated public safety systems such as TETRA and APCO25 are some wireless technologies that can be used in such scenarios [18]. But such technologies cannot deliver low-latency, flexible, on-demand, and swift adjustment to the surrounding environment in emergency situations. In this context, utilization of UAV-based flying networks, as depicted in Figure 7.1, seems to be a promising approach to allow quick, adaptable, and reliable communications in public safety scenarios. Owing to the novel qualities of UAVs such as mobility and adaptable deployment, they can efficiently set up on-demand public safety communication networks. For example, drones can be used as flying BSs for providing broadband wireless connectivity to the regions with compromised terrestrial cellular infrastructure. Thus, the usage of UAV-mounted BSs can be a viable approach for delivering quick and ubiquitous connection in public safety situations.

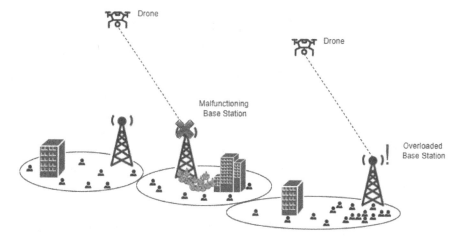

Figure 7.1   UAVs in public safety scenarios.

3. **UAV-Assisted Ground Networks for Information Dissemination:** Owing to their mobility and LoS connectivity, drones can assist ground networks in the distribution of information. For example, drones employed as flying BSs can help a D2D network or a mobile ad-hoc network disseminate information among ground devices, as shown in Figure 7.2. D2D networks can help offload cellular data traffic while also expanding the capacity and coverage of a network, but their efficiency is restricted by the low communication range of devices and the possibility for increased interference. Flying UAVs can help with speedy information distribution in this situation by intelligently broadcasting shared files between ground systems.

Drones can also play an important part in vehicular communication networks, i.e., vehicle-to-vehicle communications (V2V) by disseminating safety critical information. UAVs reduce interference by lowering the number of transmission lines required between ground devices. Mobile UAVs can help in establishing transmit diversity for increased reliability in D2D and V2V networks.

4. **Millimeter Wave and 3D MIMO Communications:** UAVs can be used for 3D MIMO and mmWave communications because of their aerial positions and ability to deploy on demand at any required location. Lately, there has been a lot of interest in using UAVs for 3D MIMO, well known as full-dimension MIMO, which exploits both the horizontal and vertical dimensions. As

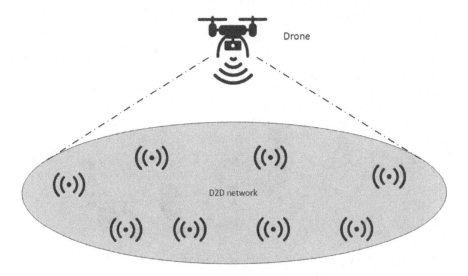

Figure 7.2 UAVs for information dissemination.

demonstrated in Figure 7.3. 3D beamforming allows for the simultaneous generation of many beams in three-dimensional space, thereby decreasing intercell interference. In general, 3D MIMO is better for circumstances when there are a lot of users and they are at varied elevation angles from the serving BS. Thus, the high

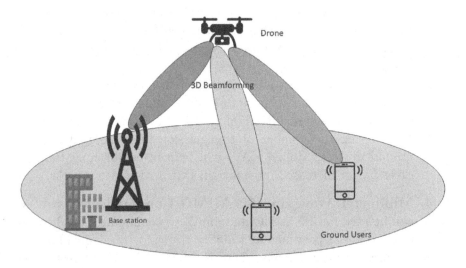

Figure 7.3 3D beamforming on drones.

altitude of these aerial BSs makes it possible to distinguish between users at different elevation angles with respect to the UAV. Moreover, in UAV-to-ground communications, the LoS channel conditions allow for productive beamforming in both the elevation and azimuth domains. As a result, drone-BSs are preferred candidates for using 3D MIMO. UAVs can potentially play an important role in mmWave communications also. UAVs with millimeter wave capability can set up LoS links with ground terminals, minimizing propagation loss and allowing them to operate at higher frequencies.

5. **UAVs for IoT Communications:** Drones can be used as aerial BSs in IoT networks to deliver dependable and energy-saving uplink IoT connectivity. In reality, because of their airborne nature and high altitude, UAVs may be used to effectively eliminate shadowing and blocking effects, which are the leading causes of signal attenuation in wireless networks. UAVs may be deployed depending on the position of IoT devices, allowing only those devices to connect to the network while consuming the least amount of transmit power possible. Furthermore, UAVs can assist large IoT systems by dynamically updating their positions based on IoT device activation patterns. This is in contrast to the use of ground small cell BSs, which may need to be significantly extended to accommodate the expected number of IoT devices. As a result, by utilizing the unique qualities of UAVs, IoT network connection and energy efficiency can be considerably increased.

6. **Cache-Enabled UAV:** Using cache-enabled UAVs in wireless networks is a potential strategy for traffic unloading. Cache-enabled drones can be optimally relocated and deployed in order to deliver required services to users by making use of user-centric information such as mobility patterns and content request distribution. Another benefit of using cache-enabled drones is that the caching complexity is minimized when compared to using static small BSs. For example, when a mobile user goes to a new cell, the required content must be saved at the new BS. For this purpose, cache-enabled drones can track users' movements and as a result, the content saved on the drones will no longer require additional caching at small BSs. In reality, in a cache-enabled drone system, a central cloud processor controls the UAV deployment by utilizing varied user-centric data such as users' mobility patterns and

content request distribution. A cloud center can learn such user-centric information from any previously available users' data. The cloud center may then calculate the mobility paths and locations of cache-enabled drones in order to aid ground users effectively. As a result, the overall overhead of refreshing the cache material can be reduced.

7.2.2.2 Cellular-Connected Drones as Aerial User Equipments

Apart from UAV as flying BSs, UAV can function as the users of the wireless cellular network, commonly known as aerial user equipments (AUEs). These AUEs can be operated for item delivery, remote sensing and surveillance, and virtual reality operations. Undoubtedly, cellular-connected drones will be the principal supporter of Internet-of-Things. The salient advantage of AUEs is that they can move swiftly and optimize their flight path to achieve their objectives quickly. However, for supporting a large-scale deployment of such aerial users, a dependable and low-latency connectivity among drones and ground BSs is necessary. In practice, a suitable wireless communication infrastructure is required to efficiently govern the drones' activities while supporting the traffic generated by their application services. Also, drone user equipments will need high-speed uplink connectivity from the ground network and from other drone-BSs when utilized for surveillance purposes. However, the current cellular network which is specifically outlined for ground-UE may not be fully ready to carry drone-UE operations, whose requirements are quite different from those of ground-UEs. In this regard, one of the main challenges is aerial ground interference due to LoS channels with non-serving BSs.

7.2.2.3 Flying Ad-Hoc Networks with UAVs

Flying ad-hoc networks (FANETs) is another important use case where in many UAVs interact in an ad-hoc fashion. FANETs can increase connectivity and communication range in places with inadequate cellular infrastructure because of their self-organizing nature, mobility, and lack of central control. They are used in a variety of applications, including disaster management, traffic monitoring, border surveillance, remote sensing, agricultural management, relay networks, and wildfire control. A relaying network of UAVs, in particular, helps in maintaining reliable communication channels between senders and receivers which are unable to interact directly due to barriers or a large separation distance. The previously mentioned applications are just a small example of the many possible applications of drones as aerial wireless platforms. Such applications, if achieved, will have profound technological as well as societal implications.

7.2.3 Network Architecture for UAVs

The general architecture of wireless communication with UAVs comprises two types of communication links: the data link and Control and Non-Payload Communication link (CNPC) [19] as shown in Figure 7.4.

1. Control and Non-Payload Communication Link: The CNPC links are necessary to enable the safe operation of UAVs. In order to exchange the safety-critical data among the UAVs or between the UAV and GCS, a reliable, secure, and low-latency communication is required. CNPC usually requires low-data rates. The CNPC information flow can be grouped into three types as follows:

   • Command and control from GCS to UAVs

   • Aircraft status report from UAVs to ground

   • Sense-and-avoid information among UAVs

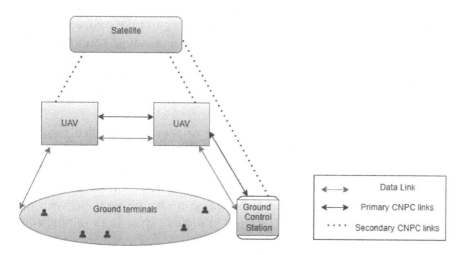

Figure 7.4   Network Architecture.

2. Data Link: Depending on the application scenario, data links are intended to provide mission-related data for ground terminals comprising terrestrial BSs, gateway nodes, mobile terminals, wireless sensors, etc. For example, the data links can be used to support direct mobile-to-UAV communication in case of BS offloading or during a BS malfunction. Also, data links maintained by the UAV can support the UAV-BS and UAV-gateway wireless backhaul. The data link capacity requirements depend on the application and may vary from several kilobits per second in UAV sensor links to dozens of gigabits per second in UAV-gateway wireless backhaul.

Compared to CNPC links, UAV data links have higher tolerance in terms of security and latency requirements.

7.3   THZ-EMPOWERED UAV COMMUNICATION

Owing to the 3D positional versatility of the UAVs, they will play a major role in various aspects of upcoming 6G wireless networks. The THz frequency, on the other hand, is one of the prospective enablers for meeting the growing demands of 6G, thus providing higher throughput, accurate localization, lower latency, and precise sensing. Recent breakthroughs in the semiconductor industry have made it possible to design more firm THz devices, which have been the key impediments to THz implementation in the past. As a result, in the envisioned 6G networks, THz-enabled UAVs may present a number of opportunities, leading the door to new applications and services. In order to take complete advantage of these opportunities, UAVs must have a steadfast communication link. Since moisture has no effect at the altitude above 16 km, THz attenuation is negligible. As a result, THz could be a good contender for establishing reliable communications in a variety of UAV application scenarios.

Moreover, to attain their self-governing or remote-controlled missions, UAVs depend on short-distance reliable links for obtaining commands or transmitting data. THz links are best considered for interchanging safety-critical data between UAV and GCS as well as among the UAVs. Furthermore, THz communication links can also be used between airplanes and UAVs to support internet in flights rather than relying completely on satellite services [4].

7.3.1 Applications of THZ-Empowered UAVs Use Cases

Here, we will discuss various applications of THZ-empowered UAVs [20]:

7.3.1.1 Interactive Aerial Telepresence

UAVs can be deployed and controlled remotely to accomplish dangerous, expensive, or time-sensitive activities for humans. Aerial telepresence is the term for this, and it gets more effective when augmented reality (AR) is employed, as it may provide 3D views and real-time tele-interaction with the surroundings. Indeed, tactile guiding improves UAV performance and most importantly allows access to experts anytime from anywhere. THz deployment is critical for such applications, especially for real-time association where it's not possible to compress AR data and high data rates are sought. UAVs equipped with THz sensors can provide accurate and instant environmental knowledge, for highly precise interactive tasks.

7.3.1.2 Reliable and Secure Functioning of UAVs

Large antenna arrays in small sizes at THz can enable cm-level localization accuracy and sensing solutions with finer range Doppler and angular resolutions [21]. Therefore, THz-enabled UAVs have accurate localization and sensor capabilities, allowing for the instantaneous perception of the environment. Such features enable THz-powered UAVs to navigate in a secure and reliable manner. It's worth noting that the safe operation of drones is critical both for compliance with airspace restrictions and for their practical implementation. Drones with THz capabilities can facilitate this goal. THz with UAVs, for example, may easily position themselves with the intended transmitters and/or receivers to increase reliability and security by utilizing accurate sensing and localization. Furthermore, they may navigate safely and avoid any obstacle by sensing the surrounding environment.

7.3.1.3 Pollution Monitoring and Tracking

Frequency scanning spectroscopy at the THz band allows air quality monitoring and detection of some chemicals since various materials at certain frequencies have vibrational absorption. The specific absorption characteristics of different gaseous mediums significantly boost THz's

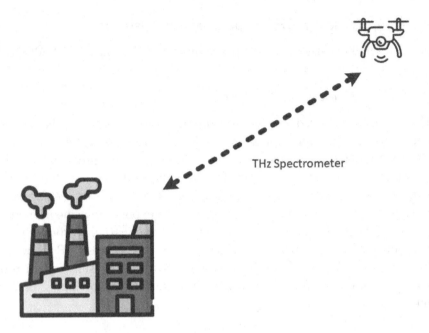

Figure 7.5   Pollution monitoring and tracking.

sensing capability [20]. THz spectroscopy, combined with the flexibility of UAVs, can provide effective pollution monitoring and tracking options. For example, drones enabled with THz spectrometers can fly near pollution sources (such as industrial chimneys) so as to detect pollutant concentrations, refer Figure 7.5. UAVs equipped with THz can also fly over the mountain top to monitor the concentration of water vapors. THz-empowered UAVs with such capabilities can aid in the fight against air pollution and the monitoring of climate change.

7.3.1.4   Time-Critical MEC-Enabled Operation

Owing to the limited amount of on-board computation capacity, drones may depend on computing (MEC) servers. These servers introduce a delay in attaining required data. Furnishing high-speed communication links enhances the time-critical remote computation capability. As shown in Figure 7.6, implementing MEC-enabled operations through THz links can reduce the payload for UAVs, thereby lowering the propulsion energy consumption and extending the operational time.

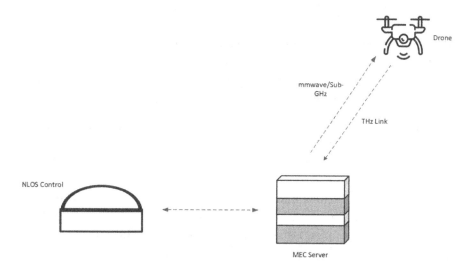

Figure 7.6 Time-critical MEC-enabled operation.

7.3.1.5 Advanced Swarm of UAVs

In UAV swarm networks, UAVs exchange a huge amount of data particularly in applications like sensing, localization, and mapping. As a result, UAVs need to enhance up the interconnection rate capacity beyond the current technology, compatible with future generations [20]. For this purpose, ultra-high-throughput UAV-2-UAV communication links can be provided with THz band in LoS propagation as shown in Figure 7.7.

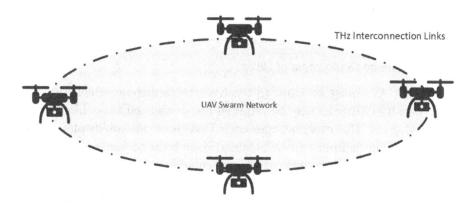

Figure 7.7 UAV swarm network.

7.3.2 Design Challenges

In the following, we present some challenges faced in the successful deployment of THz-empowered UAV communication [20]:

7.3.2.1 Wobbling of UAVs

Efficacious use of THz frequencies needs highly directional antennas on the transmitter and receiver sides for compensating the propagation losses. But UAVs wobbling because of the wind can cause misalignment of beam, thereby deteriorating the standard of communication. Furthermore, frequent rotations and tilts resulting from UAV motion make it difficult to maintain an absolute beam alignment, thereby causing handover and beam hopping issues. The frequent handovers cause delays and decrease link reliability. Intelligent reflective surface (IRS)-assisted THz communication can help in enhanced localization and sensing, thereby aiding effective beam alignment [22].

7.3.2.2 Limited Design Space

Transitions from NLoS to LOS for much favorable communication environment, which compensate for the negative effect of longer link lengths, are the primary means of achieving optimal height for UAVs in low frequencies. However, for THz, the communication quality drops substantially with the length of the link. The THz band, for instance, can provide high-speed (tbps) communication links for maintaining integrated access and backhaul for a drone-BS, but the advantages are only achievable in a narrow range of communication, limiting the space available for efficient and ideal UAV placement. Therefore, an adaptable and tactical planning for the 3D placement of UAVs is required.

7.3.2.3 Dense Deployment of UAVs

In THz, UAVs must be close to receivers/transmitters, which means a high density of drones may be required for specialized tasks like sensing or monitoring. However, because each UAV may offload/display bigger data, high throughput can be attained if each drone has a higher sum-rate capacity. Furthermore, while molecular absorption and high path loss require dense UAV deployment, it may result in significant LoS interference (which gets poor with frequent beam misalignment) and handovers.

7.3.2.4 Multi-Functionality Design

The integration of multiple features such as communication, sensing, localization, and computation benefits THz-enabled UAVs greatly. Co-designing a multi-function system is, however, a difficult task. Other than the general issues like waveform design for integrated sensing and communication, developing novel models and powerful less-complex algorithms that can be implemented on drones having limited power and processing capabilities is a big difficulty. In addition, the algorithms should account for the sensitivities of numerous UAV applications to the delay produced by the integrated design.

7.3.2.5 Uncertain Medium Condition

The THz channel is highly uncertain and unreliable due to air composition and its reliance on weather conditions, especially in an uncontrolled outdoor scenario. The proportion of water vapor in the air as well as the humidity level has a big impact on molecular absorption. As a result, offline optimal network deployment and design characteristics may only be applicable in particular situations, reducing the benefits. To some extent, online real-time communication and sensing can help solve this problem.

7.3.3 Trade-Offs

THz transmitters use massive antenna arrays with pencil-sized beams, for they require highly directional transmission. Because of the UAV's speed and wobbling, beam misalignment is common, necessitating frequent handovers. Frequent handovers can generate delays, which can negatively impact latency and reliability performance.

Large antenna gain leads to higher data rates, but it might have a negative impact on reliability and latency.

Longer flights are also desirable in THz-powered UAVs for meaningful data transfer at the expense of greater energy consumption. In order to compensate for the short-communication range, drones can be densely deployed, but this may result in significant interference, lowering reliability.

As a result, it is possible to determine the underlying trade-offs between data rate, latency, reliability, and energy consumption. Moreover, by having a good awareness of the surroundings and beam tracking, the integration of multiple features such as communication, localization, and sensing may increase the data rate and reliability. However, such multifunctionality can result in higher payload energy consumption and processing delays.

As a result, there is a trade-off between energy use and reliability. Furthermore, because multiple activities are involved, this integrated architecture may lead to some latency. As a result, higher reliability can be achieved at the expense of increased latency, leading to trade-off.

BIBLIOGRAPHY

[1] John Federici and Lothar Moeller. Review of terahertz and sub-terahertz wireless communications. *Journal of Applied Physics*, 107(11):6, 2010.

[2] Anamaria Moldovan, Michael A Ruder, Ian F Akyildiz, and Wolfgang H Gerstacker. LOS and NLoS channel modeling for terahertz wireless communication with scattered rays. In *2014 IEEE Globecom Workshops (GC Wkshps)*, pages 388–392. IEEE, 2014.

[3] Ke Su, Lothar Moeller, Robert B Barat, and John F Federici. Experimental comparison of terahertz and infrared data signal attenuation in dust clouds. *JOSA A*, 29(11):2360–2366, 2012.

[4] Hadeel Elayan, Osama Amin, Raed M Shubair, and Mohamed-Slim Alouini. Terahertz communication: The opportunities of wireless technology beyond 5G. In *2018 International Conference on Advanced Communication Technologies and Networking (CommNet)*, pages 1–5. IEEE, 2018.

[5] Xue Wu and Kaushik Sengupta. Dynamic waveform shaping with picosecond time widths. *IEEE Journal of Solid-State Circuits*, 52(2):389–405, 2016.

[6] Josep Miquel Jornet and Ian F Akyildiz. Graphene-based plasmonic nano-transceiver for terahertz band communication. In *The 8th European conference on antennas and propagation (EuCAP 2014)*, pages 492–496. IEEE, 2014.

[7] Mehdi Hasan, Sara Arezoomandan, Hugo Condori, and Berardi Sensale-Rodriguez. Graphene terahertz devices for communications applications. *Nano Communication Networks*, 10:68–78, 2016.

[8] Sebastian Priebe and Thomas Kurner. Stochastic modeling of THz indoor radio channels. *IEEE Transactions on Wireless Communications*, 12(9):4445–4455, 2013.

[9] Seunghwan Kim and Alenka Zajic. Statistical modeling of THz scatter channels. In *2015 9th European Conference on Antennas and Propagation (EuCAP)*, pages 1–5. IEEE, 2015.

[10] Seunghwan Kim and Alenka Zajic. Statistical modeling and simulation of short-range device-to-device communication channels at sub-THz frequencies. *IEEE Transactions on Wireless Communications*, 15(9):6423–6433, 2016.

[11] Danping He, Ke Guan, Alexander Fricke, Bo Ai, Ruisi He, Zhangdui Zhong, Akifumi Kasamatsu, Iwao Hosako, and Thomas Kurner. Stochastic channel modeling for kiosk applications in the terahertz band. *IEEE Transactions on Terahertz Science and Technology*, 7(5):502–513, 2017.

[12] Ali Riza Ekti, Ali Boyaci, Altan Alparslan, Ilhami Unal, Serhan Yarkan, Ali Gorccin, Huseyin Arslan, and Murat Uysal. Statistical modeling of propagation channels for terahertz band. In *2017 IEEE Conference on Standards for Communications and Networking (CSCN)*, pages 275–280. IEEE, 2017.

[13] Shahram Mollahasani and Ertan Onur. Evaluation of terahertz channel in data centers. In *NOMS 2016-2016 IEEE/IFIP Network Operations and Management Symposium*, pages 727–730. IEEE, 2016.

[14] Chunxiao Jiang, Haijun Zhang, Yong Ren, Zhu Han, Kwang-Cheng Chen, and Lajos Hanzo. Machine learning paradigms for next-generation wireless networks. *IEEE Wireless Communications*, 24(2):98–105, 2016.

[15] Matti Latva-aho, Kari Leppänen (eds.), Key Drivers and Research Challenges for 6G Ubiquitous Wireless Intelligence. 6G Research Visions, 6G Flagship, University of Oulu, September 2019, Oulu, Finland. Available at: http://urn.fi/urn:isbn:9789526223544

[16] Mohammad Mozaffari, Walid Saad, Mehdi Bennis, Young-Han Nam, and Mérouane Debbah. A tutorial on UAVs for wireless networks: Applications, challenges, and open problems. *IEEE Communications Surveys & Tutorials*, 21(3):2334–2360, 2019.

[17] Daniel Fuller. AT&T detail network testing of drones in football stadiums. Android Headlines, September 27, 2016. Available at: https://www.androidheadlines.com/2016/09/att-detail-network-testing-of-drones-in-football-stadiums.html

[18] Gianmarco Baldini, Stan Karanasios, David Allen, and Fabrizio Vergari. Survey of wireless communication technologies for public safety. *IEEE Communications Surveys & Tutorials*, 16(2):619–641, 2013.

[19] Yong Zeng, Rui Zhang, and Teng Joon Lim. Wireless communications with unmanned aerial vehicles: Opportunities and challenges. *IEEE Communications Magazine*, 54(5):36–42, 2016.

[20] M Mahdi Azari, Sourabh Solanki, Symeon Chatzinotas, and Mehdi Bennis. THz-empowered UAVs in 6G: Opportunities, challenges, and trade-offs. *IEEE Communications Magazine*, 60(5):24–30, 2022.

[21] Andre Bourdoux, Andre Noll Barreto, Barend van Liempd, Carlos de Lima, Davide Dardari, Didier Belot, Elana-Simona Lohan, Gonzalo Seco-Granados, Hadi Sarieddeen, Henk Wymeersch, et al. 6G white paper on localization and sensing. *arXiv preprint arXiv:2006.01779*, 2020.

[22] Christina Chaccour, Mehdi Naderi Soorki, Walid Saad, Mehdi Bennis, Petar Popovski, and Merouane Debbah. Seven defining features of terahertz (THz) wireless systems: A fellowship of communication and sensing. *IEEE Communications Surveys & Tutorials*, 24(2):967–993, 2022.

Performance Characterization of RSMA in THz Networks

Sadeq Bani Melhem

York University, Toronto, Canada

Hina Tabassum

York University, Toronto, Canada

CONTENTS

8.1	Introduction	178
	8.1.1 Evolution of Transmission Spectrum	178
	8.1.2 Evolution of Multiple Access Techniques	180
	8.1.2.1 Orthogonal Multiple Access (OMA)	181
	8.1.2.2 Non-orthogonal Multiple Access Schemes	182
8.2	Background Works–RSMA	184
8.3	Contribution of the Chapter	185
8.4	System Model	186
	8.4.1 Channel Model	187
	8.4.2 SINR Model	188
	8.4.3 Data Rate Computation	189
	8.4.4 User Pairing Schemes	189
8.5	THz-RSMA: Outage Analysis	190
	8.5.1 Outage – Near User	191
	8.5.2 Outage – Far User	193
8.6	Numerical Simulations	194
8.7	Summary and Open Issues	197
Bibliography		198

DOI: 10.1201/9781003282211-8

$\mathbf{F}$ UELED BY THE emergence of machine-type communications in various wireless applications, the provisioning of massive connectivity becomes instrumental. On the other hand, accommodating trillions of devices within the highly congested and limited sub-6 GHz spectrum is becoming challenging. In this context, shifting to higher frequency Terahertz (THz) communication is under consideration to obtain the data rates in the order of hundreds of gigabits per second (Gbps). Also, non-orthogonal multiple access (NOMA) schemes are becoming popular to support multiple users in the same frequency and time resource blocks while leveraging efficient interference cancellation mechanisms. This chapter will provide a comprehensive analytic framework to analyze the performance of emerging non-orthogonal channel access schemes, such as non-orthogonal multiple access (NOMA) and rate-splitting multiple access (RSMA), in sub-6 GHz and THz networks. Numerical results demonstrate a comparison between the efficiency of RSMA, orthogonal multiple access (OMA), and NOMA. The result shows that the nearest-farthest scheme performs better than the random scheme for the near user (U_1), while it is the opposite for the far user (U_2). Moreover, the outage increases with the increase in the cell radius for both users in both schemes. The increase in portion transmits power allocated for U_1 (i.e., p_1 for RSMA, a_1 for NOMA); the outage increases for RSMA and NOMA due to the increased interference from U_1 created at U_1. The gain of RSMA over NOMA and OMA improves more for low values of p_1 (a_1) for both users.

8.1 INTRODUCTION

This section explains the evolution of various frequency bands (such as sub-6 GHz, mm-wave, THz) of wireless networks and illustrates the evolution of multiple access schemes with the generations of wireless networks.

8.1.1 Evolution of Transmission Spectrum

While sub-6 GHz and mm-wave bands have already been implemented in fourth-generation (4G) and fifth-generation (5G) networks, respectively, the THz and optical frequency bands are the potential candidates for sixth-generation (6G) wireless networks [1].

In what follows, we provide a brief overview of the electromagnetic spectrum and various frequency bands (such as sub-6 GHz, mm-wave, THz):

- **Sub-6 GHz Spectrum:** Sub-6 GHz spectrum is the most commonly used spectrum for cellular communications. Lower frequency transmissions, such as sub-6 GHz transmissions, can propagate further and penetrate through buildings more effectively than higher frequency spectrum, such as mm-wave. However, the sub-6 GHz spectrum is running out of bandwidth to accommodate wireless users. Since the sub-6 GHz spectrum is extremely limited and congested, the achievable data rates can be much lower than those available in the higher spectrum. Two primary factors affect the signal, i.e., large-scale fading created by path loss and shadowing and small-scale fading caused by multi-path propagation. As a result, 5G networks rely on the mm-wave spectrum to achieve higher bandwidth and massive connectivity [2].

- **Mm-wave Spectrum:** Mm-wave transmissions (30–100 GHz) are identified as a promising technology for future-generation networks to meet the data rate requirements in the order of 10 Gbps. The key benefit of the mm-wave spectrum is its broad bandwidth and smaller form-factor which enables the deployment of massive antennas within a small area, allowing high antenna gain and beamforming to be achieved even in handsets. Unfortunately, since mm-waves experience a higher path loss than lower frequencies, their range is constrained. Also, since the wavelength is less than a centimeter, mm-wave signals are more vulnerable to environmental attenuation and absorption than sub-6 GHz signals. When mm-wave signals get into contact with large structures, they experience higher diffused scattering than sub-6 GHz signals.

- **THz Spectrum:** Following the efficient introduction of mm-waves in 5G networks, the research on THz communications is now gaining popularity. THz-band promotes ultra-reliable and low-latency applications for multiple indoor scenarios thanks to its broader bandwidth. Also, passive cyber-attacks, such as man-in-the-middle attacks, would be extremely difficult in the THz band because of incredibly short wavelengths and high-gain ultra-massive antennas producing extremely focused beams [3]. However, THz transmission generates very high propagation losses.

8.1.2   Evolution of Multiple Access Techniques

Multiple access, a critical part of wireless networks, has a significant effect on bandwidth usage, system throughput, and latency. Multiple access in cellular radio refers to a technique in which multiple users share a transmission medium to communicate with the wireless access point. Figure 8.1 shows a categorization of multiple access techniques. Multiple access schemes are instrumental to the performance of wireless networks as they describe how resources (e.g., frequency, time, antennas, power, codes, etc.) should be allocated to users [4]. For example, since bandwidth is typically limited and/or very costly, the efficient allocation between users is critical in both uplink and downlink communications. Typically, when multiple users can access wireless services through their respective dedicated resources, this is referred to as *multiple access*. Figure 8.1 illustrates the evolution of multiple access schemes with the generations of wireless networks. The multiple access schemes can be classified into two groups, i.e., OMA and NOMA schemes, which are detailed in the subsequent sections.

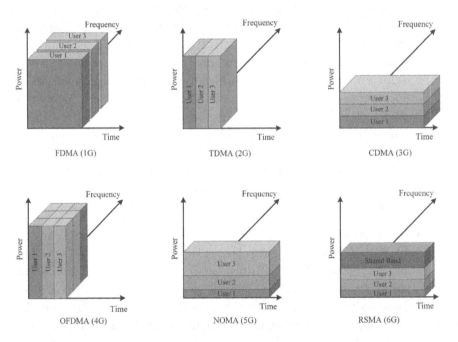

Figure 8.1   Evolution of multiple access schemes over the past decades.

8.1.2.1 Orthogonal Multiple Access (OMA)

The users' channel access in OMA is exclusive, i.e., users do not interfere with one another when sharing a networking channel. Due to the orthogonal channel access, there is no inter-user interference. Thus, low-complexity receivers can be used to identify the signal of the intended user. Nevertheless, since the amount of orthogonal resources is restricted, OMA systems cannot support massive connectivity in 5G and 6G networks. The most common OMA techniques implemented to date in wireless networks are classified as follows:

- **Frequency-Division Multiple Access (FDMA):** In FDMA, which has been used in the first generation (1G) of cellular networks, the usable spectrum is divided into non-overlapping frequency sub-bands, each providing one digital data stream.

- **Time-Division Multiple Access (TDMA):** In TDMA, the time resource is split into several time slots, each of which is assigned to users typically in a round-robin manner. TDMA has been utilized in the second generation (2G) of cellular networks. In TDMA, the network dimensions are split into non-overlapping time slots, and each user is allocated a separate time slot. Since each user in TDMA is assigned a new cyclically repeated time slot, the limitation of TDMA is that the channel features can vary drastically in different slots. As a result, receiver functions that rely on channel estimates, such as equalization, must re-estimate the channel for every period.

- **Code-Division Multiple Access (CDMA):** CDMA uses orthogonal pseudo random spreading codes to modulate the input signals of various users. The transmitted signals from different users use the same time and bandwidth. In CDMA, user-specific codes scatter the modulated symbol by the processing gain [5] such that the code signatures can be orthogonal to one another. Subsequently, the number of concurrent users that can be assisted is less than or equal to the number of orthogonal codes. The spreading code structure is used by the receiver to distinguish between various users. Third generation (3G) networks were based on wideband CDMA [6].

- **Orthogonal Frequency-Division Multiple Access (OFDMA):** OFDMA is based on the orthogonal frequency-division multiplexing (OFDM) waveform, allowing subcarriers to be packed tightly and orthogonally in the frequency domain with a subcarrier spacing inverse to the symbol length. Furthermore, OFDMA divides the time and frequency plane into two-dimensional resource blocks, each of which can be allocated to a user who transmits a modulated symbol. OMA based on OFDMA or single-carrier frequency-division multiplexing access (SC-FDMA) is used in 4G of cellular networks such as long-term evolution (LTE) and LTE-Advanced, standardized by the third Generation Partnership Project (3GPP).

In general, OMA schemes make it easier to build a transceiver and reduce co-channel interference. In these schemes, however, the number of users who may be serviced at the same time is limited. Furthermore, user scheduling and reliable feedback channels are necessary to ensure orthogonality.

8.1.2.2 Non-orthogonal Multiple Access Schemes

NOMA schemes are the ones where users can access a shared time/frequency/power resource block of the network. Some of the popular non-orthogonal schemes are classified as follows:

- **Power-Domain Non-orthogonal Multiple Access (PD-NOMA):** PD-NOMA is a crucial enabling technology for next-generation wireless networks to attain massive connectivity, higher throughput, and enhanced fairness. PD-NOMA superposes users' message signals in the power domain using linearly precoded superposition coding and leverage on successive interference cancellation (SIC) at the receivers. In particular, PD-NOMA enables the superposition of various users' message signals on a specific time-frequency resource block [7]. Then, by using SIC, the desired message signal is identified and decoded at the receiver.

- **Code-Domain Non-orthogonal Multiple Access (CD-NOMA):** CD-NOMA was inspired by CDMA, in which multiple users are served using the same time/frequency resources and distinguished by dedicated user-specific spreading sequences [8]. CD-NOMA serves multiple users by employing sparse, low-density, and low-cross-correlation sequences for each user. Multi-user detection is often performed iteratively at the receiver utilizing message passing-based techniques.

- **Space-Division Multiple Access (SDMA):** multiple input multiple output (MIMO) communication is becoming increasingly crucial in the current 5G standard and future wireless networks, thanks to the rapid development of multi-antenna technology. Compared to single-antenna communication systems, multiple antennas at transmitters and/or receivers can utilize additional spatial degrees of freedom (DoFs). As a result, numerous users/devices can be supplied in the same time/frequency/code domain while still being distinguished in the spatial domain in SDMA. Because of its simplicity, linear precoding (LP) is the most often used approach for SDMA. Inter-user interference, in particular, can be efficiently minimized by using spatial DoF to construct appropriate transmit and/or receive beamformers. SDMA, on the other hand, is only useful in underloaded and critically loaded conditions since the available spatial DoFs are used to mitigate inter-user interference.

- **Rate-Splitting Multiple Access (RSMA):** In RSMA schemes, users' messages are splitted into common and private parts, where the common parts will be encoded within one or more common streams while the private parts will be encoded into different streams. Users' messages are separated into common and private messages at the transmitter. The common messages are merged and encoded to common streams to be decoded by different users. In contrast, private messages are encoded separately into private streams to be decoded by the relevant users only [9]. At the transmitter, all streams are combined and transmitted to the users. Following that, each receiver decodes the common stream(s), conducts SIC, and then decodes its private stream. Thus, each receiver will restore its original message after combining its common and private messages. RSMA can partially decode interference using SIC and partially handle interference as noise, thus allowing for rate and quality of service (QoS) improvements [8]. This potential of RSMA to partially mitigate interference and partially address interference as noise enables rate enhancement and SIC complexity reduction [10], [8].

- **Partial-NOMA:** Users share only a portion of the resource block in partial-NOMA, thus reducing interference while allowing some spectrum reuse. Partial-NOMA bridges the gap between classic OMA and nonorthogonal multiple access (NOMA). Note that the signals of users are completely overlapped in traditional PD-NOMA, which enhances spectrum reuse while increasing the interference. On the contrary, partial-NOMA regulates the spectrum overlap ratio.

- **Delta-OMA:** D-OMA has recently been examined as a possible method for improving spectral efficiency in 6G networks. At the expense of increased interference, D-OMA allows partial overlapping of adjacent sub-channels that are assigned to two different clusters of users served by NOMA.

8.2   BACKGROUND WORKS–RSMA

This section presents the recent research on analyzing the performance of RSMA based on either optimization algorithms or analytic frameworks.

Recently, several existing studies, such as in [11–13], investigated the performance of RSMA considering a single-input-single-output (SISO) system. In [11], the authors introduced a one-dimensional search algorithm for sum-rate maximization considering a SIC constraint to optimize rate allocation and power control in a downlink SISO system with RSMA. Another study in [12] investigated the decoding order and power allocation optimization in an uplink RSMA system. Recently, a research work investigated optimizing the RSMA transmission scheme for a downlink cloud radio access network (C-RAN) in the SISO system [13]. The authors presented an efficient RSMA scheme with a hierarchical clustering mechanism to select users who decode common signals.

Several existing works such as [14–18] have recently studied RSMA considering a multiple input single output (MISO) system. In the extended version [14] of [11], the authors proposed a successive convex approximation (SCA)-based algorithm to obtain a locally optimal solution that maximizes the network sum-rate in MISO-RSMA considering the rate and SIC constraints. In [15], the authors suggested a power allocation scheme for multi-user multi-carrier MISO networks to maximize the sum-rate of users, considering random beamforming and zero-forcing beamforming for common and private messages. The energy efficiency (EE) maximization problem of RSMA in the MISO broadcast channel has been studied in [16]. The authors obtained optimal EE beamforming using an SCA-based approach considering the downlink MISO system and showed that RSMA is more energy-efficient than SDMA and NOMA. In [17], the authors presented an energy-efficient resource allocation scheme for downlink a reconfigurable intelligent surface (RIS)-assisted MISO system that optimizes the phase shifts of all RISs to maximize the system EE. In [18], the authors used a modified weighted minimum mean square error (WMMSE) approach jointly with an alternating optimization algorithm to achieve max-min fairness (MMF) in multi-beam satellite communications.

An interesting research work in [8] demonstrated that RSMA is a more generic, reliable, and efficient scheme than SDMA, OMA, and NOMA. In contrast to SDMA, which depends on totally treating any residual interference as noise, and NOMA, which depends on fully decoding interference, RSMA has the capacity to partially decode the interference and partially treating the interference as noise. In [19], the authors investigated the performance of RSMA in a unmanned aeriel vehicle (UAV) assisted multiuser downlink communication system, in which the UAV serviced many users simultaneously through RSMA. They derived closed-form outage expressions and achievable throughput at each user while considering independent and non-identical Nakagami-m small-scale fading, then used Monte-Carlo simulations to validate the accuracy of the derived analytical expressions. The outage performance of semi-grant-free (SGF) transmissions was demonstrated in [20] by deriving an accurate formula for outage probability (OP) and a high SNR approximation. They have shown that the RSMA-SGF scheme is more robust than NOMA-SGF schemes. In [21], the authors proposed two different cooperative-NOMA (C-NOMA) and cooperative-RSMA (C-RSMA) uplink user collaboration techniques. They calculated the asymptotic OP of the proposed C-NOMA and C-RSMA cooperative schemes and demonstrated that they reach a diversity order of two. Their simulation results also show that C-RSMA outperforms C-NOMA and C-OMA.

8.3 CONTRIBUTION OF THE CHAPTER

This chapter provides a comprehensive framework to analyze the outage performance of RSMA in the downlink THz network. In particular, we characterize outage expressions in the presence of various user pairing schemes, Nakagami-m channel fading, and molecular absorption noise. Numerical results validate the accuracy of the derived expressions. To the best of our knowledge, there is no research work that provides a comprehensive framework for the outage analysis of users in a THz-RSMA network. The numerical results validate the accuracy of the derived expressions.

8.4 SYSTEM MODEL

Consider an RSMA-based downlink transmission from an AP to two users, as shown in Figure 8.2. The AP is equipped with a single antenna while serving two users with a single antenna for each, U_1 and U_2. The message of user u, $\forall u \in \{1, 2\}$, denoted by W_u, is divided into common parts $W_{u,c}$ and a private message $W_{u,p}$. The common parts are combined and encoded together to form one common stream, denoted by s_c. And, s_u is the private stream of the encoded private message $W_{u,p}$, $\forall u \in \{1, 2\}$. Figure 8.3 shows the RSMA transmission model. The resulting streams are then assigned precoding weights and superposed for transmission, assuming that the AP has perfect channel knowledge. Hence, the received signal at user u can be expressed by

$$y_u = h_u x + n_u, \quad \forall u \in \{1, 2\}, \tag{8.1}$$

where n_u is the aggregate complex additive white Gaussian noise (AWGN) with zero-mean and variance σ^2 and molecular noise signal,

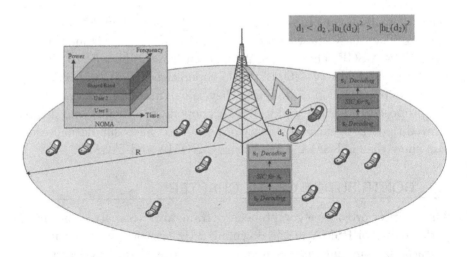

Figure 8.2 Downlink RSMA in a single cell with one access point (AP) and two users.

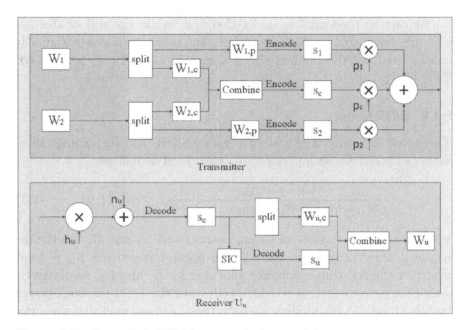

Figure 8.3 Downlink RSMA transmission model.

and x is the transmit signal which can be expressed as

$$x = \mathbf{ps},$$

where $\mathbf{s} = [s_c, s_1, s_2]^T$ is the transmitted signal vector, $\mathbf{p} = [p_c, p_1, p_2]$ is the precoding vector or power vector, and h_u is the channel gain from the AP to user u, which assumed to be perfectly known at the transmitter.

8.4.1 Channel Model

The line-of-sight (LoS) channel power between the AP and user u in THz networks is formulated as follows:

$$h_u^{\text{THz}} = \left(\frac{c}{4\pi f d_u}\right)^2 e^{-k_a(f)d_u} = \zeta d_u^{-2} e^{-k_a(f)d_u}, \quad \forall u \in \{1, 2\}, \quad (8.2)$$

where d_u represents the transmission distance from AP to user u, f is the transmitting carrier frequency, ζ represents the transfer function of the channel spreading loss and defined as $\zeta = \left(\frac{c}{4\pi f}\right)^2$, and $k_a(f)$ is the molecular absorption coefficient defined in [22].

At user u, the signal s_c is decoded first, while the rest of the signals are considered to be noise. Then, the common message signal is subtracted from the original signal using SIC. Finally, s_u is decoded, $\forall u \in \{1, 2\}$.

8.4.2 SINR Model

The signal to interference and noise ratio (SINR) for the common messages of U_1 and U_2 with perfect SIC is modeled as follows:

$$\gamma_u^c = \frac{G_t G_r p_c h_u \chi_u}{G_t G_r p_1 h_u \chi_u + G_t G_r p_2 h_u \chi_u + N_u^{\text{com}}}, \quad \forall u \in \{1, 2\}, \qquad (8.3)$$

where χ is the Nakagami-m fading channel and G_t and G_r are the directional antenna gains of the AP and users, respectively. p_1, p_2, and p_c depict the AP transmit power allocated for U_1 and U_2, respectively, such that $p_1 + p_2 + p_c = p$, where p denotes the total transmit power budget of the AP. Beam alignment strategies are assumed that align the main lobes of the users and the THz AP. The noise at the receivers of U_1 and U_2 comprises thermal noise N_0 and molecular absorption noise as defined below:

$$N_u^{\text{com}} = k_b T + G_t G_r (p_1 + p_2 + p_c) \zeta d_u^{-2} (1 - e^{-k(f)d_u}) \chi_u, \quad \forall u \in \{1, 2\},$$
$$= k_b T + \zeta d_2^{-2} (1 - e^{-k(f)d_u}) \chi_u, \quad \forall u \in \{1, 2\}, \qquad (8.4)$$

where T is the temperature of the channel and k_b is the Boltzmann constant.

The SINRs for the private messages of U_1 and U_2 with perfect cancellation of the common messages are modeled as follows:

$$\gamma_1^{\text{pr}} = \frac{G_t G_r p_1 h_1 \chi_1}{G_t G_r p_2 h_1 \chi_1 + N_1^{\text{pr}}}, \qquad (8.5)$$

and

$$\gamma_2^{\text{pr}} = \frac{G_t G_r p_2 h_2 \chi_2}{G_t G_r p_1 h_2 \chi_2 + N_2^{\text{pr}}}, \qquad (8.6)$$

where

$$N_u^{\text{pr}} = k_b T + G_t G_r (p_1 + p_2) \zeta d_u^{-2} (1 - e^{-k(f)d_u}) \chi_u, \quad \forall u \in \{1, 2\}. \quad (8.7)$$

8.4.3 Data Rate Computation

The corresponding data rates of U_1 and U_2 (in bps) for common and private messages are given, respectively, as follows:

$$R_u^c = B \log_2(1 + \gamma_u^c), \quad \forall u \in \{1, 2\}, \tag{8.8}$$

and

$$R_u^{\mathrm{pr}} = B \log_2(1 + \gamma_u^{\mathrm{pr}}), \quad \forall u \in \{1, 2\} \tag{8.9}$$

where B is the channel bandwidth. To decode the common stream s_c at both users, the common rate shall not exceed the minimum of common decoded signals, i.e.,

$$R^{\mathrm{com}} = \min\{R_1^c, R_2^c\}.$$

This common data rate R^{com} is then shared between users, where C_u^{com} is user u's portion of the common rate with

$$R^{\mathrm{com}} = C_1^{\mathrm{com}} + C_2^{\mathrm{com}}. \tag{8.10}$$

Consequently, the data rate of user u, denoted by R_u^{tot}, can be expressed as

$$R_u^{\mathrm{tot}} = C_u^{\mathrm{com}} + R_u^{\mathrm{pr}}, \quad \forall u \in \{1, 2\}. \tag{8.11}$$

8.4.4 User Pairing Schemes

We consider random and nearest-farthest user pairing schemes. In the random pairing scheme, we pick only two users randomly with independent and identically distributed distances r_1 and r_2 from AP. The near and far user's distance can thus be defined as $d_1 = \min(r_1, r_2)$ and $d_2 = \max(r_1, r_2)$, respectively. Therefore, the probability density function (PDF) cumulative distribution function (CDF) of the distances of U_1 and U_2 are given, respectively, as follows:

$$f_{d_1}(d_1) = 2f_r(d_1)(1 - F_r(d_1)), \quad F_{d_1}(d_1) = 1 - [1 - F_r(d_1)]^2, \tag{8.12}$$

$$f_{d_2}(d_2) = 2f_r(d_2) F_r(d_2), \quad F_{d_2}(d_2) = [F_r(d_2)]^2. \tag{8.13}$$

On the other hand, in the nearest-farthest scheme, we select two users out of N users with minimum and maximum distances from the AP. The near and far user's distance can thus be defined as $d_1 = \min(r_1, r_2, \ldots, r_N)$ and $d_2 = \max(r_1, r_2, \ldots, r_N)$, respectively. The PDF and CDF of d_1 and

d_2 can thus be given, respectively, as follows:

$$f_{d_1}(d_1) = N[1 - F_r(d_1)]^{N-1} f_r(d_1), \quad F_{d_1}(d_1) = 1 - [1 - F_r(d_1)]^N \tag{8.14}$$

$$f_{d_2}(d_2) = N[F_r(d_2)]^{N-1} f_r(d_2), \quad F_{d_2}(d_2) = [F_r(d_2)]^N, \tag{8.15}$$

where the PDF and CDF of r are given, respectively, as $f_r(r) = \frac{2r}{R^2}$, $F_r(r) = \frac{r^2}{R^2}$, since all users are uniformly distributed in a circular region of radius R.

8.5 THZ-RSMA: OUTAGE ANALYSIS

Considering $C_{th,u}$ is the desired data rate threshold for the common message and $R_{th,u}$ is the desired data rate of the private message for user u, we define the outage probability for user u as follows:

$$O_u^{\text{rsma}} = \Pr(C_u^{\text{com}} \le C_{th})\Pr(R_u^{\text{pr}} \le R_{th}), \quad \forall u = \{1, 2\}. \tag{8.16}$$

Using (8.9), we can rewrite the private message of (8.16) as follows:

$$\Pr(R_u^{\text{pr}} \le R_{th}) = \Pr(\gamma_u^{\text{pr}} \le \gamma_{th}^{\text{pr}}), \quad \forall u = \{1, 2\}, \tag{8.17}$$

where $\gamma_{th}^{\text{pr}} = 2^{\frac{R_{th}}{B}} - 1$. Now substituting (8.5) in (8.17) or (8.6) in (8.17) gives us the outage expression of the private messages of U_1 and U_2, respectively, as

$$\Pr(\gamma_1^{\text{pr}} \le \gamma_{th}^{\text{pr}}) = \Pr\left(h_1 \le \frac{\gamma_{th}^{\text{pr}} \kappa}{p_1 - \gamma_{th}^{\text{pr}} p_2}\right) \tag{8.18}$$

and

$$\Pr(\gamma_2^{\text{pr}} \le \gamma_{th}^{\text{pr}}) = \Pr\left(h_2 \le \frac{\gamma_{th}^{\text{pr}} \kappa}{p_2 - \gamma_{th}^{\text{pr}} p_1}\right) \tag{8.19}$$

where $\kappa = \frac{1}{G_t G_r \zeta}$. Note that the data rate of the common message depends on the minimum of common decoded signals. Using (8.10), we can rewrite the common message of (8.16) as follows:

$$\Pr(C_u^{\text{com}} \le C_{th}) = \Pr(R^{\text{com}} \le C_{th} + C_{u'}^{\text{com}}), \quad \forall u = \{1, 2\}, \forall u' = \{2, 1\}. \tag{8.20}$$

Without loss of generality, we consider the common rate is equal for both users (i.e., $C_1^{\text{com}} = C_2^{\text{com}}$), and the channel gains are sorted in ascending order, i.e., $h_1 \geq h_2$. To ensure that all both users can successfully decode the common stream, the rate of common stream should be as

$$\min\{R_1^c, R_2^c\} = R_2^c = B \, \log_2(1 + \gamma_2^c), \tag{8.21}$$

then outage probability for common message at U_1 and U_2 can be written as

$$\Pr(C_u^{\text{com}} \leq C_{\text{th}}) = \Pr\left(\frac{R^{\text{com}}}{2} \leq C_{\text{th}}\right), \quad \forall u = \{1, 2\}$$
$$= \Pr(\min\{R_1^c, R_2^c\} \leq 2C_{\text{th}})$$
$$= \Pr(R_2^c \leq 2C_{\text{th}}). \tag{8.22}$$

In the following, $\Pr(R_2^c \leq 2C_{\text{th}})$ using (8.8) can be given as follows:

$$\Pr(R_2^c \leq 2C_{\text{th}}) = \Pr\left(\gamma_2^c \leq \gamma_{\text{th}}^{\text{com}}\right), \tag{8.23}$$

where $\gamma_{\text{th}}^{\text{com}} = 2^{\frac{2C_{\text{th}}}{B}} - 1$. Now substituting (8.3) in (8.23), we get:

$$\Pr(\gamma_2^c \leq \gamma_{\text{th}}^{com}) = \Pr\left(h_2 \leq \frac{\gamma_{\text{th}}^{\text{com}} \kappa}{p_c - \gamma_{\text{th}}^{\text{com}}(p_1 + p_2)}\right). \tag{8.24}$$

8.5.1 Outage – Near User

Based on (8.5), $\Pr(\gamma_1^{\text{pr}} \leq \gamma_{\text{th}}^{\text{pr}})$ can be equivalently rewritten as

$$\Pr\left(\gamma_1^{\text{pr}} \leq \gamma_{\text{th}}^{\text{pr}}\right) = \Pr\left(h_1 \leq \frac{\gamma_{\text{th}}^{\text{pr}} N_1^{\text{pr}}}{\chi_1(p_1 - \gamma_{\text{th}}^{\text{pr}} p_2)}\right). \tag{8.25}$$

Substituting (8.2) and (8.7) in (8.25), we have the following:

$$\Pr\left(\gamma_1^{\text{pr}} \leq \gamma_{\text{th}}^{\text{pr}}\right) = \Pr\left(\chi_1 \leq \frac{\gamma_{\text{th}}^{\text{pr}} k_b T d_1^2}{\zeta e^{-k_a(f)d_1}\left(1 + p_1 - \gamma_{\text{th}}^{\text{pr}} p_2\right) - 1)}\right)$$
$$= \int_0^R \frac{\gamma\left[m, \frac{1}{\Theta} \frac{\gamma_{\text{th}}^{\text{pr}} k_b T d_1^2}{\zeta e^{-k_a(f)d_1}\left(1 + p_1 - \gamma_{\text{th}}^{\text{pr}} p_2\right) - 1)}\right]}{\Gamma(m)} f_{d_1}(d_1) \, dd_1, \tag{8.26}$$

where $\gamma(\cdot)$ is the lower incomplete Gamma function, $\Gamma(\cdot)$ is the complete Gamma function, m is the fading severity parameter, and Θ is the fading power. Since $k_b T$ is negligible compared to molecular noise, the outage probability for the private message is written as

$$\Pr\left(\gamma_1^{\mathrm{pr}} \leq \gamma_{\mathrm{th}}^{\mathrm{pr}}\right) = \Pr\left(\frac{e^{-k(f)d_1}}{\frac{p_2}{p_1} + 1 - e^{-k(f)d_1}} \leq \gamma_{\mathrm{th}}^{\mathrm{pr}}\right)$$

$$= \Pr\left(d_1 \geq \frac{1}{k(f)} \ln\left(\frac{1 + \gamma_{\mathrm{th}}^{\mathrm{pr}}}{(\frac{p_2}{p_1} + 1)\gamma_{\mathrm{th}}^{\mathrm{pr}}}\right)\right)$$

$$= 1 - F_{d_1}\left(\frac{1}{k(f)} \ln\left[\frac{1 + \gamma_{\mathrm{th}}^{\mathrm{pr}}}{(\frac{p_2}{p_1} + 1)\gamma_{\mathrm{th}}^{\mathrm{pr}}}\right]\right). \tag{8.27}$$

Now substituting (8.12) in (8.27) for the random scheme, and (8.14) in (8.27) for the nearest-farthest scheme, gives the outage probability of the private message of U_1 as shown below:

$$\text{Random}: \quad \Pr\left(\gamma_1^{\mathrm{pr}} \leq \gamma_{\mathrm{th}}^{\mathrm{pr}}\right) = \left[1 - \left(\frac{1}{Rk(f)} \ln\left[\frac{1 + \gamma_{\mathrm{th}}^{\mathrm{pr}}}{(\frac{p_2}{p_1} + 1)\gamma_{\mathrm{th}}^{\mathrm{pr}}}\right]\right)^2\right]^2$$

$$\tag{8.28}$$

$$\text{Nearest-Farthest}: \Pr\left(\gamma_1^{\mathrm{pr}} \leq \gamma_{\mathrm{th}}^{\mathrm{pr}}\right) = \left[1 - \left(\frac{1}{Rk(f)} \ln\left[\frac{1 + \gamma_{\mathrm{th}}^{\mathrm{pr}}}{(\frac{p_2}{p_1} + 1)\gamma_{\mathrm{th}}^{\mathrm{pr}}}\right]\right)^2\right]^N.$$

$$\tag{8.29}$$

By calling (8.22) and given that $\Pr(R_u^c \leq 2C_{\mathrm{th}}) = \Pr\left(\gamma_u^c \leq \gamma_{\mathrm{th}}^{\mathrm{com}}\right)$, $\forall u = \{1, 2\}$, where $\gamma_{\mathrm{th}}^{\mathrm{com}} = 2^{\frac{2C_{\mathrm{th}}}{B}} - 1$, we can calculate the outage probability of common message in RSMA-THz as

$$\Pr(C_u^{\mathrm{com}} \leq C_{\mathrm{th}}) = 1 - \Pr\left(\gamma_2^c \geq \gamma_{\mathrm{th}}^{\mathrm{com}}\right)$$

$$= 1 - \Pr\left(h_1 \geq \frac{\gamma_{\mathrm{th}}^{\mathrm{com}}\kappa}{p_c - \gamma_{\mathrm{th}}^{\mathrm{com}}(p_1 + p_2)}\right)$$

$$= 1 - \Pr\left(d_2 \leq \left(\frac{1}{k(f)} \ln\left(\frac{p_c + \gamma_{\mathrm{th}}^{\mathrm{com}}p_c}{\gamma_{\mathrm{th}}^{\mathrm{com}}}\right)\right)\right)$$

$$= 1 - F_{d_2}\left[\frac{1}{k(f)} \ln\left(\frac{p_c + \gamma_{\mathrm{th}}^{\mathrm{com}}p_c}{\gamma_{\mathrm{th}}^{\mathrm{com}}}\right)\right]. \tag{8.30}$$

By substituting (8.12) in (8.30) for the random scheme, and (8.14) in (8.30) for the nearest-farthest scheme, gives us, i.e.,

$$\text{Random}: \quad \Pr(C_u^{\mathrm{com}} \leq C_{\mathrm{th}}) = (1 - v^2) \tag{8.31}$$

$$\text{Nearest-Farthest}: \quad \Pr(C_u^{\mathrm{com}} \leq C_{\mathrm{th}}) = (1 - v^N) \tag{8.32}$$

where $v = \left[\frac{1}{Rk(f)} \ln \left(\frac{p_c + \gamma_{th}^{com} p_c}{\gamma_{th}^{com}} \right) \right]^2$. Now substituting (8.28) and (8.31) in (8.16) for the random scheme, and (8.29) and (8.32) in (8.16) for the nearest-farthest scheme, the outage probability for the near user can be given as follows: The outage of U_1 is given as

$$\text{Random}: \quad \mathcal{O}_1^{THz} = (1 - w)^2 (1 - v^2) \tag{8.33}$$

$$\text{Nearest-Farthest}: \quad \mathcal{O}_1^{THz} = (1 - w)^N (1 - v^N) \tag{8.34}$$

where $w = \left[\frac{1}{Rk(f)} \ln \left(\frac{1 + \gamma_{th}^{pr}}{(\frac{p_2}{p_1} + 1) \gamma_{th}^{pr}} \right) \right]^2$.

8.5.2 Outage – Far User

Based on (8.6), $P(\gamma_2^{pr} \leq \gamma_{th}^{pr})$ can be formulated as follows:

$$\Pr\left(\gamma_2^{pr} \leq \gamma_{th}^{pr}\right) = \Pr\left(h_2 \leq \frac{\gamma_{th}^{pr} N_2^{pr}}{\chi_2 (p_2 - \gamma_{th}^{pr} p_1)} \right). \tag{8.35}$$

Substituting (8.2) and (8.4) in (8.35), we get:

$\Pr\left(\gamma_2^{pr} \leq \gamma_{th}^{pr}\right)$

$$= \Pr\left(\zeta d_2^{-2} e^{-k(f)d_2} \leq \frac{\gamma_{th}^{pr} \left(k_b T + (p_1 + p_2)\zeta d_2^{-2}(1 - e^{-k_a(f)d_2})\chi_2 \right)}{\chi_2 (p_2 - \gamma_{th}^{pr} p_1)} \right)$$

$$= \Pr\left(\chi_2 \leq \frac{\gamma_{th}^{pr} k_b T d_2^2}{\zeta e^{-k_a(f)d_2} (1 + p_2 - \gamma_{th}^{pr} p_1) - 1)} \right)$$

$$= \int_0^R \frac{\gamma\left[m, \frac{1}{\Theta} \frac{\gamma_{th}^{pr} k_b T d_2^2}{\zeta e^{-k_a(f)d_2} (1 + p_2 - \gamma_{th}^{pr} p_1) - 1)} \right]}{\Gamma(m)} f_{d_2}(d_2) \, dd_2. \tag{8.36}$$

Since $k_b T$ is negligible compared to molecular absorption noise, then the outage probability of the private message of U_2 can be given as

$$\Pr\left(\gamma_2^{pr} \leq \gamma_{th}^{pr}\right) = \Pr\left(\frac{e^{-k(f)d_2}}{\frac{p_1}{p_2} + 1 - e^{-k(f)d_2}} \leq \gamma_{th}^{pr} \right)$$

$$= \Pr\left(d_2 \geq \frac{1}{k(f)} \ln\left[\frac{1 + \gamma_{th}^{pr}}{(\frac{p_1}{p_2} + 1)\gamma_{th}^{pr}} \right] \right)$$

$$= 1 - F_{d_2}\left(\frac{1}{k(f)} \ln\left[\frac{1 + \gamma_{th}^{pr}}{(\frac{p_1}{p_2} + 1)\gamma_{th}^{pr}} \right] \right). \tag{8.37}$$

Now substituting (8.13) in (8.37) for the random scheme, and (8.15) in (8.37) for the nearest-farthest scheme, gives the outage probability expressions for the private message as

$$\text{Random}: \quad \Pr\left(\gamma_2^{\text{pr}} \leq \gamma_{\text{th}}^{\text{pr}}\right) = 1 - \left[\frac{1}{Rk(f)} \ln\left(\frac{1 + \gamma_{\text{th}}^{\text{pr}}}{\left(\frac{p_1}{p_2} + 1\right)\gamma_{\text{th}}^{\text{pr}}}\right)\right]^4 \quad (8.38)$$

$$\text{Nearest-Farthest}: \quad \Pr\left(\gamma_2^{\text{pr}} \leq \gamma_{\text{th}}^{\text{pr}}\right) = 1 - \left[\frac{1}{Rk(f)} \ln\left(\frac{1 + \gamma_{\text{th}}^{\text{pr}}}{\left(\frac{p_1}{p_2} + 1\right)\gamma_{\text{th}}^{\text{pr}}}\right)\right]^{2N}.$$
$$(8.39)$$

The outage probability for the common message for U_2 is given in (8.31) and (8.32) for random and nearest-farthest schemes, respectively. Thus, the outage probability for U_2 can then be calculated as

$$\text{Random}: \quad \mathcal{O}_2^{\text{THz}} = (1 - w')^2(1 - v^2) \quad (8.40)$$

$$\text{Nearest-Farthest}: \quad \mathcal{O}_2^{\text{THz}} = (1 - w')^N(1 - v^N) \quad (8.41)$$

where $w' = \left(\frac{1}{k(f)} \ln\left[\frac{1 + \gamma_{\text{th}}^{\text{pr}}}{\left(\frac{p_1}{p_2} + 1\right)\gamma_{\text{th}}^{\text{pr}}}\right]\right)^2$.

8.6 NUMERICAL SIMULATIONS

In this section, we compare the performance of U_1 and U_2 in THz-RSMA networks. Unless stated otherwise, the parameters are listed herein. We consider 100 users to be uniformly distributed in a circular disc. Note that our results are general for any arbitrary value of radius in the range of 25–50 m for THz-RSMA networks. The transmission bandwidths for THz-RSMA are 0.5 MHz and 0.5 GHz, respectively. The antenna gains G_t and G_r are set as 20 dB for THz AP. The AP transmit power is 1W, The power allocation coefficients are set as $p_1 = 0.2$, $p_2 = 0.25$ and $p_c = 1 - (p_1 + p_2)$. Nakagami-m fading parameter is set as 2 and $\Omega = 1$ with frequency 1 THz and $k(f) = 0.1$ m^{-1} considering water vapour molecules for THz-RSMA.

Figure 8.4 demonstrates the outage of U_1 and U_2 as a function of the cell radius in THz-RSMA and validates (8.28), (8.38), (8.31), (8.33), and (8.40) for random scheme (8.29), (8.39), (8.32), (8.34), and (8.41) for the nearest-farthest scheme through Monte-Carlo simulations.

It is shown that the values obtained through derived expressions (shown in circles) exactly match those obtained through simulations

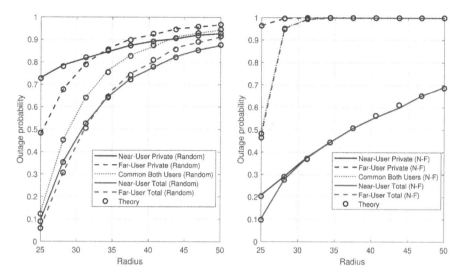

Figure 8.4 Outage probability of near and far users as a function of the cell-radius in THz spectrum, where the power factors are $p_1 = 0.15, p_2 = 0.30$, and $p_c = 0.55$.

(shown by lines). It can be seen that the outage increases with the increase in the radius for both users. As expected, for near and far users with the increase in cell-radius, the outage increases.

Figure 8.5 demonstrates the outage of U_1 and U_2 as a function of the power allocation coefficient a_1 for NOMA and p_1 for RSMA and highlights the difference gain of RSMA, NOMA, and OMA. Note that to do that comparison, p_2 is considered to have a fix value ($p_2 = 0.25$), and the portion of transmit power is allocated for U1 in NOMA $a_1 = p_1$. Observe that with the increase in a_1 and p_1, the outage increases for RSMA and NOMA due to the increased interference from U_1 observed at U_2. The gain of RSMA over NOMA and OMA improves more for low values of a_1 (p_1) for both near and far users than for high values of a_1 (p_1) for both near and far users. Figure 8.6 demonstrates the outage of U_1 and U_2 as a function of the cell-radius and highlights the difference gain of RSMA, NOMA, and OMA. Note that to do that comparison, we considered that the portion of transmit power allocated for U1 in NOMA a_1 has the same fixed value as p_1 for RSMA ($a_1 = p_1 = 0.1$), and p_2 is considered to have a fix value ($p_2 = 0.25$). Observe that with the increase in R, the outage increases for RSMA, NOMA, and OMA

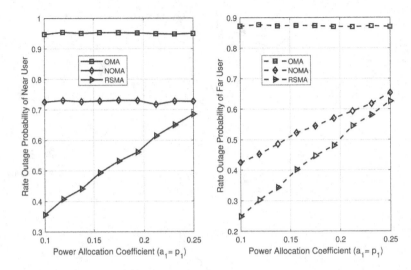

Figure 8.5 Outage probability of near and far users as a function of power allocation spectrum in the THz spectrum, for all multiple access schemes (OMA, NOMA, and RSMA), where radius $R = 60$, $p_2 = 0.25$, and $ka = 0.03\ m^{-1}$ ($f_t = 0.8$ THz).

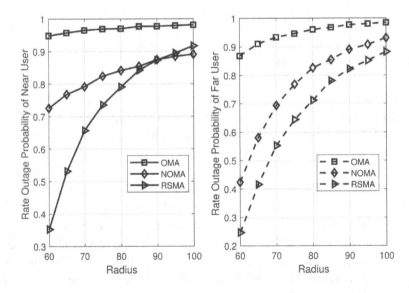

Figure 8.6 Outage probability of near and far users as a function of the radius in the THz spectrum, for all multiple access schemes (OMA, NOMA, and RSMA), where radius $a_1 = p_1 = 0.1$, $p_2 = 0.25$, and $ka = 0.03\ m^{-1}$ ($f_t = 0.8$ THz).

due to the increased interference from U_1 observed at U_2. The gain of RSMA over NOMA and OMA improves more for low values of R for near user, while the gain of RSMA over NOMA and OMA has almost the same at all values of R for far users.

8.7 SUMMARY AND OPEN ISSUES

This chapter provided an overview of the evolution of wireless network frequency bands (such as sub-6 GHz, mm-wave, THz) and multiple access techniques. The benefits and drawbacks of various OMA techniques (FDMA, TDMA, CDMA, and OFDMA) and NOMA (PD-NOMA, CD-NOMA, SDMA, Partial NOMA) were highlighted. Moreover, we present a comprehensive framework for evaluating the performance of RSMA in the downlink transmission of a single carrier in THz networks. Numerical results validate the accuracy of the derived expressions. From computing the OP, we were able to construct simplified expressions. Furthermore, these expressions can be modified to accommodate a variety of user-pairing methods. The numerical results also show that RSMA outperforms NOMA and OMA in almost all cases. The use of RSMA-THz in a next-generation (6G) environment is still in its early stages. To realize its impact on end users, a number of open issues and research problems must be solved. Some of the directions are listed in the following:

- **Multi-Carrier RSMA-THz Network Performance:** The channel statistics for each sub-channel in the THz spectrum suffers from distinct molecular absorption, as represented by the molecular absorption coefficient. As a result, the mathematical complexity of the Beer's-Lambert law-based channel propagation model on each sub-channel adds to the difficulty of outage analysis in a multi-carrier THz-RSMA network.

- **Novel User Pairing for THz-RSMA Network**: It is also worth noting that assessing the performance of multi-carrier NOMA in radio frequency (RF) networks is simple because all sub-channels have the same channel data. There is still no research that produces a low-complexity user pairing approach for the THz-RSMA network with a guaranteed gain over other multiple access techniques for each individual user and/or provides a complete framework for the outage analysis of users in a multi-carrier THz-RSMA network.

- **RSMA-THz Performance for Uplink Transmissions**: Another open research topic is analyzing the performance of RSMA in single-carrier and multi-carrier uplink transmissions in THz networks.

- **Non-line-of-sight (NLOS) propagation for THz networks**: LOS communications may not always be possible due to the existence of obstacles. Representing the coefficients for EM wave reflection, scattering, and diffraction at THz frequencies is crucial to account for NLOS propagation. These coefficients are affected by the surface's material, shape, incident electromagnetic (EM) wave frequency, and angle. If LOS is not attainable, especially indoors, NLOS propagation serves as a backup. NLOS propagation can be created by carefully positioning installed dielectric mirrors to reflect the beam to the receiver. Due to the minimal reflection loss on dielectric mirrors, the resulting path loss is tolerable. For the THz spectrum, experimental measurements are required to verify the reflection, scattering, and diffraction coefficients for typical materials used for indoor and outdoor communications.

- **Multi-bands for 6G networks**: The THz spectrum will complement mm-wave and RF frequencies in 6G to transmit/receive data. Thanks to the variety of the spectrum which will cause a traded-off in terms of coverage area, capacity, user mobility, and latency. The main challenge will be in optimizing the deployment of APs, traffic load-aware network activation methods, opportunistic spectrum selection at the users' end, and multi-connectivity solutions.

BIBLIOGRAPHY

[1] Mehdi Rasti, Shiva Kazemi Taskou, Hina Tabassum, and Ekram Hossain. Evolution toward 6G wireless networks: A resource management perspective. *arXiv preprint arXiv:2108.06527*, 2021.

[2] H. Ibrahim, H. Tabassum, and U. T. Nguyen. The meta distributions of the SIR/SNR and data rate in coexisting sub-6GHz and millimeter-wave cellular networks. *IEEE Open Journal of the Communications Society*, 1:1213–1229, 2020.

[3] Noha Hassan, Md Tanvir Hossan, and Hina Tabassum. User association in coexisting RF and terahertz networks in 6G. *2020 IEEE Canadian Conference on Electrical and Computer Engineering (CCECE)*, pages 1–5, 2020.

[4] L. Zhu, Z. Xiao, X. G. Xia, and D. Oliver Wu. Millimeter-wave communications with non-orthogonal multiple access for B5G/6G. *IEEE Access*, 7:116123–116132, 2019.

[5] O. Somekh and S. Shamai. Shannon-theoretic approach to a Gaussian cellular multiple-access channel with fading. *IEEE Transactions on Information Theory*, 46(4):1401–1425, 2000.

[6] J. De Vriendt, P. Laine, C. Lerouge, and Xiaofeng Xu. Mobile network evolution: A revolution on the move. *IEEE Communications Magazine*, 40(4):104–111, 2002.

[7] H. Tabassum, M. S. Ali, E. Hossain, M. J. Hossain, and D. I. Kim. Uplink vs. downlink NOMA in cellular networks: Challenges and research directions. *2017 IEEE 85th Vehicular Technology Conference (VTC Spring)*, pages 1–7, 2017.

[8] Yijie Mao, Bruno Clerckx, and Victor O.K. Li. Rate-splitting multiple access for downlink communication systems: Bridging, generalizing, and outperforming SDMA and NOMA. *EURASIP Journal on Wireless Communications and Networking*, 2018(1), May 2018. https://doi.org/10.1186/s13638-018-1104-7.

[9] Yijie Mao and Bruno Clerckx. *Multiple Access Techniques*, pages 63–100. Springer Int. Publishing, Cham, 2021.

[10] B. Clerckx, H. Joudeh, C. Hao, M. Dai, and B. Rassouli. Rate splitting for MIMO wireless networks: A promising PHY-layer strategy for LTE evolution. *IEEE Communications Magazine*, 54(5):98–105, 2016.

[11] Zhaohui Yang, Mingzhe Chen, Walid Saad, and Mohammad Shikh-Bahaei. Downlink sum-rate maximization for rate splitting multiple access (RSMA). *ICC 2020 – 2020 IEEE International Conference on Communications (ICC)*, pages 1–6, 2020.

[12] Zhaohui Yang, Mingzhe Chen, Walid Saad, Wei Xu, and Moham-mad Shikh-Bahaei. Sum-rate maximization of uplink rate splitting multiple access (RSMA) communication, *2019 IEEE Global Communications Conference* 2019.

[13] Daesung Yu, Junbeom Kim, and Seok-Hwan Park. An efficient rate-splitting multiple access scheme for the downlink of c-ran systems. *IEEE Wireless Communications Letters*, 8(6):1555–1558, 2019.

[14] Zhaohui Yang, Mingzhe Chen, Walid Saad, and Mohammad Shikh-Bahaei. Optimization of rate allocation and power control for rate splitting multiple access (RSMA). *IEEE Transactions on Communications*, 69(9):5988–6002, 2021.

[15] Lihua Li, Kejia Chai, Jilong Li, and Xingwang Li. Resource allocation for multicarrier rate-splitting multiple access system. *IEEE Access*, 8:174222–174232, 2020.

[16] Yijie Mao, Bruno Clerckx, and Victor O.K. Li. Energy efficiency of rate-splitting multiple access, and performance benefits over SDMA and NOMA. *2018 15th International Symposium on Wireless Communication Systems (ISWCS)*, pages 1–5, 2018.

[17] Zhaohui Yang, Jianfeng Shi, Zhiyang Li, Mingzhe Chen, Wei Xu, and Mohammad Shikh-Bahaei. Energy efficient rate splitting multiple access (RSMA) with reconfigurable intelligent surface. *2020 IEEE International Conference on Communications Workshops (ICC Workshops)*, pages 1–6, 2020.

[18] Longfei Yin and Bruno Clerckx. Rate-splitting multiple access for multibeam satellite communications. *2020 IEEE International Conference on Communications Workshops (ICC Workshops)*, pages 1–6, 2020.

[19] Sandeep Kumar Singh, Kamal Agrawal, Keshav Singh, and Chih-Peng Li. Outage probability and throughput analysis of UAV-assisted rate-splitting multiple access. *IEEE Wireless Communications Letters*, 10(11):2528–2532, 2021.

[20] Hongwu Liu, Theodoros A. Tsiftsis, Bruno Clerckx, Kyeong Jin Kim, Kyung Sup Kwak, and H. Vincent Poor. Rate splitting multiple access for semi-grant-free transmissions, 2021.

[21] Omid Abbasi and Halim Yanikomeroglu. Transmission scheme, detection and power allocation for uplink user cooperation with NOMA and RSMA, IEEE Transactions on Wireless Communications, 22(1): 471-485, 2023. doi: 10.1109/TWC.2022.3195532. 2022.

[22] Sadeq Bani Melhem and Hina Tabassum. User pairing and outage analysis in multi-carrier NOMA-THz networks. *IEEE Transactions on Vehicular Technology*, 71(5):5546–5551, 2022.

A Comprehensive Overview of Security and Privacy in the 6G Era

Shakila Zaman

School of Computer Science and Engineering, University of North Texas, Denton, Texas, USA

Faisal Tariq

James Watt School of Engineering, University of Glasgow, Glasgow, UK

Muhammad Khandaker

School of Engineering and Physical Sciences, Heriot-Watt University, Edinburgh, UK

Risala T Khan

Jahangirnagar University, Savar, Bangladesh

CONTENTS

9.1	Introduction	205
9.2	Potential Attacks/Threats in 6G Networks	205
	9.2.1 Access Control Attack	206
	9.2.2 Eavesdropping	207
	9.2.3 Jamming	208
	9.2.4 Data Leakage	208
	9.2.5 Sybil Attack	208
	9.2.6 Malicious Code Injection	209
	9.2.7 Position Tracking Attack	209
	9.2.8 Attacks on IoT Nodes	209
	9.2.9 Attacks on Molecules	210
	9.2.10 DoS/DDoS	210
	9.2.11 Pilot Spoofing Attacks	210
	9.2.12 Data Security Attack	210

DOI: 10.1201/9781003282211-9

9.2.13 Poisoning Attack 211
9.2.14 Evasion Attack 211
9.2.15 Oracle Attacks 211
9.2.16 Model Inversion Attack 211
9.2.17 Man-in-the-Middle Attack 211
9.2.18 Privilege Escalation Attacks (PEA) 212
9.3   Physical Layer Security Issues 212
    9.3.1   Security of Key Physical Layer Technologies 212
            9.3.1.1   Security of Cell-Free Massive MIMO 217
            9.3.1.2   Security of Non-orthogonal Multiple
                      Access 218
            9.3.1.3   Security of Reconfigurable Intelligent
                      Surface 219
    9.3.2   Security of mmWave 220
    9.3.3   Security of Terahertz Communication 222
    9.3.4   Security of Holographic Radio 223
    9.3.5   Security of Ultra-Reliable and Low-Latency
            Communications (URLLC) 223
9.4   Connection/Network Layer Security Issues 224
    9.4.1   Security of Connection/Network Layer Technologies 227
            9.4.1.1   Security on SDN-WAN 227
            9.4.1.2   Security on NFV 228
            9.4.1.3   Security on Deep Slicing 228
            9.4.1.4   Security on DLT 229
            9.4.1.5   Security on AI 230
            9.4.1.6   Security on Deterministic Networking ... 232
            9.4.1.7   Security on Space-Air Ground-Sea
                      Integrated Network 233
9.5   The Processing/Service Layer 234
    9.5.1   Processing/Service Layer Security 236
            9.5.1.1   Security on Container-Based
                      Virtualization 236
            9.5.1.2   Security on Zero-Touch Service
                      Orchestration 237
    9.5.2   Security on Quantum Computing 238
            9.5.2.1   Security on Distributed Computing 239
            9.5.2.2   Security on Terminals or Consumer End . 240
9.6   Conclusion 240
Bibliography 241

9.1 INTRODUCTION

Wireless communications have always been prone to higher security risk compared to it wired counterpart due to its inherent broadcast nature. Early generations of wireless or mobile communication systems suffered from several security and privacy issues such as cloning attack, eavesdropping, authentication, spoofing, and so on. With the emergence of advanced and powerful computing devices along with ever-increasing network integration, security and privacy challenges also evolved toward more sophisticated attacks which are difficult to mitigate. The recent 5th-generation (5G) systems were characterized by the widespread proliferation of Internet of Things (IoT) in all aspects of human life. The key services and technologies include vehicle to everything (V2X) communications [30], network function virtualization (NFV), cloud radio access networks. Though 5G has some weaknesses, it has been improved significantly to satisfy service-oriented architectures and fix enormous vulnerabilities for the previous version of cellular networks. However, the concept of 6G mobile communication is already envisioned with a novel research direction that includes heterogeneous physical devices and intelligent network protocols such as blockchain, visible light communications (VLC), intelligent reflecting surface (IRS) [96], Terahertz spectrum, orbital angular momentum (OAM), cell-free communications, quantum computing, etc. [102]. Therefore, cognitive security and privacy management are required to address the vulnerabilities of 6G networks. Figure 9.1 demonstrates the advancement of various security protocols from 5G to 6G network.

9.2 POTENTIAL ATTACKS/THREATS IN 6G NETWORKS

The wide adoption of diverse communication standards and technologies of 6G welcomes a plethora of conventional and cognitive threats/attacks to make a system vulnerable. This work represents three layers model to understand the operations of a network that includes physical layer, connection/network layer, and processing/service layer. However, there are different potential attacks to compromise the physical layer, connection layer, and processing layer operations of a network including access control attack, eavesdropping, jamming, data modification, attack on IoT devices, malicious code injection, location tracking, distributed denial of service (DoS/DDoS), man-in-the-middle (MITM) attack, poisoning,

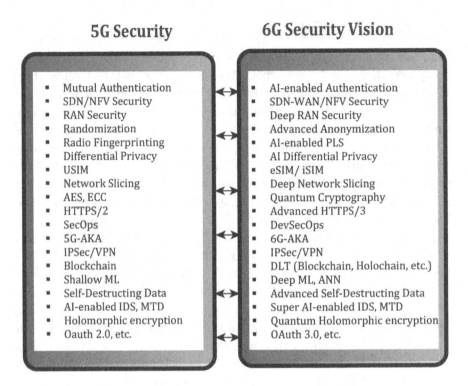

Figure 9.1 Existing 5G security vs 6G security vision.

evasion, Sybil attack, etc. This section demonstrates a brief discussion of various potential attacks/threats in 6G networks. Figure 9.2 shows some frequent attacks on 6G network.

9.2.1 Access Control Attack

Adversaries may introduce access control attacks to steal data or collect authorized users' credentials to access unauthorized resources. The multisensory users are able to communicate through different technologies (Terahertz (THz), AI, VLC, molecular, Quantum, etc.), which may seriously cause access control attacks to the smart 6G network. Existing access control methods are not adequate to mitigate the vulnerabilities of future-generation networks. For instance, conventional dual network access verification and identity management are becoming outdated methods to deal with the cognitive 6G protocols.

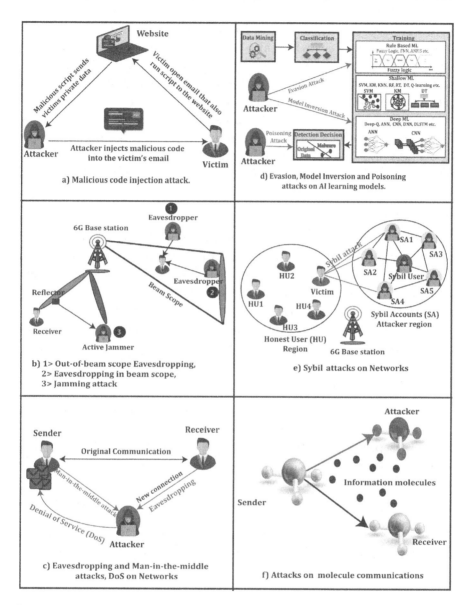

Figure 9.2 Potential frequent attacks on 6G networks.

9.2.2 Eavesdropping

Eavesdropping takes place when a malicious node tries to overhear a private communication among authorized nodes, which is continuously being considered a significant security concern from 1G to 5G beyond networks [50]. In the concept of Physical Layer Security (PLS), there are three types of eavesdropping [49], [6]. A passive eavesdropping – without taking any action an intruder only overhears the messages. An active eavesdropping – includes various malfunctions (e.g. add noise, introduce jamming, or provide fake information). The third one is potential eavesdropping – operates as an authorized user in some instances. However, new technologies are also prone to compromise eavesdropping in 6G networks like multiuser Multiple Input Multiple Output (MIMO) are considered to hide active and passive eavesdropping.

9.2.3 Jamming

Jamming still remains in IoT-based smart networks, which occurs when an attacker intentionally jams an authorized communication to make it unavailable for other connected users [108]. In 6G networks, like various methods, VLC or hybrid visible light communication-radio frequency (VLC-RF) systems can arise a successful jamming attack possibility, as the highly directed transmitter uses optical beamforming techniques which may pass the malicious activity undetected [77].

9.2.4 Data Leakage

The physical layer deals with a large degree of sensitive information which perceive and collect to transmit to the upper layer. The data leakage attack permits a network to expose sensitive information like user states, personal information, technical details, or network environment. Lack of clarity on how heterogeneous 6G technologies are governed, security risks like information leakage are becoming the main concern of various applications like biometric authentication, smart healthcare. For example, data explosion in smart healthcare will be a nightmare in case of highly sensitive information (patients records, therapy details, user-specific metadata, etc.) preservation [28].

9.2.5 Sybil Attack

During the Sybil attack, a fake node named Sybil nodes transmits a high-energy deceiving signal in the communication phase among transceivers to claim as an original one [121]. In 6G networks, most of the emerging technologies considered multipath communication systems, which make a system vulnerable in case of Sybil attacks. Though conventional cryptographic systems are used to secure routing packets against nodes' misbehavior, the methods themselves can affect energy consumption in various applications like unmanned aerial vehicle (UAV) networks using fake insider nodes [63].

9.2.6 Malicious Code Injection

Malicious code injection occurs when an intruder embeds unwanted contents or codes into an application to access remotely for possibly making a system compromised. The rapid acceptance of high resource technologies into an extensive range of applications causes targeted remote access by intruders. For example, extended reality (ER) is used in various human-computer interaction (HCI) applications including games, healthcare, and social interactions, which is also vulnerable to malicious code injection [1]. An intruder may embed malicious content into eXtended Reality (XR) systems to introduce passive attacks or expose sensitive information.

9.2.7 Position Tracking Attack

Due to the advanced wireless technology of ad hoc networks, node position can be available to third parties, which possibly introduces position tracking threats. During this attack, adversaries collect information regarding node position or mobility patterns. The key security concern of a cell-free massive MIMO in 6G technology is the exposed position of radio stripes [128]. For example, an attacker can track the location of a base station (BS) and implement malicious injection to lead various passive attacks like eavesdropping on significant keys (e.g. user-specific keys, short-term session keys, etc.) for authentication methods.

9.2.8 Attacks on IoT Nodes

Due to the resource-constrained nature of IoT devices, complex cryptographic schemes are not possible to implement, which creates immense concerns about vulnerable communication. The high-complex heterogeneous IoT networks are sometimes exposed to various attacks including DoS, eavesdropping, fake node attack, and spoofing [130]. On the other hand, 6G IoT networks incorporate new protocols like brain-computing and molecular communications that require explicit security requirements such as real-time response or trustworthy communication.

9.2.9 Attacks on Molecules

Molecular communication is an emerging technique in 5G beyond cellular networks, which practices chemical molecules or signals to communicate among nanodevices in case of extremely low-power communications (ELCP). A smart healthcare system uses nanodevices into human bodies to supply drugs or monitor patients' status continuously. These biodevices can compromise the molecular systems through unauthorized remote access to kill the molecules or take control of the system for interrupting medical information to harm or kill the host which is still a critical issue to be concerned [64].

9.2.10 DoS/DDoS

DoS and DDoS are very frequent attacks and still exist in 5G beyond networks. The attacker targets a host node and transfers overwhelming packets/messages to make it unavailable to authorized users or even can shut down the entire system. DoS/DDoS is frequently employed in almost every technology of a network. Networks such as 6G-enabled SDN-WAN are highly prone to DoS/DDoS attacks for the unique nature of SDN controller operations or flow table functionalities [92].

9.2.11 Pilot Spoofing Attacks

Pilot spoofing attacks initiate during the training phase when a spoofer broadcasts the synchronized and pilot signal as an authorized user. The original network is unable to identify the unauthorized user and provide access to the network. Therefore, spoofer can contaminate channel status, modify the authenticate beamforming design, and introduce eavesdropping. Spoofing may deploy in the 6G network through IRS, where the different uplink and downlink phases of a corresponding transmitted signal from IRS are used to incline toward the eavesdropper [126].

9.2.12 Data Security Attack

The rudimentary target of data security attacks is to destroy the data integrity or availability to produce abnormal outputs or undermine the training session of the model [124]. For example, in an IoT-based smart autopilot system, adversaries can manipulate the sensing information, which may lead to passenger death. Poisoning and evasion are two popular data security attacks on deep learning algorithms.

9.2.13 Poisoning Attack

Poisoning attacks introduce during the training phase when an intruder may inject fake data/model/algorithm to create availability issues of the outcome, which leads misprediction of the resources or misclassifying the services [120].

9.2.14 Evasion Attack

Mostly, an evasion attack takes place during the prediction stage of a learning algorithm, which creates incorrect labelling by intentionally adding noises to the original data samples. For example, an evasion attack is introduced during the prediction process in a selective audio generation system to hamper the operations of the speech recognition function [53].

9.2.15 Oracle Attacks

Oracle attacks occur when an attacker has been given the access of an Application Interface Model for creating a substitute or duplicate of the original ML model by stealing a copy. These attacks are very common in cloud-based ML services like Google, Amazon, and BigML, which offer easy-to-use API for users.

9.2.16 Model Inversion Attack

The model inversion attack takes place during training or inference time to recreate data fed through a target prediction model. In 6G networks, most of the applications utilize the advantages of shallow or deep learning algorithms for various services, which are also prone to model inversion attacks. For example, a model inversion attack model is introduced in hidden layers of a deep neural network to recreate similar inputs or stolen data during training [106]. Moreover, API-based attacks can also predict the information regarding sample features, which may introduce model inversion attacks.

9.2.17 Man-in-the-Middle Attack

MITM attack is a kind of eavesdropping attack which takes place when an attacker secretly interrupts an existing network or communication between two users. For instance, in 6G, SDN-WAN is frequently compromised by MITM attacks where an attack can take place between the

SDN controller and end user to take control over the data or network itself [92].

9.2.18 Privilege Escalation Attacks (PEA)

The fundamental vulnerability of the container-based VM is sharing kernel. The PEA may take place because of the lack of focus on permission on computing systems which increases unauthorized access. A successful PEA is launched inside the Linux container mechanism, which is able to disable the protection methods of the system [60]. Table 9.1 represents the impact of the above attacks and corresponding layers.

Three layers such as physical layer, connection/network layer, and processing or service layer on 6G security issues are considered to demonstrate the involved technologies along with the vulnerabilities and countermeasures. Figure 9.3 represents layer-based fundamental technologies of 6G networks.

9.3 PHYSICAL LAYER SECURITY ISSUES

The 6G of wireless cellular networks are designed to integrate heterogeneous devices and protocols in the physical layer. The rudimentary functionalities of this layers include intelligent modulation techniques, manage enormous nodes (macrodevices to nanodevices), design the capacity of communication links, implement appropriate topologies, and deal with immense sensitive information. In 6G networks, the physical layer will have introduced various advanced technologies like ultra-massive MIMO, non-orthogonal multiple access (NOMA), millimeter-wave (mmWave)/molecular/Terahertz Communications, and multiuser LDPC (low-density parity-check code) to provide a robust fast network than ever before. Therefore, with the rapid growth of intelligent and autonomous technologies in the physical layer, ensuring secure communication becomes paramount against various conventional and novel cognitive threats/attacks.

9.3.1 Security of Key Physical Layer Technologies

The existing 5G and beyond networks bring challenges of having high energy efficiency, full spectral productivity, low-latency, trustworthy communication experience through the emerging concept of Internet of Intelligence with connected things and people. There are immense technological breakthroughs in the physical layer to deal with the

TABLE 9.1   Potential Attacks with Their Impact on 6G Networks

Attacks	Description/Impact	Compromised Layer	Reference
Access control	Gain access to unauthorized resources	Physical layer, connection layer, service layer	[65]
Eavesdropping	Overhear the messages and networks	Physical layer, connection layer, service layer	[6]
Jamming	Make authorized user unavailable to others	Physical layer, service layer	[77]
Data leakage	Explore sensitive data intentionally	Physical layer, service layer	[28]
Sybil attack	Fake user take control the network decisively	Physical layer, connection layer	[63]
Malicious code injection	Injects malicious code to access networks/devices remotely	Physical layer, service layer	[1]
Position tracking	Collects node position or mobility patterns to harm the network	Physical layer, service layer	[128]
Attacks on IoT nodes	Compromised physically or virtually through various attacks	Physical layer	[130]
Attacks on molecules	Use chemical molecules or signals to initiate attacks on nanodevices	Physical layer	[64]
DoS/DDoS	Make authorized user unavailable to others	Physical layer, connection layer, service layer	[92]
Pilot spoofing	Broadcast fake pilot signal	Physical layer	[126]
Data security	Reduce the data integrity and availability to produce abnormal outputs	Physical layer, service layer	[124]
Poisoning	Inject fake models to modify the training result	Connection layer, service layer	[120]
Evasion	Add noise to incorrect labeling amid prediction stage	Connection layer, service layer	[53]
Oracle	Create duplicate Machine Learning (ML) model by accessing the Application Programming Interface (API)	Connection layer, service layer	[139]
Model inversion	Duplicate data sample of a target production model	Connection layer, service layer	[120]
Man-in-the-middle	intercept communications between two users to take the role of a proxy	Physical layer, connection layer, service layer	[120]
PEA	Take unauthorized access on Virtual Machine (VM)	Service layer	[60]

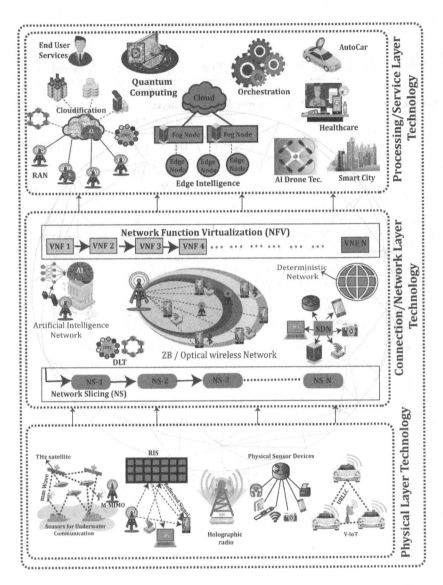

Figure 9.3 Layer-based fundamental technologies of 6G networks.

unprecedented requirements to boost network performances in various perspectives. However, these fast-ever 6G networks are also vulnerable toward various attacks/threats, which demands adoptable countermeasure to have the better throughput of a network. Table 9.2 lists the frequent attacks and the countermeasures on physical layer technologies.

TABLE 9.2 Frequent Attacks and Countermeasures on Physical Layer Key Technologies

Technology	Attacks/Threats	Countermeasure	Reference
CFM-MIMO	Pilot spoofing	Channel estimation	[81]
		GRFIM and AQNM	[135]
		Power allocation	[39]
		MMSE	[105]
	Eavesdropping	AOA and DFT	[81]
		MMSE	[105]
		Fingerprints and K-means	[39]
NOMA	Multidisciplinary attacks on physical layer security (MA-PLS)	Cooperative communication	[35]
		Beamforming	[35]
		Channel-based PA	[35]
		SIC	[90]
		Beamforming and PA	[136]
		OSRS, TSRS, ODRS	[136]
		IRS/VLC/THz-based NOMA	[75]
	Eavesdropping	IRS	[101]
		Multi-dimensional directional modulation method (MDDMM), channel dependent interleaving (CDI), cyclic feature suppression-based schemes (CFSS)	[29]
RIS	MA-PLS	Optimized transmitted power	[118]
		Phase shifting	[118]
		Fisher-Snedecor framework	[67]
	Eavesdropping	Noise beamforming	[137]
		Power allocation	[137]

(*continued*)

TABLE 9.2 Frequent Attacks and Countermeasures on Physical Layer Key Technologies (*Continued*)

Technology	Attacks/Threats	Countermeasure	Reference
	Jamming	Enhanced signal strength	[112]
		Randomness of modulations	[112]
	Eavesdropping	Artificial noise	[40]
		Friendly jamming	[40]
		AI/ML-based learning	[9, 47]
	Pilot spoofing	Enhanced signal strength	[112]
mmWave		Phase shifting	[118]
		Beamforming	[118]
		PA and beamforming	[69]
	MA-PLS	Exchanging physical keys	[100]
		Directional modulation	[109]
		Hybrid beamforming	[55]
		Frequency hopping	[93]
		Lightweight encryption	[76], [65]
	MA-PLS	Signal power variation	[93]
THz Communication		Fingerprint-based authentication	[82]
		ES-based authentication	[82]
	Eavesdropping	Creating multiple paths	[93]
	Data leakage	RIS-based multiple paths	[93]
Holographic radio	Eavesdropping	Directional beamforming	[138]
		MIMO-based holographic radio	[93]
	MA-PLS	CIPC	[24]
		Location-based beamforming	[125]
URLLC	Eavesdropping	friendly jamming	[56]
	Malicious attack	NFV/Blockchain-based SA	[138]

9.3.1.1 Security of Cell-Free Massive MIMO

Cell-free massive MIMO is a kind of network technology where a large degree of distributed low-cost, low-energy access points (APs) antennas are traditionally used to form a heavy and expensive massive MIMO BS to accelerate the magnitude of cell radius where APs are responsible for ensuring services to the users in the same time-frequency resource [71]. The great advantage of having a distributed framework is releasing the pressure of a central CPU processing through APs. On the other hand, cell-free massive MIMO utilizes the low path loss and high diversity gain to boost the quality of services (QoS) of cell edge users.

However, due to the distributed antenna structure and having multiple APs, cell-free massive MIMO is more complex in terms of signal processing and being compromised by malicious eavesdroppers. Moreover, a single BS of this technology consisting of a small number of antennas causes poor directionality of beamforming leads to eavesdropping and interference. Therefore, providing security on cell-free massive MIMO through PLS is crucial.

In cell-free Massive MIMO networks, pilot spoofing attacks are the most frequent attacks. To harm the estimation of authorized user channels, active eavesdroppers forward the uplink pilot sequence to create interference with the pilot training phase to affect the downlink transmission for stealing the authorized information [81].

Various strategies have been introduced to deal with this kind of PLS on cell-free massive MIMO networks. To measure the level of pilot contamination, the Gaussian RF impairment model (GRFIM) and additive quantization noise model (AQNM) are considered where the linear minimum mean square error (MMSE) is used to find the irregularity of the affected channel [135]. To reduce the impact of pilot spoofing attacks, enormous power allocation schemes are used to maximize the achievable data rate or secrecy rate of the targeted users and minimize energy consumption as long as security requirements are ensured [39]. Another way of dealing with the attacks is to downlink pilot optimization through derivative closed-form expression of secrecy rate and various channel estimation parameters [134]. In addition, using MMSE to limit the active pilot contamination is another popular way by which download link precoders are constructed to examine the impact of unplanned beamforming of hidden signals toward the active eavesdropper through the distributed APs [105]. However, the key disadvantage of using the above-mentioned approaches is high complexity. Therefore, the combination of a channel estimation method with user location measurement through fingerprint positioning is used to mitigate the effect of the active pilot attack [81]. In this technique, fingerprint positioning and K-means clustering are considered to collect the position status of users and eavesdroppers. Moreover, the non-overlapping angle of arrival (AOA) method and Discrete Fourier transform (DFT) are utilized to separate the uplink channels of legitimate users from eavesdroppers and eliminate the pilot contamination.

9.3.1.2 Security of Non-orthogonal Multiple Access

NOMA is a potential physical layer technology that allocates resources nonorthogonally to support multicast/unicast transmission for 6G massive networks. The NOMA distributes channel resources to the receiver in a fair manner where more energy is allocated for the users who have weak signals compared to strong signal users to enhance the number of simultaneously served users [46]. The key advantages of using NOMA over orthogonal multiple access (OMA) are low-signal processing cost, less latency, high-quality channel feedback, and accelerated throughput.

However, in spite of having various benefits, NOMA still faces multiple attacks through both active (internal) with known channel state information (CSI) and passive (external) with unknown CSI eavesdroppers due to its exclusive structure. For instance, an internal eavesdropper can introduce jamming attacks not only on the beam scope of NOMA clusters but also reflected beams from IRS-based NOMA networks [72].

There are various conventional techniques that are popular to secure communication from active eavesdroppers including cooperative communication, beamforming, and channel-based power allocation. For example, cooperative communication security scheme is used in a large-scale downlink system of full-duplex NOMA users with the presence of jammers and artificial noise to enhance the output of PLS [35]. Implementation of a jammer in cooperative NOMA is a popular method of improving the secrecy performance of a network [129]. The NOMA uses successive interference cancellation (SIC) processes to measure the channel health status of authorized users, which can be beneficial for CSI knowledge-based security protocols [90]. Moreover, a combination of beamforming and power allocation technique is also considered to enhance the performance of the PLS system in case of known/unknown CSI of eavesdroppers [136]. Various relay selection methods like optimal single relay selection (OSRS), two-step single relay selection (TSRS), and optimal dual selection (ODRS) are used to decode authorized messages to forward to NOMA users [54].

Nevertheless, some eavesdroppers may harm the general functions of NOMA. For example, eavesdroppers can interface the transmitter signal by sending a large degree of identical pilot signals to consume the transmitter power for allocating to legitimate users. The amalgamation of NOMA and IRS/VLC/THz techniques are becoming emerging solutions for the above-mentioned scenario [75]. During communication between cell-edge users and BS, IRS can be implemented to improve the quality of the downlink in terms of various fadings like Nakagami-m [101].

However, most of the above uses technologies require the CSI of the eavesdropper, which is not possible for cognitive networks. Therefore, multidimensional directional modulation method, channel-dependent interleaving, or cyclic feature suppression-based schemes can be the potential solution for real-time NOMA [29].

9.3.1.3 Security of Reconfigurable Intelligent Surface

Reconfigurable intelligent surface (RIS) is a revolutionary 6G physical layer technology that is capable of controlling the challenges of the cognitive environment through software-based solutions. To accelerate the propagation performance of a channel, RIS uses a large degree of low-cost reflecting components in a planar array which can control the reflected signals' amplitude/phase shift by adjusting the reflective coefficient.

In contrast to traditional communication technology, RIS can have a unique feature of controlling the propagation environment that ensures secure transmission. However, in an adverse propagation situation, an eavesdropper may reduce secrecy performance even with the presence of artificial intelligence (AI). RIS technology uses the rudimentary concept of multipath propagation. Therefore, position exposure attack is introduced in RIS communication [25].

Ensuring PLS on a RIS-assisted framework is significant as this technology has become a predominant choice for the 6G network to enhance data rate and efficiency. Generally, optimizing transmitted power and controlling phase shift can enhance the security and the QoS of the perceived signal [118]. In a RIS-assisted network, a Fisher-Snedecor framework is employed to investigate composite fading and shadowing channels in the presence of eavesdroppers [67]. The average secrecy performance showed that deploying the RIS-based access point for the above-mentioned scenario has a beneficial impact on the impact of eavesdropping. However, the RIS technique is also united with other 6G physical layer technologies including NOMA. RIS-assisted NOMA networks are commonly prone to both external and internal eavesdropping. Noise beamforming is utilized to resist external eavesdropping with/without the presence of CSI of the attacker. Moreover, a power allocation sub-optimal system is employed for limiting the attacks in case of known CSI [137].

9.3.2 Security of mmWave

The mmWave technology refers to a communication system where users can transmit the messages through a 30–300 GHz frequency band with the wavelength ranging from 1 to 10 mm. Enhanced security, less interference, small components sizes, and huge bandwidth are key advantages of using mmWave, which are enabled to outweigh the disadvantages like blockage losses or higher penetration. With the explosive development of diverse high-end applications like UAV, IoT, virtual reality (VR), satellite communication, object imaging/detection with mmWave is intended to use mmWave to meet the demand of fulfilling a large degree of low-latency and high-speed connected devices in 6G networks [18].

The physical layer mmWave technology is mostly affected by three common adversarial behaviors such as jamming, eavesdropping, and pilot spoofing attack. In mmWave communication, eavesdropping can be introduced by interfering and wiretapping in unsecured wireless communication. Eavesdropper utilizes the internal domain of beam scope or a reflector to monitor the channel. According to the transmitted signal information, the eavesdropper sometimes generates jamming or pilot spoofing attacks on the network to harm the legitimate user [72]. In extreme cases, a jammer can form DoS attack by sending the same frequency radio signal to occupy the shared resources of the network. To generate pilot spoofing in mmWave, the intruder deliberately sends identical pilot signals to equivocate regarding authorized user detection and channel assessment phase of the transmitter.

The key technology of PLS is to analyze and maximize the performance of the security measurements such as outage probability and secrecy rate to enlarge the signal strength for mitigating jamming and pilot spoofing attack. On the other hand, most signal strength-based approaches require CSI, which informs regarding the status of the communication channel by a transmitter. Therefore, the fundamental idea of PLS technology is to use randomness in the modulation techniques to create random wireless channels to avoid being predicted by the eavesdropper [112]. Many researchers have focused on beamforming design for PLS. During the design stage, artificial noise or friendly jamming is added to the free space of authorized users to confuse the intruder on the original transmission channel [40]. For instance, the artificial noise-based optimal beamforming framework can be utilized for reducing the transmit energy of a BS while satisfying the adequate requirements of secrecy rate for all secret messages [132]. Moreover, a unique beamforming technique is implemented for escalating the secrecy rate of a user to improve the PLS in a dual-polarized mmWave system [95]. The integration of optimal power allocation algorithms and zero-forcing beamforming technique at the transmitter is another way to enhance the PLS for the NOMA-based MIMO system [69]. Another emerging concept of ensuring security on mmWave communication is to use a pair of physical keys, which needs to be exchanged among authorized users and transmitter to verify the legitimate transmission [100].

On the other hand, conventional encryption and decryption algorithms in the precoding can experience a bad impact on a network where eavesdropper is one of the authorized users. In such cases, knowing the CSI status of every user is crucial. Therefore, AI/ML-based learning methods are employed to improve the CSL knowledge for applying the appropriate resistance strategies like channel hopping. Due to the repeated changes of transmission frequency using the emerging concept of frequency hopping, legitimate users will be less prone to attackers [47]. Though AI/ML-based learning methods are significant to have a better impact on security, the learning algorithms themselves are vulnerable to various attacks and produce faulty output by manipulating the learning schemes. Therefore, adversarial training-based learning techniques are utilized to mitigate the attack on training algorithms [26].

However, the directional modulation technique is another security solution where a definite constellation diagram is transmitted toward the targeted user, while a noise pattern is transmitted in another direction. For example, a simple directional modulation scheme is used to modulate the radiation pattern through a randomly selected antenna subset for each symbol [109]. To formulate a more artificial noise, a new programmable weight phased array-enabled optimized weight antenna subset method is considered to secure mmWave wireless communication [41]. Moreover, to enhance the security in the multiuser PLS modulation technique, hybrid beamforming is also introduced where the cross-entropy iteration method is used to select the optimal antenna combination that reduces the eavesdroppers' signal power to improve the security performance [55].

9.3.3 Security of Terahertz Communication

THz communication collectively works on optical waves and microwaves in the ranges between 0.1 and 10 THz to ensure high-speed communication with less interference and less-complex combination methods of sensing and transmission. THz communication technology supports enormous applications like VR gaming, ultra-HD video conferencing, nano-technology-based communications (e.g. The Internet of nano thing (IoNT) or bio-nano-technology), and molecular communication that require high bandwidth and low latency [88]. Higher data rate (100 Gbps or higher), narrow beam structure, and low attenuations are rudimentary benefits of THz technology. However, despite of having immense advantages, there are various drawbacks of THz communications. THz has low penetration capability regarding particular obstacles like thick walls. Therefore, the network coverage domain is comparatively small. Moreover, high absorption resonance by the surrounding environment, highly directional signals, and energy consumption are other disadvantages [12].

Though the characteristics of being highly directional and low communication coverage ensure more security in an indoor/outdoor application against common attacks like eavesdropping and jamming [113], THz still is not entirely protected by various attacks in special conditions like line-of-sight (LOS) transmission [78]. THz transmissions are compromised to access control attacks, eavesdropping, and data transmission exposure. For example, an unauthorized user may deploy a reflector in the LOS transmission path to scatter the radiation from the legitimate receiver to eavesdropper antennas to leakage or collect specific information [65].

There are various conventional ways to enhance the THz communication security including signal power variation, frequency hopping, the strategic position of access points, lightweight cryptography, and beam encryption and create a network more robust for the attackers [93]. The concept of transmitting messages through multiple paths is a popular way to accelerate security against eavesdropping not only saves sensitive information but also secures key exchange in THz communication [76]. Moreover, to combat the problem of eavesdropping, an artificial noise-aided self-interface cancellation free receiver is also proposed where an efficient DNN algorithm is used to achieve a high secrecy rate with low hardware complexity [31]. On the other hand, IRS technology can also assist the multipath nature of the network to boost the signal energy and resist information leakage [80].

Enormous authentication schemes are considered to ensure the PLS. A distance-dependent-path loss-based authentication method is used for a THz technology-enabled vivo nanonetwork [82]. Moreover, this authentication method applied path loss as a device fingerprint with a THz time-domain spectroscopic structure. In THz communication, the electromagnetic signature of the frequencies is also formed to provide authentication at the physical layer [111].

9.3.4 Security of Holographic Radio

Holographic signal/radio/beamforming is an innovative step for 6G communications where software/phonetically defined antenna arrays are used instead of traditional-phased array or MIMO technology [42]. The key advantages of holographic signals are enhanced spectrum efficiency, escalated network capacity, and smoothly integrated of imaging with wireless transmission [87]. Holographic communication requires large bandwidth, which also rises equitably with the increasing number of users. Therefore, ensuring security with less complex and lightweight techniques is desired [78].

Eavesdropping is the major concern of holographic communication. In wireless communication, highly directional beamforming is a good idea to deal with eavesdropping. Holographic communication is integrated with large intelligent surface (LIS) technology to create perfect directional beamforming through recording and rebuilding [138]. On the other hand, MIMO-assisted holographic communication can also ensure PLS security against eavesdropping by canceling out the reflection of the BS to the attacker [42]. However, uncontrollably scattered electromagnetic waves cause the issues of being interference to take the software control over the electromagnetic behavior of a wireless environment. Therefore, the software-based solution is used to take control of the scattered behavior of the waves of wireless communication [59].

9.3.5 Security of Ultra-Reliable and Low-Latency Communications (URLLC)

URLLC is envisioned to start real-time cooperative transmission by exchanging the packets with less error probability and ultra-low latency. The unique features of extreme reliability and ultra-low latency encourage to design of time-sensitive and life-threatening applications including autonomous vehicular networks, remote surgery, augmented/virtual reality, and tactile internet [15]. Due to the leakage of confidential and significant information, ensuring security in URLLC has got paramount attention. For instance, an eavesdropper can place between two authorized vehicles in a vehicular internet of things (VIoT) network and extract transmitted messages through a random sequence and replace them by misleading information which may cause a serious accident.

There are enormous conventional security solutions exist that require CSI information to ensure security for wireless communication. However, due to the short blocklength nature, adequate CSI status estimation is not available in URLLC [27]. Channel inversion power control (CIPC) based on channel reciprocity is considered to enhance security instead of CSI for various applications which use URLLC for downlink and uplink [24]. On the other hand, location-based beamforming is also provided PLS in LOS-based transmission on URLLC environments [125]. Though this technology cannot ensure intense security like CSI-enabled beamforming, it can achieve the significant feature of being an ultra-low-latency communication tool. By contrast to traditional cryptographic methods, distributed networks apply friendly jamming methods for URLLC to accelerate the performances of the PLS [56]. Deploying jamming signals is a considerable technique that mitigates the decoding capability of an eavesdropper. To deal with the developing security challenges in URLLC, NFV and blockchain based integrated method is proposed where considered network devices have to sign the contract before spectrum sharing [43]. This feature of the method ensures authorized access to the network which is able to fight against malicious attacks. The framework also used the parallelized NVF concept, which helps to execute firewall and intrusion detection methods simultaneously for improving security with the low-latency output.

9.4 CONNECTION/NETWORK LAYER SECURITY ISSUES

The connection layer is an amalgamation of network and transport layers, which is responsible for perceiving/sensing information from the physical layer entities for being processed and transmit it to the end user through the appropriate route. In 6G networks, this layer consists of enormous advanced technologies including SDN-WAN, deep slicing, network function virtualisation (NFV), distributed ledger technology (e.g. blockchain, holochain), advanced AI, deterministic networking, space-air ground-sea integrated network, and quantum communication. Table 9.3 shows significant attacks on the Connection/Network layer, and the countermeasures.

TABLE 9.3 Frequent Attacks and Countermeasures on Connection/Network Layer Key Technologies

Technology	Attacks/Threats	Countermeasure	Reference
	MITM	Open flow protocols	[19]
		IDS	[19]
SDN-WAN	DDoS/DoS	Flood tracker	[20]
		ReCON-based MTD	[34]
	ARP flooding	suppression algorithm	[22]
		SASE framework	[117]
NFV	DoS/DDoS	Back-propagation NN	[86]
		Shuffling-based algorithm	[61]
	Side channel attack	External noise	[35]
	Jamming	ML algorithms	[91]
		key-based authentication	[73]
Deep slicing	DoS/DDoS	Boolean logic	[83]
		Deep learning	[104]
	Spoofing	Deep learning	[104]
	DDoS	Nash learning-based gaming	[119]
		Secure smartcontract	[7]
		Coin-based protocol	[4]
	Routing attacks	Bitcoin relay network	[8]
	Eclipse attacks	Add more buckets	[38]
DLT		Add outgoing links	[38]
		Directional channel	[38]
	Sybil attacks	Trustchain and CTI	[55]
		Coin-based protocol	[4]
	51% attack	ML and gaming	[21]
		Proof of work (PoW)-based consensus	[127]

(*continued*)

TABLE 9.3 Frequent Attacks and Countermeasures on
Connection/Network Layer Key Technologies (*Continued*)

Technology	Attacks/Threats	Countermeasure	Reference
		Detection methods	[52]
	Poisoning attacks	Sphere and slab defense	[97]
		Shared key-based protocol	[123]
		TFP system	[3]
		Safetynet	[33]
	Evasion attack	Fingerprint	[37]
AI		ML algorithms	[51], [74]
		SecDefender	[14]
	API-based attack	Recurrent Neural Network (RNN) Algorithm	[140]
		CBA framework	[84]
		Additive noise	[106]
	Inversion attacks	additive Laplacian noise	[106]
		Hash method	[79]
Deep Network (DN)	MITM	Redundant path	[68]
	DDoS	Redundant path	[68]
		Hash-based authentication	[68]
	Spoofing	Redundant path	[68]
SAGSIN	DoS	ML classifier	[11]
		Half-duplex-based PT	[44]
	Eavesdropping	Cooperative beamforming	[23]
		Artificial noise	[23]
		Multipoint transmission	[110]
		Particle Filter (PF)-based GPS	[57]
	Spoofing	Two-stage validation	[66]
		Deep learning	[107]
	Jamming	Game theory	[122]

NN–Neural Network, TFP–Tamper-free provenance, CBA–Characteristic-based Automated framework, DN–Deterministic Networking, PT–Packet Transmission.

9.4.1 Security of Connection/Network Layer Technologies

Due to the broad range of technology, network layers have been subject to various treats/attacks. For instance, a spoofing attack may introduce for the lack of integrity protection of transmitted signal traffic. DoS is introduced by injecting millions of mobile UEs to overload the authentication servers to make them unavailable to the authorized user [45]. Therefore, ensuring security on this layer becomes crucial, specially for the network access points like signal units, gNodeB (gNB), and ground station or for the core networks such as endpoint gateways, validation server, and edge/fog servers.

9.4.1.1 Security on SDN-WAN

SDN-WAN is a new paradigm of research in 6G network, which offers automated or programmed networking architecture through WAN connection. The emerging SDN-WAN is not only a recognized cognitive technology in 6G but also expends the scope of new vulnerabilities. Therefore, to implement this technology in a framework, it needs to ensure that all the connected resources and information preserve the security requirements including availability, integrity and privacy.

There are various threats/attacks landscape in SDN-WAN architecture. For instance, MITM attacks may introduce by inserting an intercepting node between the SDN switch and the controller to tamper with the communication without being detected. On the other hand, flow table or flow buffer, or controller can be compromised by DoS/DDoS attacks [92]. For example, in the case of an updated rule insertion to the switch, an attacker creates a large degree of packets to transmit to the unknown host within a short period of time, which causes an overflow of flow table storage capacity.

To protect SDN/SD-WAN, various detection and mitigation techniques for MITM and eavesdropping have been developed. Most of the traditional ways of dealing the security in SDN-WAN is changing the rules of open flow protocol including FlowVisor, FlowChecker, and Flow-Guard and implementing an Intrusion Detection System (IDS) on the network [19]. A novel lightweight SDN-WAN flood tracker is suggested to trace the attack flows and remove them without having an impact on legitimate traffic of the network [20]. This scheme is applicable to detect attacks in both internal and external sources of a controller. Due to the distinctive nature of in-band control and centralized control architecture of SDN-WAN, an attacker may launch DDoS on the SDN controller through the open flow agents or by sharing a link. A REsilent COntrol Network Architecture (ReCON) is proposed as a Moving Target Defense (MTD) method to minimize the DDoS attack impact by reducing transmission of the critical resources between data and control traffic and increasing the control agent's capacity [34]. In a multicontroller SDN network, a two-stage "storing-blocking" suppression algorithm is used against address resolution protocol (ARP) flooding [22]. The result shows that this algorithm can mitigate the impact of ARP flooding in terms of attack time, scope, and controller overhead in the multicontroller SDN network. However, to ensure security for a collection of cloud-native integrated services, a secure access service-edge (SASE) framework is introduced in SDN-WAN mobile networks to reduce the impact of the above-mentioned attacks [117].

9.4.1.2 Security on NFV

The 6G networks are intended to include NFV for limiting the resource cost, improving functional efficiency, and shorter lifespan to market. In spite of having immense benefits of using NFV in a network, vulnerable landscapes remain to be a significant concern. For instance, the functionality of the virtual layer of NFV may be undermined by the presence of attackers [85]. Moreover, the communication among NFV systems, Orchestra frameworks, and NFV managers is also vulnerable in terms of eavesdropping, flooding, side channel attacks, and DDoS attacks.

The conventional AI is a common method of protecting NFV networks. For instance, the back-propagation neural network training module is used to design a novel architecture named Cloud Trace Back to identify the attack traffic for reducing DDoS on the NFV network [86]. A cost-effective shuffling-based algorithm is proposed in SDN/NFV networks to solve DDOS attacks, where an HTTP request is sent to create flooding and slow the network performance [61]. However, adding external noise to the signal or mitigating information from being exposed from the channel can be a significant idea for handling side-channel attacks in NFV networks [5].

9.4.1.3 Security on Deep Slicing

In 5G network, slicing has become an interesting technique for getting the reliable resource allocation with a high data rate, which is intended to implement through the deep learning mechanism that refers as deep slicing in 6G architecture. Therefore, enormous learning methods are considered as a decision-making tool for slicing the network to enhance the network performance, which is also prone to attacks. For instance, the reinforcement learning algorithm is deployed in a radio network for slicing the network load balancing where a novel over-the-air attack and jamming attacks are employed through adversarial ML to manipulate the decision matrix of the learning algorithm [91].

There are three defense mechanisms against jamming attacks on RL-based network slicing [91]. Firstly, an update of the RL algorithm needs to be avoided whenever an attack is detected. Secondly, ensure the randomness of the learning algorithms' decisions. Finally, manipulate the feedback (NACK) method to protect reliable information from being hacked. On the other hand, an efficient session key-based authentication architecture is proposed for securing network slicing in IoT networks where Fog systems and IoT servers are used to ensure the secure authentication for the data of fog cache and remote servers without considerable delay [73]. During the implementation of deep learning on network slicing and resource allocation, DDoS is a frequent attack which may also introduce in SDN/NFV-based frameworks like advanced mobile broadband, URLLC, and massive machine-type communication [83]. To secure the SDN/NFV network, Boolean logic is integrated with a password-enabled key generation function to produce traffic-aware scheduling for protecting the network against DDoS attacks. However, a deep learning-based mechanism is proposed to preserve the network slicing process against various attacks including DDoS [104]. The degree of flooding and spoofing attacks was considered to reflect the performance of the proposed learning algorithm and got nearly 98% accuracy in terms of DDoS detection.

9.4.1.4 Security on DLT

Key features of the 6G network include trustworthy information sharing by mitigating source's misbehavior/fakeness and single-point failure, which is a challenge to achieve. Therefore, various emerging distributed technologies (DLT) like blockchain [103] and Holochain [131] are being used in time-sensitive/trust-significant transactions such as cryptocurrency or finance for real-time applications including 6G-enabled smart healthcare, UAV, and autonomous driving. Furthermore, DLT is capable of ensuring secure networks by introducing data integrity, distributed security, transparency, immutability, traceability, and availability.

Since the implementation of blockchain is in the early stage, there are potential threat/attack landscapes that need to work on to provide secure communication. For instance, the transactions, chain of blocks, and smart contracts left vulnerabilities that cause various attacks including 51%/majority attacks, double spending attacks, Sybil attacks, DDoS, routing attacks, and eclipse attacks in the network layer.

The deployment of numerous security mechanisms can mitigate the attack possibilities on blockchain networks. Therefore, ensuring security on the smart contract is inevitable. Performing intensive formal verification [99], employing security check tools [48], and detecting semantic flows [116] are popular mechanisms to protect the smart contracts of a blockchain network. The Nash learning-based gaming approach is utilized to solve the DDoS attacks which are introduced through the dynamic mining pool [119], while penalty-based and age-based mechanisms are considered to limit the flooding of the transactions by filtering spam ones [114]. However, a smart contract-based collaboration method is proposed in SDN networks to enhance the detection rate of DDoS by monitoring the attack information among autonomous systems [7]. A software-hardware integrated bitcoin relay network is suggested to protect routing against attacks in SDN networks where the relay nodes are responsible for inter-domain routing protocols, ensuring security on the attack positions and selecting secure paths for the users [8]. There are various countermeasures suggested for resolving the eclipse attacks on blockchain by setting up more buckets, adding additional outgoing links, diversifying incoming channels, and detecting abnormal behavior [38]. However, blockchain-enabled trustchain and cyber threat intelligent (CTI) architecture are proposed to defend against Sybil attacks. Furthermore, coin/currency-based multiparty hybrid protocols are designed to address Sybil attacks and DoS [4]. Machine learning and gaming methods are recommended to mitigate the 51% attack, which is also known as a double-spending attack [21]. By contrast, a hybrid approach with PoW consensus is utilized to collect minor's past information and ensure 51% attack-free networks [127].

9.4.1.5 Security on AI

Due to the vast implementation scopes, AI/ML has become a significant technology in the 6G network. Designing cognitive physical layer, dynamic resource allocation, optimization, providing security on networks/users, resource management, and processing information are the popular using areas of AI/ML. Undoubtedly, this technology has attracted huge attention and is also used to detect or defense from dynamic/real-time threats/attacks. However, AI/ML itself can be compromised by internal and external intruders during the training or testing phase. AI/ML technologies are prone to various attacks including poisoning attacks, evasion attacks, API-based attacks/Oracle attacks, and model inversion attacks.

The frequent countermeasure against poisoning attacks is to build efficient detection methods amid the training session. One of the detection mechanisms employed influence functions to map the correlation among prediction and training sample sets, and the result shows that it is capable of detecting malicious data detection against poisoning attacks [52]. On the other hand, two workable defense strategies like sphere defense and slab defense are utilized to eliminate the injected malicious data sample on a DNN model [97]. To resolve the integrity problem of the calculated outcomes, a shared key-based protocol is designed to protect the training data against poisoning attacks [123]. In addition, emerging tamper-free provenance systems are utilized which use contextual information regarding the real and transformation of the data during training to detect poisoning attacks [3]. This work considered both untrusted and partially trusted data sets and showed that machine learning models can be protected from the adversarial environment in terms of reliable provenance data sets.

A safety-net is proposed as a verification framework to preserve the data integrity of a cloud server against evasion attacks [33]. The work used an interactive proof protocol to achieve end-to-end verifiability. Moreover, fingerprint technology is utilized as sensitive samples for DNN models to verify the return results from a server [37]. In contrast, the advantages of various ML techniques are also employed to protect the system from adversarial activities. For instance, an support vector machine (SVM) classifier based on a secure and robust learning model is proposed to enhance the malware detection performance against evasion attacks [51]. The feature of neuron activation is used during test time in a 3-layer ANN-based IDS architecture, where various ML classifiers like Random forest, Nearest Neighbor classifier, and SVM are selected to train and test to detect the evasion attacks [74]. A robust learning paradigm named SecDefender is proposed for malware detection, which considered both classifier retraining technique and evasion cost for feature influence by adversarial attacks as security regularization to accelerate the system security [14].

An RNN-based algorithm is proposed to detect API-based malicious activities, where generative RNN is used to produce irrelevant or malicious API requests and add them to the original sequence of API. On the other hand, a modified RNN is deployed to train the API request dataset for detecting adversarial activities [141]. However, a novel characteristic-based automated framework is utilized to identify new attacks on cloud servers or applications [84]. The fundamental advantage of using this framework is to identify the application traffic pattern along with various attacks with a minimal computational load.

A straight and less complex additive noise method is imposed into an intermediate tensor on one-dimensional ECG data to ensure security against model inversion attacks [106]. However, additive noise technology generates a negative impact on the accuracy of performance. Therefore, additive Laplacian noise is employed in intermediate data to make it harder on the sender side before transformation to deal with model inversion attacks by maintaining a good level of accuracy. However, to ensure the security against model inversion attacks in a face recognition system, the concept of hash is implemented perceptually to the part of the selected training images, which provide a countermeasure of preventing original image recreation with high classification accuracy [79].

9.4.1.6 Security on Deterministic Networking

In 5G and 5G beyond technology, instead of having conventional networking services, there are multiple industries that require deterministic network services with extremely low-latency and ultra-low packet loss within a specified user domain.

Since for faster communication, packet delay is as sensitive and significant as the packet itself, providing security on deterministic networking architecture has added a new dimension. The time protocol can be attacked by enormous internal or external attackers and causes significant negative consequences. For instance, a MITM attack may adjust additional latency into a link which may harm a real-time application without even cracking any encryption model. Moreover, DoS attacks may be introduced by IP/ARP/Medium Access Control (MAC) spoofing, which affects time protocol along with others. Deterministic networking is compromised by various types of attacks such as MITM attacks, spoofing, DoS, replay attack, and packet manipulation [68].

Though countermeasures against enormous attacks vary with the nature of the specified deterministic networking environment, there are some traditional ways to mitigate the attacks. For instance, path redundancy methodology is considered to mitigate various deterministic networking attacks including DoS, MITM attacks, and spoofing. The concept of path replication or redundancy enhances the strength against failures and attacks. Hash-based Message Authentication Code is used to reduce data modification attacks and spoofing on deterministic networking [68]. On the other hand, the idea of source authentication may improve the detection performance of malicious activities like spoofing and data modification. Moreover, while integrity protection helps to secure intermediary nodes, authentication verifies the source information.

9.4.1.7 Security on Space-Air Ground-Sea Integrated Network

In 6G networks, Space-air-ground-sea integrated network (SAGSIN) is an emerging technology that integrates heterogeneous communications like satellite networks, aerial networks, terrestrial networks, and marine networks to offer universal internet services. Various 6G communication technologies are used to communicate in SAGSIN including Mm-wave communication, Terahertz Communication, VLC, and optical fiber communication. Due to the network's unique characteristics both the nodes and the network itself can be exposed by active or passive attacks that introduce DoS, spoofing, eavesdropping, and jamming attacks [36].

However, a jammer harms communication links or signals or nodes in the physical layer, injecting false messages or taking routing control packets to reduce throughput and increase satellite services in the network/connection layer. To identify the vulnerable nodes against jamming, a novel packet classifier is adopted in an encrypted wireless ad hoc network [11]. A half-duplex-based packet transmission method and long propagation delay are used by the receiver to produce interference with the eavesdropper. The receiver sends jamming packets to the adversarial nodes to keep busy with receiving, which prevents malicious signal transmission [44]. However, the concept of game theory is also considered in underwater sensor networks to mitigate jamming attacks by selecting the transmitted energy in the presence of an interfering signal [122].

Cooperative beamforming technology and artificial noise are designed to secure satellite communication against eavesdropping [23]. Another countermeasure of eavesdropping in SAGSIN is proposed where coordinated multipoint transmission and signal alignment are used by applying the nature of water velocity and antenna's spatial diversity [110].

Nowadays, the Monte Carlo simulation-based enhanced particle filter has become a popular solution against spoofing in SAGSIN. For instance, to detect spoofing attack, a maximum particle weight monitoring method is employed where the position of the receiver of a global navigation satellite system. This advanced PF-based GPS anti-spoofing handles the attacks through both spoofing detection and spoofing suppression steps [57].

To prevent DoS attacks, a two-step efficient validation system is proposed where the system uses a nonce value that needs to be equal to the packet sequence number, and the private key is also compared with the measured MAC address value. However, not having protection for the sequence number may help to compromise DoS attacks too. Therefore, a one-way Rabin function is employed to encrypt the sequence number to intensify the level of security against DoS attacks [66]. Moreover, a deep reinforcement learning-based attack mitigation strategy is imposed to reduce the abnormal traffic of SDSN generated by DDOS attacks [107].

9.5 THE PROCESSING/SERVICE LAYER

The processing layer closely works with the business or application layer that consists of edge/fog/cloud technologies and processes the big data regarding specified application requirements. Currently, the processing layer has become an independent research domain to transform the network from 5G to 6G through the utilization of emerging AI and intelligent cloud-based technologies. Therefore, protecting the layer from adversarial activities demand various security aspects including authentication, transformation of encrypted data, access control, hardware/software security, and operation/kernel system reinforcement. While in 5G, the service layer ensures the performance of the network using cloud computing, edge computing, or massive IoT services, 6G networks introduce container-based virtualization, zero-touch service orchestration, distributed/autonomous computing, Quantum computing for getting faster and safer communication. However, the layer is also prone to various conventional and cognitive attacks/threats due to the nature of automation networks in 6G communication. Therefore, security needs to be ensured from the host, operating systems, applications, virtual machine, and containers to API. Table 9.4 lists the common attacks and countermeasures of the processing layer.

TABLE 9.4 Frequent Attacks and Countermeasures on Processing/Service Layer Key Technologies

Technology	Attacks/Threats	Countermeasure	Reference
CBV	Unauthorized access	Setting privilege functions	[98]
	Data tampering	Scrambling	[98]
	Data leakage	Two-stage defense	[32]
		Permission setup	[58]
	DoS	Scanning malicious events	[98]
		Hybrid method	[58]
ZTSO	Deception attacks	Limit VM resources	[10]
		DLT	[89]
	DoS	AI-based learning	[10]
	MITM	Data encryption and VPN	[10].
	Data leakage	Implementation of IBN	[115]
		Fuzzy-based AC	[78]
	Malicious attack	Implementation of IBN	[115]
		AI-based prediction	[78]
Quantum computing	Quantum cloning	Post-quantum cryptography	[62]
		Lattice-based encryption	[70]
	Quantum collision	QkD-enabled fiber links	[83]
		Post-quantum cryptography	[62]
DC	Data tampering	Softwarization	[72]
		Secure Smartcontract	[7]
		Coin-based protocol	[4]
	DoS	Probabilistic mechanism	[94]
	Malicious attacks	Softwarization	[72]
		Probabilistic mechanism	[94]

CBV – container-based virtualization, ZTSO – zero-touch service orchestration, AC – access control, DC – distributed computing.

9.5.1 Processing/Service Layer Security

There are various attacks that take place on the processing/service layer technologies to degrade the performance of the network. Some frequent and popular attacks, along with their countermeasures, are discussed here.

9.5.1.1 Security on Container-Based Virtualization

In 5G beyond networks, container-based VM introduces a lightweight alternative of VM where containers share the same kernel to limit the time and resources for microservice applications. There are various containers such as Docker, Linux-Vserver, OpenVZ, and LXC that are being used in heterogeneous 6G applications like IoT, smart cars/healthcare, fog/cloud computing, and so on. The characteristics of considering limited resources of a container, hosts are vulnerable to enormous adversarial activities such as unauthorized access, data tempering, DoS, ARP spoofing, MITM, and PEA [98]. In container-based VM applications, the container itself or inter-containers or the communication between host and container may attack by malicious events and can be able to access confidential data, modify data, gain resource usage patterns, and breach the integrity of the information.

There are various attacks employed on a container through malicious applications within the container. Executing an application in a container that provides unnecessary root privilege causes unauthorized users to take control of the container. Therefore, providing root privilege to the least privilege-specific functionalities can limit unauthorized access [98]. Containers store clear test which can be compromised by the attacker to disclose the information or data tampering. The mitigation technique for this type of tampering is to store secrets outside the image or data [98]. Moreover, a two-stage defense mechanism is proposed to protect the containers from information leakage. The first approach is masking the channel and accelerating the container resource isolation method [32]. On the other hand, a vulnerable application within a container may introduce a DoS attack to create the unavailability of the container. Creating root file system read-only and scanning applications within an interval can reduce the impact of DoS attacks. However, a dynamic DDoS alleviation method is used where the number of instances of a container and the coordination of resource allocation are checked regularly for different users [58].

Attackers can also generate syn flooding and internet control message protocol (ICMP) flooding to introduce DoS on neighboring containers that create a negative impact on container operation. Therefore, scanning malicious activities on images or applications periodically can be a countermeasure against DoS attacks [98]. On the other hand, MITM attack is placed in poorly separated containers that introduce potential ARP spoofing, DoS, MITM, and MAC flooding. Limiting the container resources and configuring the network as the least-necessary privilege is used to provide security [13]. However, instead of having various popular container-based protection mechanisms, kernel security mechanisms such as Capability, Seccomp, and MAC are considered for preventing PEA [60].

9.5.1.2 Security on Zero-Touch Service Orchestration

The rudimentary advantages of 6G networks such as extremely low latency, very high reliability, high degree of machine-to-machine communication, and the adaptability to the required demands for high capacity are looking for major changes in network services and management including automation in orchestration. Due to the advanced emerging technologies in 6G frameworks, end-to-end automation has been introduced to ensure an intelligent networking system. For instance, network slicing is a significant issue in 6G architecture which can be automated by featuring an AI-driven automated orchestration framework [17]. However, intelligence and zero-touch orchestration also deploy various security challenges including threats and attacks such as deception attacks, DoS, MITM, information leakage, and malicious activity.

The closed-loop network automation system for the 6G architecture allows a zero-touch management system for monitoring the error and congestion identification that may create a feedback loop of communication among various steps of the given task such as detecting, observing, modifying, and optimizing the output to ensure self-optimization. Therefore, the closed-loop automation system is compromised by various threats. DoS attacks may introduce by employing fake loads on VNF-enable networks to escalate the capacity of virtual machines. One of the countermeasures of these attacks is to limit the resources for virtual machines. In addition, how many resources are required for a network can be predicted by AI-based learning algorithms to resist DoS attacks [10]. MITM attacks are hosted in the closed-loop network by triggering fault events and rerouting the message in the incorrect path through a malicious switch. Scramble the information using various encryption methods and using VPNs such as internet protocol security (IPSec), secure sockets layer (SSL), transport layer security (TLS), and host identity protocol (HIP) is considered to limit the impact of the MITM attacks. Moreover, communication may prone to deception attacks to tamper the transmitted data [89]. Various DLTs like blockchain and holochain are used to ensure the integrity validation of a transaction.

Another concept of automation in network orchestration is intent-based networking (IBN) which includes AI technologies in 6G mobile networks that are capable of monitoring, configuring, and operating through AI methods. The fundamental security concerns of implementing IBN include information leakage and malicious activity [115]. An attacker may intercept authorized users/information to compromise the security aims and publish the authorized data to the public. Authentication mechanisms (OpenID connect, Signed JWD tokens) between attacker and consumer, authorized access control (fuzzy-based access control, OAuth 2.0), and secure transport layer protocols like TLS are used to limit the information leakage/exposure security vulnerabilities [78]. Moreover, a malicious user causes a network outage by changing the behavior of the transmission. AI-based prediction models can be used to detect malicious activities.

9.5.2 Security on Quantum Computing

Quantum computing-based communication is an emerging technology that is expected to be commercially available for enormous 6G-enabled applications including sea communication, terrestrial network, and satellite/TeraHertz communication. To achieve high reliability, quantum communication proposes quantum channels instead of using noiseless classical channels. In terms of security, 6G networks use quantum computing to identify, mitigate, and prevent security vulnerabilities. For instance, quantum key distribution (QKD) is employed in various applications to create a secret key pair for authorized users. However, adversaries are also capable of utilizing quantum power to generate threats/attacks on IoT devices or networks.

Quantum cloning attack is a common attack in quantum communications where an intruder selects random quantum information status to copy without any modification of the original information [2]. Quantum collision is another possible attack that may introduce in a quantum setting where two different hash function produces the identical result.

Though considerable advancement in security protocols has been achieved, quantum communication is still struggling to provide intense security. Various post-quantum cryptographic solutions (code-based, lattice-based, hash-based, and multivariate-based) are provided to ensure secure communication [62]. Since IoT devices require lightweight security solutions, lattice-based encryption is appropriate for small-length keys. The QKD is employed in extremely sensitive applications as a quantum-safe cryptographic mechanism [70]. By contrast, without generating secret keys, transmitting an encrypted message through a direct channel is another method for securing quantum communication [133]. Moreover, with a large degree of distributed long-distance users, traditional QKD is sometimes not adequate. Therefore, an integrated space-to-ground quant communication network is combined with huge QkD-enabled fiber links to provide a stable and secure connection for more than 2600 km [16].

9.5.2.1 Security on Distributed Computing

6G is intended to involve immense heterogeneous technologies and a massive number of users. Moreover, datasets that use through the networks become more complex and giant in size, which demands intelligent processing methods over traditional approaches. Therefore, distributed computing introduces the concepts of parallelizing the computation in a distributed manner by dividing the tasks into many machines in a network to ensure large-scale data processing. However, the modern design of the distributed computing system is still concerned with security issues. The system is compromised by various types of attacks including data tampering, DoS, malicious attacks, etc.

The distributed networking system requires a significant security solution that will be less complex as well as capable of delivering the security updates among distributed nodes/devices with extremely low latency. To ensure intensive security for this kind of distributed architecture, softwarization and cloudization-based security solutions are intended to implement [72]. In a distributed computing environment, a coding strategy is adopted to resist message tampering attacks and malicious activity from collaborative attackers. The coding strategy also provides some advantages such as reducing the impact of the attacks, alleviating the communication cost and frequency range, and providing data confidentiality. Another strategy called group strategy is presented to efficiently detect a set of malicious activities in distributed networks. However, a probabilistic mechanism is proposed to successfully detect malicious nodes in a distributed environment where the attackers are non-communicating/non-colluding [94].

9.5.2.2 Security on Terminals or Consumer End

Advanced portable communication still depends on the physical engagement of symmetric keys in the Subscriber Identity Module (SIM). Undoubtedly, the nanoscale technology of SIM cards has replaced the conventional ones, but they still require to be inserted in physical devices or gadgets, which resists the flexibility of Internet of Everything (IoE) architecture and introduces security vulnerabilities such as unauthorized access and fake node insertion in 5G beyond networks. Therefore, the emerging technology of eSIMs and iSIMs are injected as a part of system-on-chip in upcoming devices. However, the conventional symmetric key is not adequate for the advanced end-user identity (SIM). So, asymmetric public/private keys and post-quantum keys, TLS and elliptic curve cryptography are significantly used to ensure user and device authentication.

9.6 CONCLUSION

Since 6G will potentially offer new services with enabling technologies, challenges will, therefore, be unique in nature. The chapter provides a thorough and comprehensive overview of security and privacy issues in 6G networks. The security and privacy issues related to individual layers of a communication systems such as physical layer, link and network layer, application layer, etc., are analyzed. Potential solutions to those security and privacy challenges are also discussed in detail. Challenges related to new technologies including but not limited to mmWave, NOMA, cell-free massive MIMO, DLT, NFV, quantum communications and computing are discussed in detail. Special consideration is given to processing and service layer techniques such as security issues in zero-touch orchestration and container-based virtualization. Although a broad range of security challenges for emerging technologies are discussed, new applications and technologies will certainly emerge in the coming years that are unthinkable now. Those new technologies will also come with their own unique security and privacy challenges. Therefore, sustained effort and new solutions will be needed to tackle those future security challenges.

BIBLIOGRAPHY

[1] Jassim Happa, Mashhuda Glencross, and Anthony Steed. Cyber Security Threats and Challenges in Collaborative Mixed-Reality *Frontiers in ICT* 6:5, 2019.

[2] Frédéric Bouchard, Robert Fickler, Robert W. Boyd, and Ebrahim Karimi. High-dimensional quantum cloning and applications to quantum hacking. *Science Advances*, 3(2):e1601915, 2017.

[3] Nathalie Baracaldo, Bryant Chen, Heiko Ludwig, and Jaehoon Amir Safavi. Mitigating Poisoning Attacks on Machine Learning Models : A Data Provenance Based Approach. In *Proceedings of the 10th ACM Workshop on Artificial Intelligence and Security*, pages 103–110, 2017.

[4] George Bissias, A. Pinar Ozisik, Brian N. Levine, and Marc Liberatore. Sybil-Resistant Mixing for Bitcoin. In *Proceedings of the 13th Workshop on Privacy in the Electronic Society*, pages 149–158, 2014.

[5] Mohammed Adil Abbas, Hojin Song, and Jun-Pyo Hong. Opportunistic scheduling for average secrecy rate enhancement in fading downlink channel with potential eavesdroppers. *IEEE Transactions on Information Forensics and Security*, 14(4):969–980, 2014.

[6] Mohammed Adil Abbas, Hojin Song, and Jun-Pyo Hong. Opportunistic scheduling for average secrecy rate enhancement in fading downlink channel with potential eavesdroppers. *IEEE Transactions on Information Forensics and Security*, 14(4):969–980, 2019.

[7] Zakaria Abou El Houda, Abdelhakim Senhaji Hafid, and Lyes Khoukhi. Cochain-SC: An intra- and inter-domain DDoS mitigation scheme based on blockchain using SDN and smart contract. *IEEE Access*, 7:98893–98907, 2019.

[8] Maria Apostolaki, Aviv Zohar, and Laurent Vanbever. Hijacking bitcoin: Routing attacks on cryptocurrencies. In *2017 IEEE Symposium on Security and Privacy (SP)*, pages 375–392, 2017.

[9] Youness Arjoune and Saleh Faruque. Smart jamming attacks in 5G new radio: A review. In *2020 10th Annual Computing and Communication Workshop and Conference (CCWC)*, pages 1010–1015, 2020.

[10] Chafika Benzaid and Tarik Taleb. Ai-driven zero touch network and service management in 5G and beyond: Challenges and research directions. *IEEE Network*, 34(2):186–194, 2020.

[11] Timothy X. Brown, Jesse E. James, and Amita Sethi. Jamming and sensing of encrypted wireless ad hoc networks. In *Proceedings of the 7th ACM International Symposium on Mobile Ad Hoc Networking and Computing*, MobiHoc '06, pages 120–130, New York, NY, May 2006. Association for Computing Machinery.

[12] Emilio Calvanese Strinati, Sergio Barbarossa, Jose Luis Gonzalez-Jimenez, Dimitri Ktenas, Nicolas Cassiau, Luc Maret, and Cedric Dehos. 6G: The next frontier: From holographic messaging to artificial intelligence using subterahertz and visible light communication. *IEEE Vehicular Technology Magazine*, 14(3):42–50, 2019.

[13] Jeeva Chelladhurai, Pethuru Raj Chelliah, and Sathish Alampalayam Kumar. Securing docker containers from denial of service (dos) attacks. In *2016 IEEE International Conference on Services Computing (SCC)*, pages 856–859, 2016.

[14] Lingwei Chen, Yanfang Ye, and Thirimachos Bourlai. Adversarial machine learning in malware detection: Arms race between evasion attack and defense. In *2017 European Intelligence and Security Informatics Conference (EISIC)*, pages 99–106, 2017.

[15] Riqing Chen, Chunhui Li, Shihao Yan, Robert Malaney, and Jinhong Yuan. Physical layer security for ultra-reliable and low-latency communications. *IEEE Wireless Communications*, 26(5):6–11, 2019.

[16] Yu-Ao Chen, Qiang Zhang, Teng-Yun Chen, Wen-Qi Cai, Sheng-Kai Liao, Jun Zhang, Kai Chen, Juan Yin, Ji-Gang Ren, Zhu Chen, Sheng-Long Han, Qing Yu, Ken Liang, Fei Zhou, Xiao Yuan, Mei-Sheng Zhao, Tian-Yin Wang, Xiao Jiang, Liang Zhang, Wei-Yue Liu, Yang Li, Qi Shen, Yuan Cao, Chao-Yang Lu, Rong Shu, Jian-Yu Wang, Li Li, Nai-Le Liu, Feihu Xu, Xiang-Bin Wang, Cheng-Zhi Peng, and Jian-Wei Pan. An integrated space-to-ground

quantum communication network over 4,600 kilometres. *Nature*, 589(7841):214–219, 2021.

[17] Hatim Chergui, Adlen Ksentini, Luis Blanco, and Christos Verik-oukis. Toward zero-touch management and orchestration of massive deployment of network slices in 6G. *IEEE Wireless Communications*, 29(1):86–93, 2022.

[18] Gayatri Chittimoju and Usha Devi Yalavarthi. A comprehensive review on millimeter waves applications and antennas. *Journal of Physics: Conference Series*, 1804(1):012205, 2021.

[19] Juan Camilo Correa Chica, Jenny Cuatindioy Imbachi, and Juan Felipe Botero Vega. Security in SDN: A comprehensive survey. *Journal of Network and Computer Applications*, 159:102595, 2020.

[20] Neelam Dayal and Shashank Srivastava. SD-WAN Flood Tracer: Tracking the entry points of DDoS attack flows in WAN. *Computer Networks*, 186:107813, 2021.

[21] Somdip Dey. Securing majority-attack in blockchain using machine learning and algorithmic game theory: A proof of work. In *2018 10th Computer Science and Electronic Engineering (CEEC)*, pages 7–10, 2018.

[22] Jinhua Du, Xianming Gao, Jingchao Wang, Shaohua Liu, Weipeng Cao, Yanbo Song, and Shanqing Jiang. Research on an approach of ARP flooding suppression in multi-controller SDN networks. In *2021 IEEE Intl Conf on Parallel Distributed Processing with Applications, Big Data Cloud Computing, Sustainable Computing Communications, Social Computing Networking (ISPA/BDCloud/SocialCom/SustainCom)*, pages 1159–1166, 2021.

[23] Jun Du, Chunxiao Jiang, Haijun Zhang, Xiaodong Wang, Yong Ren, and Mérouane Debbah. Secure satellite-terrestrial transmission over incumbent terrestrial networks via cooperative beamforming. *IEEE Journal on Selected Areas in Communications*, 36(7):1367–1382, 2018.

[24] Hesham ElSawy and Ekram Hossain. On stochastic geometry modeling of cellular uplink transmission with truncated channel inversion power control. *IEEE Transactions on Wireless Communications*, 13(8):4454–4469, 2014.

[25] Hesham ElSawy and Ekram Hossain. On stochastic geometry modeling of cellular uplink transmission with truncated channel inversion power control. *IEEE Transactions on Wireless Communications*, 13(8):4454–4469, 2014.

[26] Evren, Ferhat Ozgur Catak, and Arild Moldsvor. Adversarial machine learning security problems for 6G: mmwave beam prediction use-case. In *2021 IEEE International Black Sea Conference on Communications and Networking (BlackSeaCom)*, pages 1–6, 2021.

[27] Daquan Feng, Lifeng Lai, Jingjing Luo, Yi Zhong, Canjian Zheng, and Kai Ying. Ultra-reliable and low-latency communications: applications, opportunities and challenges. *Science China Information Sciences*, 64(2):120301, 2021.

[28] Talon Flynn, George Grispos, William Glisson, and William Mahoney. Knock! Knock! Who Is There? Investigating Data Leakage from a Medical Internet of Things Hijacking Attack. 2020. 10.24251/HICSS.2020.791. Accepted: 2020-01-04T08:31:35Z.

[29] Haji M. Furqan, Jehad M. Hamamreh, and Huseyin Arslan. Physical Layer Security for NOMA: Requirements, Merits, Challenges, and Recommendations. *arXiv:1905.05064 [eess]*, July 2020. arXiv preprint arXiv:1905.05064.

[30] Jin Gao, Muhammad R. A. Khandaker, Faisal Tariq, Kai-Kit Wong, and Risala T. Khan. Deep neural network based resource allocation for V2X communications. In *2019 IEEE 90th Vehicular Technology Conference (VTC2019-Fall)*, pages 1–5, 2019.

[31] Weijun Gao, Chong Han, and Zhi Chen. Receiver artificial noise aided terahertz secure communications with eavesdropper in close proximity. In *GLOBECOM 2020 – 2020 IEEE Global Communications Conference*, pages 1–6, 2020.

[32] Xing Gao, Zhongshu Gu, Mehmet Kayaalp, Dimitrios Pendarakis, and Haining Wang. Containerleaks: Emerging security threats of information leakages in container clouds. In *2017 47th Annual IEEE/IFIP International Conference on Dependable Systems and Networks (DSN)*, pages 237–248, 2017.

[33] Zahra Ghodsi, Tianyu Gu, and Siddharth Garg. SafetyNets: Verifiable execution of deep neural networks on an untrusted cloud. Advances in Neural Information Processing Systems, 30:4673–4682, 2017.

[34] Fida Gillani, Ehab Al-Shaer, and Qi Duan. In-design Resilient SDN Control Plane and Elastic Forwarding Against Aggressive DDoS Attacks. In *Proceedings of the 5th ACM Workshop on Moving Target Defense*, MTD '18, pages 80–89, New York, NY, January 2018. Association for Computing Machinery.

[35] Caihong Gong, Xinwei Yue, Zhenyu Zhang, Xiyuan Wang, and Xiaoming Dai. Enhancing physical layer security with artificial noise in large-scale noma networks. *IEEE Transactions on Vehicular Technology*, 70(3):2349–2361, 2021.

[36] Hongzhi Guo, Jingyi Li, Jiajia Liu, Na Tian, and Nei Kato. A survey on space-air-ground-sea integrated network security in 6G. *IEEE Communications Surveys & Tutorials*, 24(1):53–87, 2022.

[37] Zecheng He, Tianwei Zhang, and Ruby B. Lee. VerIDeep: Verifying Integrity of Deep Neural Networks through Sensitive-Sample Fingerprinting. Technical Report arXiv:1808.03277, arXiv, August 2018. arXiv:1808.03277 [cs] type: article.

[38] Ethan Heilman, Alison Kendler, Aviv Zohar, and Sharon Goldberg. Eclipse attacks on bitcoin's peer-to-peer network. Cryptology ePrint Archive, Report 2015/263, 2015. https://ia.cr/2015/263.

[39] Tiep M. Hoang, Hien Quoc Ngo, Trung Q. Duong, Hoang Duong Tuan, and Alan Marshall. Cell-free massive mimo networks: Optimal power control against active eavesdropping. *IEEE Transactions on Communications*, 66(10):4724–4737, 2018.

[40] Sheng Hong, Cunhua Pan, Hong Ren, Kezhi Wang, and Arumugam Nallanathan. Artificial-noise-aided secure mimo wireless communications via intelligent reflecting surface. *IEEE Transactions on Communications*, 68(12):7851–7866, 2020.

[41] Yuanquan Hong, Xiaojun Jing, and Hui Gao. Programmable weight phased-array transmission for secure millimeter-wave wireless communications. *IEEE Journal of Selected Topics in Signal Processing*, 12(2):399–413, 2018.

[42] Chongwen Huang, Sha Hu, George C. Alexandropoulos, Alessio Zappone, Chau Yuen, Rui Zhang, Marco Di Renzo, and Merouane Debbah. Holographic mimo surfaces for 6g wireless networks: Opportunities, challenges, and trends. *IEEE Wireless Communications*, 27(5):118–125, 2020.

[43] Haojun Huang, Wang Miao, Geyong Min, Jialin Tian, and Atif Alamri. Nfv and blockchain enabled 5G for ultra-reliable and low-latency communications in industry: Architecture and performance evaluation. *IEEE Transactions on Industrial Informatics*, 17(8):5595–5604, 2021.

[44] Yi Huang, Peng Xiao, Shengli Zhou, and Zhijie Shi. A half-duplex self-protection jamming approach for improving secrecy of block transmissions in underwater acoustic channels. *IEEE Sensors Journal*, 16(11):4100–4109, 2016.

[45] Bilal Hussain, Qinghe Du, Bo Sun, and Zhiqiang Han. Deep learning-based ddos-attack detection for cyber–physical system over 5G network. *IEEE Transactions on Industrial Informatics*, 17(2):860–870, 2021.

[46] SM Riazul Islam, Jae Moung Kim, and Kyung Sup Kwak. On non-orthogonal multiple access (NOMA) in 5G systems. *The Journal of Korean Institute of Communications and Information Sciences*, 40(12):2549–2558, 2015. Publisher: The Korean Institute of Commucations and Information Sciences.

[47] Furqan Jameel, Shurjeel Wyne, Georges Kaddoum, and Trung Q. Duong. A comprehensive survey on cooperative relaying and jamming strategies for physical layer security. *IEEE Communications Surveys Tutorials*, 21(3):2734–2771, 2019.

[48] Bo Jiang, Ye Liu, and W.K. Chan. Contractfuzzer: Fuzzing smart contracts for vulnerability detection. In *2018 33rd IEEE/ACM International Conference on Automated Software Engineering (ASE)*, pages 259–269, 2018.

[49] Dzevdan Kapetanovic, Gan Zheng, and Fredrik Rusek. Physical layer security for massive MIMO: An overview on passive eavesdropping and active attacks. *IEEE Communications Magazine*, 53(6):21–27, 2015.

[50] Rabia Khan, Pardeep Kumar, Dushantha Nalin K. Jayakody, and Madhusanka Liyanage. A survey on security and privacy of 5G technologies: Potential solutions, recent advancements, and future directions. *IEEE Communications Surveys Tutorials*, 22(1):196–248, 2020.

[51] Zeinab Khorshidpour, Sattar Hashemi, and Ali Hamzeh. Learning a secure classifier against evasion attack. In *2016 IEEE 16th International Conference on Data Mining Workshops (ICDMW)*, pages 295–302, 2016.

[52] Pang Wei Koh and Percy Liang. Understanding black-box predictions via influence functions. In *Proceedings of the 34th International Conference on Machine Learning*, pages 1885–1894. PMLR, July 2017. ISSN: 2640-3498.

[53] Hyun Kwon, Yongchul Kim, Hyunsoo Yoon, and Daeseon Choi. Selective audio adversarial example in evasion attack on speech recognition system. *IEEE Transactions on Information Forensics and Security*, 15:526–538, 2020.

[54] Hongjiang Lei, Zixuan Yang, Ki-Hong Park, Imran Shafique Ansari, Yongcai Guo, Gaofeng Pan, and Mohamed-Slim Alouini. Secrecy outage analysis for cooperative noma systems with relay selection schemes. *IEEE Transactions on Communications*, 67(9):6282–6298, 2019.

[55] Jianguo Li, Neng Ye, Siqi Ma, Xiangyuan Bu, and Jianping An. Multi-user hybrid beamforming design for physical layer secured mmWave LOS communications. *Electronics*, 10(21):2635, 2021.

[56] Xuran Li, Hong-Ning Dai, Mahendra K. Shukla, Dengwang Li, Huaqiang Xu, and Muhammad Imran. Friendly-jamming schemes to secure ultra-reliable and low-latency communications in 5G and beyond communications. *Computer Standards & Interfaces*, 78:103540, 2021.

[57] Yibing Li, Xiaochen Guo, Taige Zhang, and Qian Sun. GPS anti-spoofing algorithm based on improved particle filter. In *2018 USNC-URSI Radio Science Meeting (Joint with AP-S Symposium)*, pages 17–18, 2018.

[58] Zhi Li, Hai Jin, Deqing Zou, and Bin Yuan. Exploring new opportunities to defeat low-rate DDoS attack in container-based cloud environment. *IEEE Transactions on Parallel and Distributed Systems*, 31(3):695–706, 2020.

[59] Christos Liaskos, Shuai Nie, Ageliki Tsioliaridou, Andreas Pitsillides, Sotiris Ioannidis, and Ian Akyildiz. A new wireless communication paradigm through software-controlled metasurfaces. *IEEE Communications Magazine*, 56(9):162–169, 2018.

[60] Xin Lin, Lingguang Lei, Yuewu Wang, Jiwu Jing, Kun Sun, and Quan Zhou. A Measurement Study on Linux Container Security: Attacks and Countermeasures. In *Proceedings of the 34th Annual Computer Security Applications Conference*, ACSAC '18, pages 418–429, New York, NY, USA, December 2018. Association for Computing Machinery.

[61] Yi-Hui Lin, Jian-Jhih Kuo, De-Nian Yang, and Wen-Tsuen Chen. A cost-effective shuffling-based defense against HTTP DDoS attacks with SDN/NFV. In *2017 IEEE International Conference on Communications (ICC)*, pages 1–7, 2017.

[62] Ankur Lohachab, Anu Lohachab, and Ajay Jangra. A comprehensive survey of prominent cryptographic aspects for securing communication in post-quantum IoT networks. *Internet of Things*, 9:100174, 2020.

[63] Martin Andreoni Lopez, Michael Baddeley, Willian T. Lunardi, Anshul Pandey, and Jean-Pierre Giacalone. Towards secure wireless mesh networks for UAV swarm connectivity: Current threats,

research, and opportunities. In *2021 17th International Conference on Distributed Computing in Sensor Systems (DCOSS)*, pages 319–326, 2021.

[64] Valeria Loscrì, César Marchal, Nathalie Mitton, Giancarlo Fortino, and Athanasios V. Vasilakos. Security and privacy in molecular communication and networking: Opportunities and challenges. *IEEE Transactions on NanoBioscience*, 13(3):198–207, 2014.

[65] Jianjun Ma, Rabi Shrestha, Jacob Adelberg, Chia-Yi Yeh, Zahed Hossain, Edward Knightly, Josep Miquel Jornet, and Daniel M. Mittleman. Security and eavesdropping in terahertz wireless links. *Nature*, 563(7729):89–93, 2018.

[66] Ting Ma, Yee Hui Lee, and Maode Ma. Protecting satellite networks from disassociation DoS attacks. In *2010 IEEE International Conference on Communication Systems*, pages 662–666, 2010.

[67] Abubakar Makarfi, Khaled Rabie, Omprakash Kaiwartya, Osamah Badarneh, Galymzhan Nauryzbayev, and Rupak Kharel. Physical layer security in ris-assisted networks in Fisher-Snedecor composite fading. In *2020 12th International Symposium on Communication Systems, Networks and Digital Signal Processing (CSNDSP)*, pages 1–6, 2020.

[68] Tal Mizrahi. Security Requirements of Time Protocols in Packet Switched Networks. Request for Comments RFC 7384, Internet Engineering Task Force, October 2014. Pages: 36.

[69] Nibedita Nandan, Sudhan Majhi, and Hsiao-Chun Wu. Beamforming and power optimization for physical layer security of MIMO-NOMA based CRN over imperfect CSI. *IEEE Transactions on Vehicular Technology*, 70(6):5990–6001, 2021.

[70] Syed Junaid Nawaz, Shree Krishna Sharma, Shurjeel Wyne, Mohammad N. Patwary, and Md. Asaduzzaman. Quantum machine learning for 6G communication networks: State-of-the-art and vision for the future. *IEEE Access*, 7:46317–46350, 2019.

[71] Elina Nayebi, Alexei Ashikhmin, Thomas L. Marzetta, and Hong Yang. Cell-free massive MIMO systems. In *2015 49th Asilomar*

Conference on Signals, Systems and Computers, pages 695–699, 2015.

[72] Van-Linh Nguyen, Po-Ching Lin, Bo-Chao Cheng, Ren-Hung Hwang, and Ying-Dar Lin. Security and privacy for 6G: A survey on prospective technologies and challenges. *IEEE Communications Surveys Tutorials*, 23(4):2384–2428, 2021.

[73] Jianbing Ni, Xiaodong Lin, and Xuemin Sherman Shen. Efficient and secure service-oriented authentication supporting network slicing for 5G-enabled IoT. *IEEE Journal on Selected Areas in Communications*, 36(3):644–657, 2018.

[74] Marek Pawlicki, Michał Choraś, and Rafał Kozik. Defending network intrusion detection systems against adversarial evasion attacks. *Future Generation Computer Systems*, 110:148–154, 2020.

[75] Hao Peng, Zixiong Wang, Shiying Han, and Yang Jiang. Physical layer security for MISO NOMA VLC system under eavesdropper collusion. *IEEE Transactions on Vehicular Technology*, 70(6):6249–6254, 2021.

[76] Vitaly Petrov, Dmitri Moltchanov, Josep Miquel Jornet, and Yevgeni Koucheryavy. Exploiting multipath terahertz communications for physical layer security in beyond 5G networks. In *IEEE INFOCOM 2019 – IEEE Conference on Computer Communications Workshops (INFOCOM WKSHPS)*, pages 865–872, 2019.

[77] Pawani Porambage, Gürkan Gür, Diana Pamela Moya Osorio, Madhusanka Livanage, and Mika Ylianttila. 6G security challenges and potential solutions. In *2021 Joint European Conference on Networks and Communications 6G Summit (EuCNC/6G Summit)*, pages 622–627, 2021.

[78] Pawani Porambage, Gürkan Gür, Diana Pamela Moya Osorio, Madhusanka Liyanage, Andrei Gurtov, and Mika Ylianttila. The roadmap to 6G security and privacy. *IEEE Open Journal of the Communications Society*, 2:1094–1122, 2021.

[79] Pavana Prakash, Jiahao Ding, Hongning Li, Sai Mounika Errapotu, Qingqi Pei, and Miao Pan. Privacy preserving facial

recognition against model inversion attacks. In *GLOBECOM 2020 – 2020 IEEE Global Communications Conference*, pages 1–6, 2020.

[80] Jingping Qiao and Mohamed-Slim Alouini. Secure transmission for intelligent reflecting surface-assisted mmWave and terahertz systems. *IEEE Wireless Communications Letters*, 9(10):1743–1747, 2020.

[81] Jiahua Qiu, Kui Xu, Xiaochen Xia, Zhexian Shen, Wei Xie, Dongmei Zhang, and Meng Wang. Secure transmission scheme based on fingerprint positioning in cell-free massive MIMO systems. *IEEE Transactions on Signal and Information Processing over Networks*, 8:92–105, 2022.

[82] Muhammad Mahboob Ur Rahman, Qammer H. Abbasi, Nishtha Chopra, Khalid Qaraqe, and Akram Alomainy. Physical layer authentication in nano networks at terahertz frequencies for biomedical applications. *IEEE Access*, 5:7808–7815, 2017.

[83] Ali J. Ramadhan. T-s3ra: traffic-aware scheduling for secure slicing and resource allocation in SDN/NFV enabled 5g networks. *International Journal of Engineering Trends and Technology*, 69(7):215–232, 2021.

[84] P Ravinder Rao and Dr. V.Sucharita. A framework to automate cloud based service attacks detection and prevention. *International Journal of Advanced Computer Science and Applications*, 10(2):241–250, 2019.

[85] François Reynaud, François-Xavier Aguessy, Olivier Bettan, Mathieu Bouet, and Vania Conan. Attacks against network functions virtualization and software-defined networking: State-of-the-art. In *2016 IEEE NetSoft Conference and Workshops (NetSoft)*, pages 471–476, 2016.

[86] François Reynaud, François-Xavier Aguessy, Olivier Bettan, Mathieu Bouet, and Vania Conan. Attacks against network functions virtualization and software-defined networking: State-of-the-art. In *2016 IEEE NetSoft Conference and Workshops (NetSoft)*, pages 471–476, 2016.

[87] Walid Saad, Mehdi Bennis, and Mingzhe Chen. A vision of 6G wireless systems: Applications, trends, technologies, and open research problems. *IEEE Network*, 34(3):134–142, 2020.

[88] Nithin V. Sabu and Abhishek K. Gupta. Analysis of diffusion based molecular communication with multiple transmitters having individual random information bits. *IEEE Transactions on Molecular, Biological and Multi-Scale Communications*, 5(3):176–188, 2019.

[89] Ignacio Sanchez-Navarro, Pablo Salva-Garcia, Qi Wang, and Jose M. Alcaraz Calero. New immersive interface for zero-touch management in 5G networks. In *2020 IEEE 3rd 5G World Forum (5GWF)*, pages 145–150, 2020.

[90] Mahdi Shakiba-Herfeh, Arsenia Chorti, and H. Vincent Poor. Physical Layer Security: Authentication, Integrity, and Confidentiality. In Khoa N. Le, editor, *Physical Layer Security*, pages 129–150. Springer International Publishing, Cham, 2021.

[91] Yi Shi, Yalin E. Sagduyu, Tugba Erpek, and M. Cenk Gursoy. How to Attack and Defend 5G Radio Access Network Slicing with Reinforcement Learning. *arXiv:2101.05768 [cs]*, January 2021. arXiv: 2101.05768.

[92] Zhaogang Shu, Jiafu Wan, Di Li, Jiaxiang Lin, Athanasios V. Vasilakos, and Muhammad Imran. Security in software-defined networking: Threats and countermeasures. *Mobile Networks and Applications*, 21(5):764–776, 2016.

[93] Rohit Singh and Douglas Sicker. THz Communications – a Boon and/or Bane for Security, Privacy, and National Security. SSRN Scholarly Paper ID 3750493, Social Science Research Network, Rochester, NY, December 2020.

[94] Atnav Solanki, Martina Cardone, and Soheil Mohajer. Noncolluding attacks identification in distributed computing. In *2019 IEEE Information Theory Workshop (ITW)*, pages 1–5, 2019.

[95] Jiho Song, Byungju Lee, Juho Park, Moon-Sik Lee, and Jong-Ho Lee. Beamformer design for physical layer security in dual-polarized millimeter wave channels. *IEEE Transactions on Vehicular Technology*, 69(10):12306–12311, 2020.

[96] Yizhuo Song, Muhammad R. A. Khandaker, Faisal Tariq, Kai-Kit Wong, and Apriana Toding. Truly intelligent reflecting surface-aided secure communication using deep learning. In *2021 IEEE 93rd Vehicular Technology Conference (VTC2021-Spring)*, pages 1–6, 2021.

[97] Jacob Steinhardt, Pang Wei W Koh, and Percy S Liang. Certified Defenses for Data Poisoning Attacks. In *Advances in Neural Information Processing Systems*, volume 30. Curran Associates, Inc., 2017.

[98] Sari Sultan, Imtiaz Ahmad, and Tassos Dimitriou. Container security: Issues, challenges, and the road ahead. *IEEE Access*, 7:52976–52996, 2019.

[99] Tianyu Sun and Wensheng Yu. A formal verification framework for security issues of blockchain smart contracts. *Electronics*, 9(2):255, 2020.

[100] Jie Tang, Long Jiao, Kai Zeng, Hong Wen, and Kai-Yu Qin. Physical layer secure MIMO communications against eavesdroppers with arbitrary number of antennas. *IEEE Transactions on Information Forensics and Security*, 16:466–481, 2021.

[101] Zhiqing Tang, Tianwei Hou, Yuanwei Liu, Jiankang Zhang, and Lajos Hanzo. Physical Layer Security of Intelligent Reflective Surface Aided NOMA Networks. *arXiv:2011.03417 [eess]*, November 2020. arXiv: 2011.03417.

[102] Faisal Tariq, Muhammad R. A. Khandaker, Kai-Kit Wong, Muhammad A. Imran, Mehdi Bennis, and Merouane Debbah. A speculative study on 6G. *IEEE Wireless Communications*, 27(4):118–125, 2020.

[103] Pinyaphat Tasatanattakool and Chian Techapanupreeda. Blockchain: Challenges and applications. In *2018 International Conference on Information Networking (ICOIN)*, pages 473–475, 2018.

[104] Anurag Thantharate, Rahul Paropkari, Vijay Walunj, Cory Beard, and Poonam Kankariya. Secure5G: A deep learning framework towards a secure network slicing in 5G and beyond. In *2020 10th*

Annual Computing and Communication Workshop and Conference (CCWC), pages 0852–0857, 2020.

[105] Santosh Timilsina, Dhanushka Kudathanthirige, and Gayan Amarasuriya. Physical layer security in cell- free massive MIMO. In *2018 IEEE Global Communications Conference (GLOBECOM)*, 1–7, 2018.

[106] Tom Titcombe, Adam J. Hall, Pavlos Papadopoulos, and Daniele Romanini. Practical Defences Against Model Inversion Attacks for Split Neural Networks. Technical Report arXiv:2104.05743, arXiv, April 2021. arXiv:2104.05743 [cs] type: article.

[107] Zhe Tu, Huachun Zhou, Kun Li, Man Li, and Aleteng Tian. An energy-efficient topology design and DDoS attacks mitigation for green software-defined satellite network. *IEEE Access*, 8:211434–211450, 2020.

[108] B. Upadhyaya, S. Sun, and B. Sikdar. Machine learning-based jamming detection in wireless IoT networks. In *2019 IEEE VTS Asia Pacific Wireless Communications Symposium (APWCS)*, pages 1–5, 2019.

[109] Nachiappan Valliappan, Angel Lozano, and Robert W. Heath. Antenna subset modulation for secure millimeter-wave wireless communication. *IEEE Transactions on Communications*, 61(8):3231–3245, 2013.

[110] Chaofeng Wang and Zhaohui Wang. Signal alignment for secure underwater coordinated multipoint transmissions. *IEEE Transactions on Signal Processing*, 64(23):6360–6374, 2016.

[111] Minghao Wang, Tianqing Zhu, Tao Zhang, Jun Zhang, Shui Yu, and Wanlei Zhou. Security and privacy in 6G networks: New areas and new challenges. *Digital Communications and Networks*, 6(3):281–291, 2020.

[112] Ning Wang, Weiwei Li, Amir Alipour-Fanid, Long Jiao, Monireh Dabaghchian, and Kai Zeng. Pilot contamination attack detection for 5G mmwave grant-free iot networks. *IEEE Transactions on Information Forensics and Security*, 16:658–670, 2021.

[113] Rong Wang, Yu Mei, Xiangzhu Meng, and Jianjun Ma. Secrecy performance of terahertz wireless links in rain and snow. *Nano Communication Networks*, 28:100350, 2021.

[114] Yingying Wang and Guoqiang Li. Detect triangle attack on blockchain by trace analysis. In *2019 IEEE 19th International Conference on Software Quality, Reliability and Security Companion (QRS-C)*, pages 316–321, 2019.

[115] Yiming Wei, Mugen Peng, and Yaqiong Liu. Intent-based networks for 6G: Insights and challenges. *Digital Communications and Networks*, 6(3):270–280, 2020.

[116] Maximilian Wohrer and Uwe Zdun. Smart contracts: Security patterns in the ethereum ecosystem and solidity. In *2018 International Workshop on Blockchain Oriented Software Engineering (IWBOSE)*, pages 2–8, 2018.

[117] Michael Wood. How SASE is defining the future of network security. *Network Security*, 2020(12):6–8, 2020.

[118] Qingqing Wu and Rui Zhang. Intelligent reflecting surface enhanced wireless network via joint active and passive beamforming. *IEEE Transactions on Wireless Communications*, 18(11):5394–5409, 2019.

[119] Shuangke Wu, Yanjiao Chen, Minghui Li, Xiangyang Luo, Zhe Liu, and Lan Liu. Survive and thrive: A stochastic game for DDoS attacks in bitcoin mining pools. *IEEE/ACM Transactions on Networking*, 28(2):874–887, 2020.

[120] Huang Xiao, Battista Biggio, Gavin Brown, Giorgio Fumera, Claudia Eckert, and Fabio Roli. Is feature selection secure against training data poisoning? In *Proceedings of the 32nd International Conference on Machine Learning*, pages 1689–1698. PMLR, June 2015. ISSN: 1938-7228.

[121] Liang Xiao, Larry J. Greenstein, Narayan B. Mandayam, and Wade Trappe. Channel-based detection of sybil attacks in wireless networks. *IEEE Transactions on Information Forensics and Security*, 4(3):492–503, 2009.

[122] Liang Xiao, Qiangda Li, Tianhua Chen, En Cheng, and Huaiyu Dai. Jamming games in underwater sensor networks with reinforcement learning. In *2015 IEEE Global Communications Conference (GLOBECOM)*, pages 1–6, 2015.

[123] Guowen Xu, Hongwei Li, Sen Liu, Kan Yang, and Xiaodong Lin. Verifynet: Secure and verifiable federated learning. *IEEE Transactions on Information Forensics and Security*, 15:911–926, 2020.

[124] Guowen Xu, Hongwei Li, Hao Ren, Kan Yang, and Robert H. Deng. Data security issues in deep learning: Attacks, countermeasures, and opportunities. *IEEE Communications Magazine*, 57(11):116–122, 2019.

[125] Shihao Yan and Robert Malaney. Location-based beamforming for enhancing secrecy in Rician wiretap channels. *IEEE Transactions on Wireless Communications*, 15(4):2780–2791, 2016.

[126] Jie Yang, Xinsheng Ji, Feihu Wang, Kaizhi Huang, and Lin Guo. A novel pilot spoofing scheme via intelligent reflecting surface based on statistical CSI. *IEEE Transactions on Vehicular Technology*, 70(12):12847–12857, 2021.

[127] Xinle Yang, Yang Chen, and Xiaohu Chen. Effective scheme against 51history weighted information. In *2019 IEEE International Conference on Blockchain (Blockchain)*, pages 261–265, 2019.

[128] Mika Ylianttila, Raimo Kantola, Andrei Gurtov, Lozenzo Mucchi, Ian Oppermann, Zheng Yan, Tri Hong Nguyen, Fei Liu, Tharaka Hewa, Madhusanka Liyanage, Ahmad Ijaz, Juha Partala, Robert Abbas, Artur Hecker, Sara Jayousi, Alessio Martinelli, Stefano Caputo, Jonathan Bechtold, Ivan Morales, Andrei Stoica, Giuseppe Abreu, Shahriar Shahabuddin, Erdal Panayirci, Harald Haas, Tanesh Kumar, Basak Ozan Ozparlak, and Juha Räining. 6G White paper: Research challenges for Trust, Security and Privacy. *arXiv:2004.11665 [cs]*, April 2020. arXiv: 2004.11665.

[129] Chao Yu, Hak-Lim Ko, Xin Peng, Wenwu Xie, and Peng Zhu. Jammer-aided secure communications for cooperative NOMA systems. *IEEE Communications Letters*, 23(11):1935–1939, 2019.

[130] Shakila Zaman, Khaled Alhazmi, Mohammed A. Aseeri, Muhammad Raisuddin Ahmed, Risala Tasin Khan, M. Shamim Kaiser, and Mufti Mahmud. Security threats and artificial intelligence based countermeasures for internet of things networks: A comprehensive survey. *IEEE Access*, 9:94668–94690, 2021.

[131] Shakila Zaman, Muhammad R. A. Khandaker, Risala T. Khan, Faisal Tariq, and Kai-Kit Wong. Thinking out of the blocks: Holochain for distributed security in iot healthcare. *IEEE Access*, 10:37064–37081, 2022.

[132] Wei Zhang, Jian Chen, Yonghong Kuo, and Yuchen Zhou. Transmit beamforming for layered physical layer security. *IEEE Transactions on Vehicular Technology*, 68(10):9747–9760, 2019.

[133] Wei Zhang, Dong-Sheng Ding, Yu-Bo Sheng, Lan Zhou, Bao-Sen Shi, and Guang-Can Guo. Quantum secure direct communication with quantum memory. *Physical Review Letters*, 118(22):220501, 2017.

[134] Xianyu Zhang, Daoxing Guo, Kang An, Zhiguo Ding, and Bangning Zhang. Secrecy analysis and active pilot spoofing attack detection for multigroup multicasting cell-free massive mimo systems. *IEEE Access*, 7:57332–57340, 2019.

[135] Xianyu Zhang, Tao Liang, Kang An, Gan Zheng, and Symeon Chatzinotas. Secure transmission in cell-free massive mimo with RF impairments and low-resolution ADCs/DACs. *IEEE Transactions on Vehicular Technology*, 70(9):8937–8949, 2021.

[136] Zhe Zhang, Chensi Zhang, Chengjun Jiang, Fan Jia, Jianhua Ge, and Fengkui Gong. Improving physical layer security for reconfigurable intelligent surface aided NOMA 6G networks. *IEEE Transactions on Vehicular Technology*, 70(5):4451–4463, 2021.

[137] Zhe Zhang, Chensi Zhang, Chengjun Jiang, Fan Jia, Jianhua Ge, and Fengkui Gong. Improving physical layer security for reconfigurable intelligent surface aided NOMA 6G networks. *IEEE Transactions on Vehicular Technology*, 70(5):4451–4463, 2021.

[138] Zhengquan Zhang, Yue Xiao, Zheng Ma, Ming Xiao, Zhiguo Ding, Xianfu Lei, George K. Karagiannidis, and Pingzhi Fan. 6G wireless

networks: Vision, requirements, architecture, and key technologies. *IEEE Vehicular Technology Magazine*, 14(3):28–41, 2019.

[139] Florian Tramèr, Fan Zhang, Ari Juels, Michael K. Reiter, and Thomas Ristenpart. Stealing Machine Learning Models via Prediction APIs. In *USENIX Security Symposium*, vol. 16, pages 601–618, 2016.

[140] P. Vinod, Akka Zemmari, and Mauro Conti. A machine learning based approach to detect malicious android apps using discriminant system calls. *Future Generation Computer Systems* 94(2019):333–350.

[141] Weiwei Hu and Ying Tan. Black-Box Attacks against RNN based Malware Detection Algorithms. AAAI Workshops, 2017.

CHAPTER 10

Hybrid Massive-MIMO and Its Practical Beamforming Implementation

Kai Xu

School of Engineering and Physical Sciences, Institute of Sensors Signals and Systems, Heriot-Watt University, Edinburgh, UK
School of Engineering, Institute of Digital Communications, The University of Edinburgh, Edinburgh, UK

Jiayu Hou

School of Engineering and Physical Sciences, Institute of Sensors Signals and Systems, Heriot-Watt University, Edinburgh, UK

Yuan Ding

School of Engineering and Physical Sciences, Institute of Sensors Signals and Systems, Heriot-Watt University, Edinburgh, UK

CONTENTS

10.1   Introduction 260
10.2   Three Hybrid m-MIMO Transmitter Variants 262
     10.2.1   System Model of Hybrid m-MIMO 262
10.3   Fully Connected m-MIMO 262
     10.3.1   Hybrid Beamformer Design Algorithm 263
     10.3.2   Implementations 264
10.4   Fixed Subarray m-MIMO 266
     10.4.1   Fixed Subarray Beamformer Design Algorithm 267
     10.4.2   Implementations 267

DOI: 10.1201/9781003282211-10

10.5   Dynamic Subarray m-MIMO 269
          10.5.1 Fully Dynamic Subarray m-MIMO 270
                    10.5.1.1 Partially Dynamic Subarray m-MIMO ... 271
10.6   Reconfigurable Power Divider (RPD)-Based Dynamic
          Subarray m-MIMO 271
          10.6.1 Hardware Implementation of N-Way (4-Way as an
                    Example) RPDs 272
          10.6.2 System Simulation Results 275
10.7   Conclusion and Future Works 278
Bibliography 278

10.1 INTRODUCTION

The proliferation of data and wireless devices promotes the demand on higher capacity of wireless communication networks. Multiple-input multiple-output (MIMO) becomes a prevailing technology to significantly strengthen the channel capacity by employing multiple antennas at both transmitter (Tx) and receiver (Rx) to explore multipath radio propagation channels. This high dimension of Tx and Rx propagation pairs creates an extra multiplexing domain, namely spatial multiplexing, in addition to those in the conventional time and frequency domains. MIMO systems can be used to achieve spatial diversity gain and/or spatial multiplexing gain through different precoding techniques in various network architecture and wireless channel conditions. As for the 6G cellular system, higher millimeter wave (mmWave) and even Terahertz (THz) communication are reckoned to be a key technology, which provide abundant unused vacant spectrum. The expected huge path losses will be compensated by more antenna elements which can provide sufficiently high beamforming and multiplexing gain [1, 2]. Therefore, the massive-MIMO (m-MIMO) [3], where hundreds of antenna elements are equipped at Tx and/or Rx ends, is becoming a reality to increase Rx signal power and extend the communication distance.

A major challenge of traditional m-MIMO transmitter architecture, as depicted in Figure 10.1(a), is its high cost and complexity since every antenna is equipped with a full radio frequency (RF) chain consisting of power-hungry power amplifiers, mixers, digital-to-analogue converters (DACs), etc. Furthermore, channel estimation, equalization, pre-coding, and baseband processing are all heavily burdened because of channel matrices of large dimensions. For practical implementations, RF amplifiers

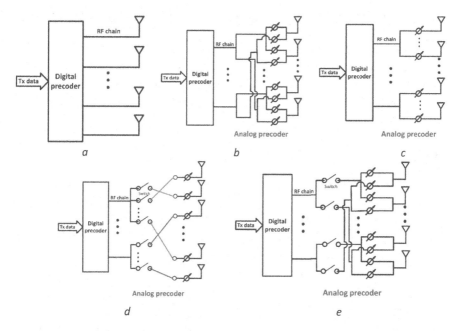

Figure 10.1 Diagrams of different m-MIMO transmitter architectures. (a) Fully digital m-MIMO, (b) fully connected hybrid m-MIMO, (c) fixed subarray m-MIMO, (d) fully dynamic subarray m-MIMO, and (e) partially dynamic subarray m-MIMO.

may not be readily integrated and implemented to be equipped before each antenna element, let alone the complexity involved in biasing and heat dissipation [4]. In order to alleviate these issues, various simplified m-MIMO architectures have been proposed and extensively studied [5–7]. They can be categorized into:

- Fully connected (FC) m-MIMO, studied in Section 10.3,

- Fixed subarray m-MIMO, studied in Section 10.4, and

- Dynamic subarray m-MIMO, studied in Sections 10.5 and 10.6.

These three architectures all follow the same concept, i.e., employing analog RF precoding networks so as to reduce the number of RF chains and thus the dimension of equivalent baseband channels. Hence, they are also labeled as analog-digital hybrid m-MIMO.

10.2   THREE HYBRID m-MIMO TRANSMITTER VARIANTS

In this section, the system model of hybrid m-MIMO is first presented, which followed with a discussion on the three hybrid m-MIMO architectures and their associated design algorithms for seeking optimized hybrid precoders. Recently, reported implementations are particularly highlighted.

10.2.1   System Model of Hybrid m-MIMO

Consider a hybrid m-MIMO system in which the transmitter employs N_t Tx antennas, served by N_{RF} RF chains, to communicate N_s data streams s to a single receiver equipped with N_r Rx antennas. To guarantee the multi-stream propagation, $N_s \leq N_{RF} \leq \min(N_t, N_r\}$ has to be satisfied. To pre-process data streams, the transmitter first applies an $N_{RF} \times N_s$ baseband precoder P_B, followed by an analog precoder P_A with a dimension of $N_t \times N_{RF}$. As for the architecture of the receiver, some of the works employ the hybrid combining structure [5, 8] and the others use a fully digital or analog combiner [9]. To facilitate the discussion and place emphasis on the works on a hybrid precoder at the transmitter end, the combiner is generically represented as W, which can be a hybrid combiner or a fully digital/analog combiner.

Under a multi-path massive MIMO channel, and assuming it can be represented as a matrix H, the received signal y after combining can be mathematically expressed as

$$y = \sqrt{\rho} W H P_A P_B s + W n, \tag{10.1}$$

where ρ is the average received power and n refers to the channel noise vector.

10.3   FULLY CONNECTED m-MIMO

In [5], the FC hybrid m-MIMO structure, reproduced in Figure 10.1 (b), was introduced. Here, each RF chain, the number of which is less than the number of Tx antennas, is routed to all antennas via RF beamforming networks consisting of power dividers (PDs), power combiners, and phase shifters (PSs). This is commonly categorized as FC m-MIMO, which has been extensively investigated.

10.3.1 Hybrid Beamformer Design Algorithm

When designing a conventional fully digital m-MIMO transmitter, the optimal precoder $\boldsymbol{P}^{\text{opt}}$ and combiner $\boldsymbol{W}^{\text{opt}}$ can be calculated using Singular Vector Decomposition (SVD) method, as shown in (2).

$$\boldsymbol{H} = \boldsymbol{U}\boldsymbol{\Sigma}\boldsymbol{V}^H$$

$$\boldsymbol{P}^{\text{opt}} = \boldsymbol{V}$$

$$\boldsymbol{W}^{\text{opt}} = \boldsymbol{U}^H \tag{10.2}$$

Here, $(\bullet)^H$ is the conjugate transpose operation and the matrix Σ contains eigenvalues, determining gains of each virtual parallel channels, in its trace.

In the FC hybrid m-MIMO, when analog precoders are implemented using analog PSs, each element of the analog precoder matrix $\boldsymbol{P}_A$ is constrained to have an equal norm [5] and be non-zero. And for the scenario where the PSs can only have discrete values, e.g., digital PSs of finite bits, the phases of $\boldsymbol{P}_A$ elements are constrained and belong to a certain set $\mathcal{A}$ according to the resolution of PSs. By taking analog precoder constraints into consideration, some useful methods to design an FC m-MIMO in reported literatures are summarized as below:

- **Maximizing the mutual information of the m-MIMO channel**: In [5], the m-MIMO system is studied for mmWave propagations. By leveraging the sparse property of mmWave channels, the design of the FC hybrid precoder and combiner is formulated as a sparse reconstruction problem to obtain maximum mutual information of the channel. Based on the optimal precoder $\boldsymbol{P}^{\text{opt}}$, orthogonal matching pursuit (OMP) algorithm is exploited to construct optimal digital and FC analog precoders by iterating N_{RF} times. Whereby the proposed hybrid precoders are able to achieve spectral efficiency performance that is close to the theoretical limit.

- **Maximizing the spectral efficiency**: Beamsteering codebook is used in this method. It is a set of pre-defined analog beamforming vectors spatially matched for a selected single path channel. As a result, each of them has the same form of the array response vector and can be parameterized by a simple spatial direction [9]. By properly selecting the analog precoder and combiner from the RF codebook, the FC m-MIMO system can obtain an acceptable performance in terms of spectral efficiency. A further extension to the fixed codebook-based method, the proposed iteration methods in [6, 10] are able to achieve higher spectral efficiency. The codebook-based method is limited in the scenarios where there is a dominant beam in the wireless channel. The case of multiple beams will lead to low efficiency due to the low beamforming gain.

- **Minimizing the Euclidean distance between FC precoder $P_A P_B$ and optimal precoder P^{opt}**: When the channel state information (CSI) is assumed to be perfectly known, the optimal precoder P^{opt} in a corresponding fully digital m-MIMO system can be calculated using (2). FC precoder $P_A P_B$ can thus be optimized by finding a solution that minimizes the problem formulation in (3):

$$\min_{P_A, P_B} \left\| P^{\text{opt}} - P_A P_B \right\|_F . \tag{10.3}$$

In [11], the authors proposed a manifold optimization-based effective alternating minimization (MO-AltMin) algorithm and a low-complexity PE-AltMin algorithm to solve the objective function and calculate the FC hybrid precoder. The essence of the MO-AltMin algorithm is that by inputting the optimal precoder P^{opt} and constructing the analog precoder with random phases, the analog precoder P_A is iteratively optimized using manifold optimization-based Conjugate Gradient Algorithm [11] and keep digital precoder P_B updated by multiplying the complex conjugate of P_A with P^{opt}. While the PE-AltMin algorithm employs two closed-form solutions for the analog and digital precoders with low complexity. The resulting FC precoders can approach the performance of the fully digital precoder for $N_{RF} > N_s$. Besides, a coordinate descent method (CMD)-based algorithm is proposed in [12], and has better performance than the AltMin algorithm and shows good tolerance to low-resolution PSs.

10.3.2 Implementations

The FC analog precoder is commonly implemented with the analog high-precision PSs, PDs, and power combiners. And it is proven that the performance of this FC m-MIMO system is capable of approaching the unconstrained fully digital m-MIMO system [5, 11]. On the other hand, a vast array of high-resolution PSs to realize the analog beamformer will be prohibitively costly and they demand highly complicated control and biasing networks. Consequently, some works are investigating FC m-MIMO transmitters which employ more efficient low-resolution B-bit PSs, for example $B = 1$ in [8].

A work attempting to replace PSs of infinite resolution with the ones of finite and low resolution was presented in [9]. The authors found out that the codebook-based FC precoder only shows a slight degradation in system performance when using 4-bit or 5-bit PSs. While the gap between the optimal fully digital precoder is quite large. Subsequently, the author Jung-Chieh presented a study on FC m-MIMO with low-resolution PSs (for $B = 2, 3, 5$) and proposed a CMD-based algorithm to design the hybrid beamformer in [12]. It was corroborated that the FC precoder implemented by low-resolution PSs is able to get a similar performance as the FC precoder based on infinite-resolution PSs. And the gap between the optimal fully digital precoder becomes negligible. Furthermore, authors in [10] investigated the design of FC beamformers with low-resolution ($B = 1, 2$) PSs and proposed an efficient iterative algorithm. In the single-user m-MIMO system, FC beamformers with binary PSs ($B = 1$) deliver the same spectral efficiency as a fully digital one and obtain excellent performance in multi-user m-MIMO when $B = 2$.

Some other works employ lens-based analog beamforming networks in order to remove a large number of power splitting, power combining, and phase shifting devices. In [13, 14], a Fourier Rotman lens, see Figure 10.2(a), is used to realize the beam codebooks. Its planar

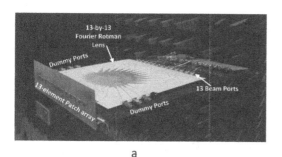
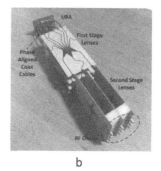

a   b

Figure 10.2 (a) Photograph of the fabricated 13-by-13 Fourier Rotman lens [13]. (b) Two-stage stacked Rotman lens and uniform rectangular array (URA) based beamformer for azimuth and elevation zone coverage [14].

structure and its scalability to any N_t and N_{RF} greatly simplify the analog beamform network designs. For 2-dimensional beam codebook, the cascaded Rotman lenses were proposed and implemented, see Figure 10.2(b).

10.4 FIXED SUBARRAY m-MIMO

The beamforming networks that enable FC m-MIMO, however, are still extremely complicated and costly when hardware implementation is considered, even with a reduced resolution of PSs. To further simplify required RF beamforming hardware, another hybrid structure in Figure 10.1(c) was proposed [6, 15], named the fixed subarray m-MIMO, in which each RF chain is connected to a fixed subset of Tx antenna elements, referred to as the subarray. For simplicity of exposition, N_t is assumed as a multiple of N_{RF} and each RF chain serves the same dimension of subarray $N_{\text{sub}} = N_t/N_{RF}$.

An example of an adjacent-positioned subarray is given here. Let the antenna indices be $\{1, \ldots, N_t\}$ and S_r denotes the partitioned subset of antenna indices connected to the rth RF chain such as [7]

$$S_1 = \{1, \ldots, N_{\text{sub}}\},$$

$$S_2 = \{N_{\text{sub}} + 1, \ldots, 2N_{\text{sub}}\},$$

$$\vdots$$

$$S_{N_{\text{RF}}} = \{(N_{\text{RF}} - 1)N_{\text{sub}} + 1, \ldots, N_{\text{RF}}N_{\text{sub}}\}. \tag{10.4}$$

With this architecture, the analog precoder $\boldsymbol{P}_A$ has a form of block diagonal matrix as

$$\boldsymbol{P}_A = \begin{pmatrix} \boldsymbol{p}_{A,S_1} & \cdots & 0 \\ \vdots & \ddots & \vdots \\ 0 & \cdots & \boldsymbol{p}_{A,S_{N_{RF}}} \end{pmatrix}, \tag{10.5}$$

where $\boldsymbol{p}_{A,Sr}$ is an $N_{\text{sub}} \times 1$ analog precoding vector associated with the rth RF chain. This is a distinct property compared to the FC case whose analog precoding matrix takes the form $\boldsymbol{P}_A = [\boldsymbol{P}_{A,1}, \boldsymbol{P}_{A,2}, \ldots, \boldsymbol{P}_{A,NRF}]$, with $\boldsymbol{P}_{A,r}$ an $N_t \times 1$ vector for the rth RF chain.

10.4.1 Fixed Subarray Beamformer Design Algorithm

Fixed subarray m-MIMO aims to maximize energy efficiency with reduced hardware complexity while suffering lower spectral efficiency. The fixed subarray structure is perdominantly applied to the transmitter (e.g. base stations), and the receiver (e.g. mobile devices) is assumed to employ the fully digital structure. To optimize the design, Gao et al. [15] propose an efficient successive interference cancellation (SIC)-based algorithm to find a hybrid precoder by maximizing the sum-rate of the system. The optimization is iteratively conducted for each subarray with the help of an auxiliary matrix T, which is optimized to maximize the sum-rate of each subarray and update the resulting auxiliary matrix T for the next subarray. The auxiliary matrix T is defined as

$$T = I_{N_r} + \frac{\rho}{N_r \sigma^2} H P_A P_B (P_A P_B)^H H^H \qquad (10.6)$$

where σ^2 is the average noise power. It is proved that the system sum-rate maximization problem can be decomposed into N_{sub} sum-rate optimization problems of each subarray system. This method avoids the need for the complex SVD and matrix inversion operation, raising the prospect of high energy efficiency in processing. Nevertheless, the SIC hybrid precoder performs poorly with regard to spectral efficiency Yu et al. [11] make a comparison between their proposed Altmin algorithm and the SIC-based algorithm. Under the same scenario, Altmin algorithm outperforms the SIC algorithm for about 4 bits/s/Hz in spectral efficiency simulation.

10.4.2 Implementations

The aforementioned works on fixed subarray m-MIMO only present a prototype of the system where analog PSs are deployed to implement the analog precoding network and emphasise more on the hybrid precoder design algorithm.

The work in [16] demonstrated that the fabrication complexity of the beam-steerable-phased 80-element array systems can be drastically reduced by using fixed subarray techniques. Using a multitude of innovative approaches, a simple two-layer realization is obtained, as depicted in Figure 10.3(a),(b), where the top layer is the antenna array and the bottom is the phase shifting and feeding network layer. Grouping the elements of a phased array into a number of partially overlapped subarrays and using a single PS for each subarray, generally results in a

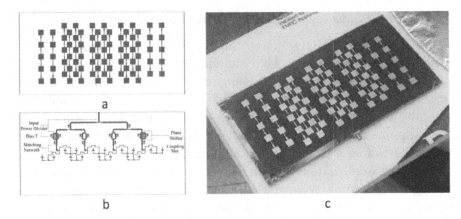

Figure 10.3 Layout of the two-layer phased array: (a) antenna layer and (b) feed layer. And (c) X band prototype of the 80-element array [16]. (© 2003 IEEE. Reprinted with permission from [16].)

considerable reduction in array size and manufacturing costs, as depicted in Figure 10.3(c). This design uses a combination of series and parallel feeding schemes to achieve the desired array coefficients in three beam-steering with a scanning width of ±10°. This can be used in applications where a narrow scanning range is required.

Another work in [17] presents an implementation of a fixed subar-ray analog beamformer using simple phase over-samplers (POSs) and a switch (SW) network, as illustrated in Figure 10.4. In this architecture, each RF chain is associated with one POS, and all the antennas are con-nected with these POSs through a switch network. POS can generate several signals, simultaneously, with different phases, depending on the resolution. For example, a 2-bit POS can generate signals with phase shifting $(0, \pi/2, \pi, 3\pi/2)$. By controlling the switches, this structure can provide the function the same as the traditional PS structure in an ideal situation, but with low hardware complexity. On the contrary, there ex-ist phase nonlinearity issues, potentially more serious to multi-carrier signals. At the output port of the POS, a variable amplifier is com-monly employed to compensate for signal strength variation for different phases. Finally, the simulation results in [17] show that the performance gap between a traditional structure with high-resolution PSs is small by exploring a low-resolution POS, and the gap tends to be ignorable when the PS resolution increases to 8 bits.

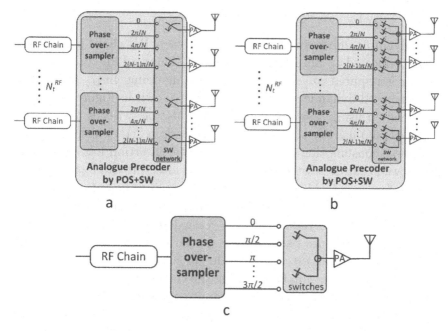

Figure 10.4 (a) Partially connected structure with N_t switches, (b) partially connected structure with more than N_t switches, and (c) an example of a dynamic connected structure with a quaternary POS [17].

10.5 DYNAMIC SUBARRAY m-MIMO

In order to recover some of the performance (i.e., spectral efficiency) loss suffered in the fixed subarray m-MIMO, the dynamic subarray structure is study which adapts the subarray configuration to better 'match' wireless channel environments [7]. It aims to strike a better trade-off between energy efficiency and spectral efficiency. The dynamic subarray can be constructed with various degrees of reconfigurability of connecting RF chains and Tx antenna elements/subarrays. For example, in Figure 10.1(d), a fully dynamic subarray [7, 18, 19] permits each RF chain to be adaptively linked to an arbitrary subset of the array antennas. On the other hand, a partially dynamic subarray [20] allows RF chains to dynamically connect to any subarrays where the antenna partition is fixed in each subarray, as illustrated in Figure 10.1(e). Inspired by these two structures, we propose a novel practical dynamic subarray m-MIMO based on RPDs, which is elaborated later in this section.

10.5.1 Fully Dynamic Subarray m-MIMO

Fully dynamic subarray m-MIMO allows each RF chain to be adaptively connected to an arbitrary subset of the array antennas where a mapper is required to achieve the dynamic mapping. In [7], a dynamic-mapper-based dynamic subarray m-MIMO, shown in Figure 10.1(d), is implemented using a switch network. The use of switches at mmWave and m-MIMO has been motivated by the work in [21], which have much less power consumption at the cost of greater insertion loss. Authors in [22] showed that the switch-based hybrid architecture can be a promising solution for large MIMO systems. When designing the optimal dynamic subarray precoder, the authors find that the performance of the hybrid precoding in a fully dynamic subarray case depends on how to configure the subarray structure based on the spatial channel covariance matrix R. Therefore, the largest singular value of covariance matrix R is optimized using an algorithm by changing the configuration of the subarray. On the other hand, a low-complexity approximation of the largest singular value of R is used to reduce the computational complexity which is normalized Minkowski L1-norm. This chapter first developed a criterion for constructing these subarrays, and it is used to design an antenna partitioning algorithm. The simulation results showed that the achievable spectral efficiency in dynamic subarrays outperforms that of fixed subarray architecture [7]. One limitation of this covariance matrix-based method is its high computation complexity, which requires a lot of iterations in order to find the optimal dynamic subarray beamformer.

A further development in [19] presents a simple criterion and a corresponding low-complexity algorithm to design the optimal partition of antennas using statistical CSI. It derives the lower bound and its approximation for the average mutual information as hybrid precoder design criterion. With these treatments, the computational complexity is greatly reduced. In addition, it also shows that the lower bound with a constant shift offers a very accurate approximation to the average mutual information. This chapter further proposes utilizing the lower bound approximation as a low-complexity and accurate alternative for developing a manifold-based gradient ascent algorithm to find near-optimal analog and digital precoders. It was proved that their proposed algorithm performs very effectively and efficiently.

In [18], the fully dynamic subarray beamformer is applied to a wideband mmWave MIMO orthogonal frequency division multiplexing (MIMO-OFDM) system. The hybrid precoder and combiner are jointly designed to maximize the average spectral efficiency of the mmWave MIMO-OFDM system. When designing the hybrid precoder and combiner, the spectral efficiency maximization problem is first converted to a mean square error (MSE) minimization problem. Then, an efficient iterative hybrid beamformer algorithm is developed based on classical block coordination descent (BCD) methods. Extensive simulation results demonstrate the superiority of the proposed hybrid beamforming algorithm with dynamic subarrays and low-resolution PSs.

10.5.1.1 Partially Dynamic Subarray m-MIMO

In [20], Park et al. propose a partially dynamic subarray m-MIMO system where the adaptive connection between RF chains and antennas in the hybrid precoding architecture has been investigated, as reproduced in Figure 10.1(e). RF chains are permitted to dynamically connect to any subarrays where the antenna partition is fixed. However, these studies [7, 18, 19] consider inserting a switch between each RF chain and each antenna, which requires the use of thousands of switches and imposes prohibitively high hardware and computational complexity. To solve the intractable partially dynamic subarray hybrid precoding problem, the element-by-element (EBE) and vectorization-based (VEC) algorithms are derived. In addition, various algorithms were developed and tested, e.g., near-optimal progressive stage-by-stage (PSBS), low-complexity alternating-selection (AS), and block-diagonal-search (BDS) algorithms, so that the connections of switches can be determined.

10.6 RECONFIGURABLE POWER DIVIDER (RPD)-BASED DYNAMIC SUBARRAY m-MIMO

A practical dynamic subarray hybrid m-MIMO constructed is proposed here using low-complexity RPDs and low-resolution (3-bit) PSs, taking into account all RF hardware constraints. Following the antenna structure in [23], we study a transmitter employing a planar antenna panel consisting of 12 rows and 8 columns of antenna elements. Each antenna element has two input ports exciting signals of two orthogonal linear polarizations, see the illustration of right and left tilted lines in the panel in Figure 10.5(a). The analog beamforming network of the proposed dynamic subarray is depicted in Figure 10.5(b), where the RF signals for each subarray are fed into two low-resolution 3-bit PSs via a 2-way PD. Each PS is then connected, via a port-selectable 3-way PDs, to a fixed

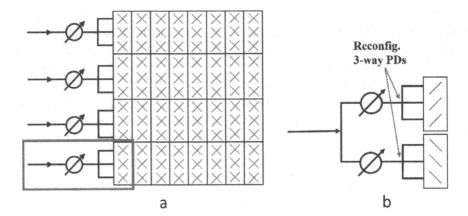

Figure 10.5 Typical m-MIMO Tx antenna [23]: (a) full antenna panel and (b) subarray.

sub-array consisting of three adjacent antenna elements of each polarization, see Figure 10.5(b). It is pointed out that the RF chain connections for only one column and one polarization are depicted for illustration purpose. Besides, the dimension of the antenna panel is extended to 16 rows and each subarray consists of four adjacent antenna elements of each polarization, where a 4-way RPD is implemented and discussed in Section 10.6.1.

10.6.1 Hardware Implementation of N-Way (4-Way as an Example) RPDs

An N-way RPD proposed in [24] is considered as practicable reconfigurable beamforming network to be applied in this dynamic subarray hybrid m-MIMO architecture. This N-way RPD has two key features which enables the dynamic connection between RF chains and antennas in the preselected subarray. First, an N-way RPD, consisting of N different output ports, can be configured to N transmission modes. All output ports combinations can be achieved in mode n $(n = 1, 2, \ldots, N)$ when n out of N output ports are selected. It is then clear that there are C_n^N configurations/states in the mode n, and $(2^N - 1)$ configurations can be achieved with an N-way RPD, indicating that each antenna in the subarray can be activated/deactivated by correctly configuring the PD. Second, the RPD introduced in [24] can achieve low insertion loss at 'ON'

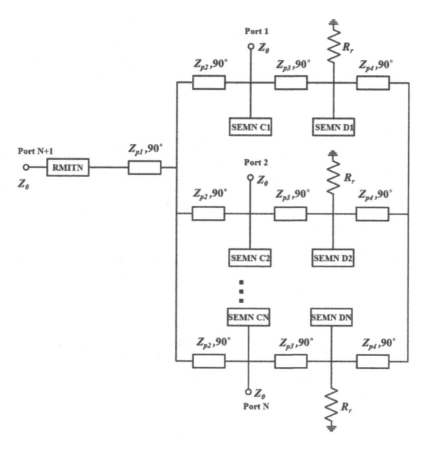

Figure 10.6   Sketch of an N-way RPD.

state ports and high isolation at 'OFF' state ports in all transmission states. This is preferred as the antennas can be well matched/isolated as expected in all transmission states, and low insertion loss will be achieved in the system.

The structure of an RPD proposed in [24] is redrawn in Figure 10.6. The p-i-n diode-based switching element with matching network (SEMN) is designed for controlling the 'ON/OFF' state, where the corresponding output ports will be activated in 'ON' state, while in 'OFF' state, the corresponding output ports will be deactivated and the signal will be reflected then can be recycled and rerouted to the activated ports. The reconfigurable multiple-impedance transformer network

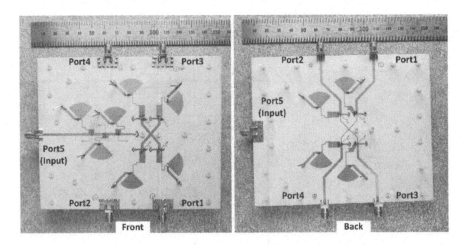

Figure 10.7   Photographs of a fabricated 4-way RPD operating at 3.7 GHz.

(RMITN) shown in Figure 10.6 is for impedance matching purpose, in order to alleviate the mismatching issue caused by the transmission mode change. Thanks to p-i-n diodes, the SEMN and RMITN could provide different configurations, and RPD modes and states act as switches and impedance converters.

One 4-way RPD following the principle in [24] for 3.7 GHz operation was designed and fabricated, as shown in Figure 10.7. It is used as an example to evaluate its performance and verify its feasibility of employing it in the dynamic subarray hybrid m-MIMO architecture. The measured S-parameter results are listed in Table 10.1.

The numerical values in the shaded cells in Table 10.1 indicate that the insertion loss for Mode 1 is no more than 1.5 dB, and is about 1.6 dB for both Mode 2 and Mode 3, and is about 1.65 dB for Mode 4. The ports in 'OFF' conditions are well isolated, with isolations all below −23 dB and the return loss at the input port (Port5) are no larger than −9.80 dB.

The measured results again verified that the aforementioned N-way RPDs could provide the flexible connection/disconnection between antennas and RF chains by adaptively configuring RPDs to desired transmission modes and states, and it is a suitable beamforming network for dynamic connections to the antennas in the subarray.

TABLE 10.1   Measured S-Parameters of the 4-Way RPD Shown in Figure 10.7 at 3.7 GHz

	States	S_{15} (dB)	S_{25} (dB)	S_{35} (dB)	S_{45} (dB)	S_{55} (dB)
Mode 1	(1)	−1.50	−26.99	−23.72	−25.78	−17.94
	(2)	−23.46	−1.54	−23.85	−25.35	−17.77
	(3)	−23.97	−26.28	−1.55	−26.95	−16.88
	(4)	−23.81	−25.80	−23.51	−1.55	−16.83
Mode 2	(1,2)	−4.59	−4.74	−26.85	−27.50	−18.42
	(1,3)	−4.66	−28.46	−4.47	−28.10	−17.40
	(1,4)	−4.53	−28.17	−26.41	−4.58	−19.49
	(2,3)	−26.78	−4.78	−4.40	−27.80	−18.67
	(2,4)	−27.12	−4.68	−26.92	−4.54	−20.86
	(3,4)	−26.98	−27.57	−4.48	−4.75	−19.29
Mode 3	(1,2,3)	−6.45	−6.62	−6.05	−30.82	−10.27
	(1,2,4)	−6.27	−6.58	−27.65	−6.23	−10.87
	(1,3,4)	−6.30	−30.84	−6.20	−6.43	−10.54
	(2,3,4)	−28.39	−6.47	−6.06	−6.49	−10.83
Mode 4	(1,2,3,4)	−7.60	−7.86	−7.52	−7.71	−9.80

10.6.2 System Simulation Results

The proposed practical dynamic subarray hybrid m-MIMO system is illustrated in Figure 10.8, where a 0.5-ns-delayed clustered delay line (CDL)-B channel [25] with 15 clusters and 20 rays per cluster is considered. The system is simulated in wideband scenario and orthogonal frequency division multiplexing (OFDM) signals with 64 subcarriers are employed whose subcarrier spacing is 15 kHz. Each subcarrier is quadrature phase shift keying (QPSK) modulated. To evaluate the performance of the proposed dynamic subarray hybrid m-MIMO system, another three m-MIMO systems are simulated for comparison purpose.

- *Fully digital m-MIMO*: In this structure, the 2-way PDs between RF chains and Tx antenna panels are removed in Figure 10.5(b). Instead, each RF chain connects to a {3, 4}-element antenna subarray via a {3, 4}-way RPD.

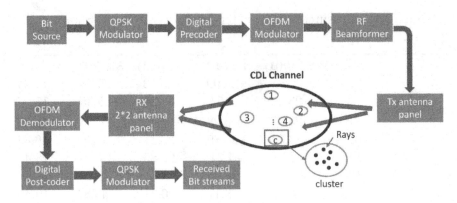

Figure 10.8 The diagram of a hybrid m-MIMO system.

- *Real fully digital m-MIMO*: Differ from the fully digital m-MIMO, each RF chain is routed to one antenna array in the real fully digital m-MIMO system.

- *Fixed subarray hybrid m-MIMO*: Here, the Tx structure is similar to that in Figure 10.5, but all the PDs are fixed, not reconfigurable.

The bit error rate (BER) results are plotted in Figure 10.9. In addition, the 4-way RPDs are adapted to the dynamic subarray hybrid

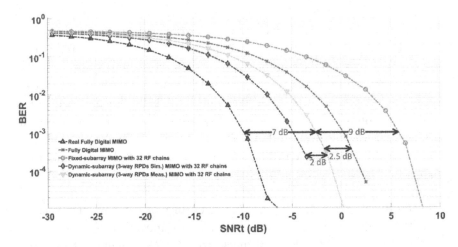

Figure 10.9 BER simulation results of the 3-way RPD-based dynamic subarray m-MIMO. (Sim. refers to simulated; Meas. represents measured.)

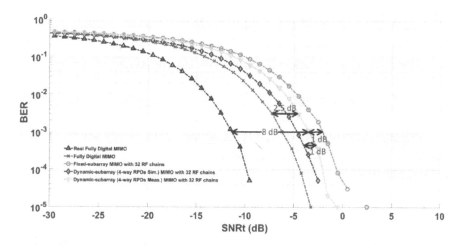

Figure 10.10 BER simulation results of the 4-way RPD-based dynamic subarray m-MIMO. (Sim. refers to simulated, Meas. Represents measured.)

m-MIMO and the results are shown in Figure 10.10. The performances or SNR gaps among the curves are summarized in Table 10.2.

The implementation of the RPDs improves the performance of the fixed subarray. In particularly, a 3-way RPD-based dynamic subarray improves the system performance significantly by about 9 dB and it achieves even better performance than the fully digital system by about 2.5 dB. In contrast, a 3-way RPD-based dynamic subarray m-MIMO system offers less gain. Compared with the real fully digital m-MIMO system, these two dynamic subarray systems suffer around 8 dB.

TABLE 10.2 A Summary of the Performance Gaps

	Gap 1	Gap 2	Gap 3
3-way RPD-based dynamic subarray	+9 dB	−7 dB	+2.5 dB
4-way RPD-based dynamic subarray	+1 dB	−8 dB	−2.5 dB

(Gap 1: between dynamic subarray and fixed subarray; Gap 2: between dynamic subarray and real fully digital; Gap 3: between dynamic subarray and fully digital.)

10.7 CONCLUSION AND FUTURE WORKS

This chapter reviewed three types of hybrid m-MIMO up to date. Sections 10.3, 10.4, and 10.5 elaborated the FC m-MIMO, fixed subarray m-MIMO, and dynamic subarray m-MIMO, respectively. Section 10.6 described a practical dynamic subarray m-MIMO implemented using RPDs.

Although the hybrid m-MIMO has been extensively studied, some challenges remain to be solved in order to pave the way for their practical field applications:

- Some efforts have been made on finding a better trade-off between the system performances and algorithm complexity based on the three hybrid m-MIMO structures. There exist no best solutions for all wireless channel conditions. In this sense, a generic evaluation metric should be developed in order for fair and meaningful comparison across different systems in various channel scenarios.

- Most previous works were theoretical models and simulation-based analysis, where the insertion loss and phase errors introduced by the beamforming networks were largely ignored. However, it is essential to consider the applications and their practical hardware implementation. For example, the low loss and low-cost hybrid beamformer architecture for dynamic subarrays and low-resolution PSs could be a direction to be devoted to.

- The current hybrid m-MIMO system was mainly explored in the mmWave band. Higher frequency communication, like THz in 6G communication, with higher propagation losses should be further studied.

BIBLIOGRAPHY

[1] W. Saad, M. Bennis, and M. Chen, "A vision of 6G wireless systems: Applications, trends, technologies, and open research problems," *IEEE Network*, vol. 34, no. 3, pp. 134–142, 2019.

[2] Z. Zhang et al., "6G wireless networks: Vision, requirements, architecture, and key technologies," *IEEE Vehicular Technology Magazine*, vol. 14, no. 3, pp. 28–41, 2019.

[3] T. L. Marzetta, "Noncooperative cellular wireless with unlimited numbers of base station antennas," *IEEE Transactions on Wireless Communications*, vol. 9, no. 11, pp. 3590–3600, 2010.

[4] T. S. Rappaport et al., "Wireless communications and applications above 100 GHz: Opportunities and challenges for 6G and beyond," *IEEE Access*, vol. 7, pp. 78729–78757, 2019.

[5] O. E. Ayach, S. Rajagopal, S. Abu-Surra, Z. Pi, and R. W. Heath, "Spatially sparse precoding in millimeter wave MIMO systems," *IEEE Transactions on Wireless Communications*, vol. 13, no. 3, pp. 1499–1513, 2014, doi: 10.1109/twc.2014.011714.130846.

[6] S. Han, I. Chih-Lin, Z. Xu, and C. Rowell, "Large-scale antenna systems with hybrid analog and digital beamforming for millimeter wave 5G," *IEEE Communications Magazine*, vol. 53, no. 1, pp. 186–194, 2015.

[7] S. Park, A. Alkhateeb, and R. W. Heath, "Dynamic subarrays for hybrid precoding in wideband mmWave MIMO systems," *IEEE Transactions on Wireless Communications*, vol. 16, no. 5, pp. 2907–2920, 2017, doi: 10.1109/twc.2017.2671869.

[8] F. Sohrabi and W. Yu, "Hybrid digital and analog beamforming design for large-scale antenna arrays," *IEEE Journal of Selected Topics in Signal Processing*, vol. 10, no. 3, pp. 501–513, 2016, doi: 10.1109/jstsp.2016.2520912.

[9] A. Alkhateeb, G. Leus, and R. W. Heath, "Limited feedback hybrid precoding for multi-user millimeter wave systems," *IEEE Transactions on Wireless Communications*, vol. 14, no. 11, pp. 6481–6494, 2015.

[10] Z. Wang, M. Li, Q. Liu, and A. L. Swindlehurst, "Hybrid precoder and combiner design with low-resolution phase shifters in mmWave MIMO systems," *IEEE Journal of Selected Topics in Signal Processing*, vol. 12, no. 2, pp. 256–269, 2018, doi: 10.1109/jstsp.2018.2819129.

[11] X. Yu, J.-C. Shen, J. Zhang, and K. B. Letaief, "Alternating minimization algorithms for hybrid precoding in millimeter wave MIMO systems," *IEEE Journal of Selected Topics in Signal Processing*, vol. 10, no. 3, pp. 485–500, 2016.

[12] J.-C. Chen, "Hybrid beamforming with discrete phase shifters for millimeter-wave massive MIMO systems," *IEEE Transactions on Vehicular Technology*, vol. 66, no. 8, pp. 7604–7608, 2017.

[13] Y. Ding, V. Fusco, A. Shitvov, Y. Xiao, and H. Li, "Beam index modulation wireless communication with analog beamforming," *IEEE Transactions on Vehicular Technology*, vol. 67, no. 7, pp. 6340–6354, 2018.

[14] M. A. B. Abbasi, H. Tataria, V. F. Fusco, and M. Matthaiou, "Performance of a 28 GHz two–stage Rotman lens beamformer for millimeter wave cellular systems," in *2019 13th European Conference on Antennas and Propagation (EuCAP)*, 2019: IEEE, pp. 1–4.

[15] X. Gao, L. Dai, S. Han, I. Chih-Lin, and R. W. Heath, "Energy-efficient hybrid analog and digital precoding for MmWave MIMO systems with large antenna arrays," *IEEE Journal on Selected Areas in Communications*, vol. 34, no. 4, pp. 998–1009, 2016, doi: 10.1109/jsac.2016.2549418.

[16] A. Abbaspour-Tamijani and K. Sarabandi, "An affordable millimeter-wave beam-steerable antenna using interleaved planar subarrays," *IEEE Transactions on Antennas and Propagation*, vol. 51, no. 9, pp. 2193–2202, 2003.

[17] M. Li, Z. Wang, H. Li, Q. Liu, and L. Zhou, "A hardware-efficient hybrid beamforming solution for mmWave MIMO systems," *IEEE Wireless Communications*, vol. 26, no. 1, pp. 137–143, 2019, doi: 10.1109/mwc.2018.1700391.

[18] H. Li, M. Li, Q. Liu, and A. L. Swindlehurst, "Dynamic hybrid beamforming with low-resolution PSs for wideband mmWave MIMO-OFDM systems," *IEEE Journal on Selected Areas in Communications*, vol. 38, no. 9, pp. 2168–2181, 2020, doi: 10.1109/jsac.2020.3000878.

[19] J. Jin, C. Xiao, W. Chen, and Y. Wu, "Channel-statistics-based hybrid precoding for millimeter-wave MIMO systems with dynamic subarrays," *IEEE Transactions on Communications*, vol. 67, no. 6, pp. 3991–4003, 2019, doi: 10.1109/tcomm.2019.2899628.

[20] L. Yan, C. Han, and J. Yuan, "A dynamic array-of-subarrays architecture and hybrid precoding algorithms for terahertz wireless communications," *IEEE Journal on Selected Areas in Communications*, vol. 38, no. 9, pp. 2041–2056, 2020, doi: 10.1109/jsac.2020.3000876.

[21] R. Méndez-Rial, C. Rusu, N. González-Prelcic, A. Alkhateeb, and R. W. Heath, "Hybrid MIMO architectures for millimeter wave communications: Phase shifters or switches?" *IEEE Access*, vol. 4, pp. 247–267, 2016.

[22] A. Alkhateeb, Y.-H. Nam, J. Zhang, and R. W. Heath, "Massive MIMO combining with switches," *IEEE Wireless Communications Letters*, vol. 5, no. 3, pp. 232–235, 2016.

[23] V. Aue. "The Open RAN System Architecture and mMIMO," *Microwave Journal* vol. 64, no. 11, pp. 70-82, 2021.

[24] H. Fan, X. Liang, J. Geng, L. Liu, and R. Jin, "An N-way reconfigurable power divider," *IEEE Transactions on Microwave Theory and Techniques*, vol. 65, no. 11, pp. 4122–4137, 2017, doi: 10.1109/tmtt.2017.2702115.

[25] 3GPP, "Study on channel model for frequencies from 0.5 to 100 GHz," *TR 38.900 Release 14*, 2017.

CHAPTER 11

Blockchain Technology for 6G-Oriented IoT Systems

Principle, Applications and Challenges

Tianqi Yu

School of Electronic and Information Engineering, Soochow University, Suzhou, China

Yongxu Zhu

Dept. of Computer Science and Informatics, London South Bank University, London, UK

Xianbin Wang

Dept. of Electrical and Computer Engineering, Western University, London, Ontario, Canada

CONTENTS

11.1 Fundamentals of Blockchain 285
    11.1.1 Structure of Blockchain 287
        11.1.1.1 Overview of Block Structure 287
        11.1.1.2 Block Header 288
        11.1.1.3 Transaction 289
    11.1.2 Decentralized Consensus Mechanism for Blockchain 291
    11.1.3 Technical Advantages of Blockchain 293
11.2 Applications of Blockchain Technology in 6G-Oriented IoT
    Systems 294
    11.2.1 Data Privacy and Security Management 294

DOI: 10.1201/9781003282211-11

11.2.2 Distributed Data Sharing and Storage 296
11.2.3 Smart Contract-Based Property Trading 298
11.2.4 Intelligent Manufacturing and Asset Tracking 299
11.2.5 Intelligent and Autonomous Vehicular IoT 300
11.3 Future Research Directions 303
11.3.1 Malicious Attacks 303
11.3.2 Heterogeneous Devices and Scalability 305
11.3.3 Parallel Transactions and Side-Chains 305
11.3.4 System Performance and Overhead 307
11.4 Summary 308
Bibliography 308

T HE FORTHCOMING SIXTH-GENERATION (6G) [39] era will empower
the massive deployment of the intelligent and autonomous Internet
of Things (IoT) applications, by integrated use of many emerging technologies including integrated sensing and communications (ISC), space-
air-ground-sea communications, Terahertz (THz) communications, edge
artificial intelligence (AI), and reconfigurable intelligent surfaces (RIS).
However, 6G-oriented IoT systems still face many critical challenges.
Due to the inevitable use of distributed data in IoT applications, data
privacy and protection, secure data sharing and storage, and massive device management have become more important. Blockchain technology
is among the most promising solutions to address the related challenges,
which has been gradually applied to information and communication
technology (ICT) systems since the fifth-generation (5G) era. Specifically, a blockchain is a decentralized database that can record and validate all the peer-to-peer (P2P) transactions without central authority or
third-party verification. The decentralized, secure, and autonomous characteristics of blockchain technology make it possible to improve the performance of 6G-oriented IoT systems, by introducing many new mechanisms including data privacy and security management, distributed data
sharing and storage, smart contracts, asset tracking, and mobile management. This chapter provides a comprehensive overview of how blockchain
technology could boost 6G-oriented IoT systems. The fundamental principle of blockchain technology is first presented. Existing applications of
blockchain technology on enhancing the performance of the 6G-oriented
IoT systems are then surveyed. Finally, the main challenges of applying

blockchain technology to the 6G-oriented IoT systems and future research directions are highlighted.

11.1 FUNDAMENTALS OF BLOCKCHAIN

In 2008, Satoshi Nakamoto published a paper entitled "Bitcoin: A Peer-to-Peer Electronic Cash System [31]", where the concept of blockchain was proposed for the first time. Specifically, in the paper, Satoshi Nakamoto conceived a purely peer-to-peer (P2P) digital-encrypted currency "Bitcoin". Bitcoin allowed online payments to be made directly from one participant to another without the operation or supervision of a third-party financial institution. The participant involved in the online payment was termed a node in the Bitcoin network.

In the Bitcoin network, the blockchain was proposed as the public ledger to record all the executed transactions. As the public ledger, the blockchain autonomously shares the recorded transactions among all the nodes in the network for decentralized consensus. Thus in blockchain technology, the decentralized consensus mechanism was used to verify and authorize the recorded transactions instead of the central authority or third-party verification.

After rapid development for around 10 years, blockchain technology has been extensively investigated and applied to many different areas beyond the original Bitcoin. According to Statista [37], the size of the blockchain technology market worldwide has grown from 1.2 to 7 billion US dollars from 2018 to 2021, expected to reach 39.7 billion in 2025. Except for the usage as the public ledgers for cryptocurrencies similar to the originated Bitcoin, blockchain technology has also been exploited as a secure and distributed database to overcome the technical challenges in many evolving non-financial applications, such as decentralized proof of the existence of documents in Proof-of-Existence, distributed data storage in Storj, anti-counterfeit product verification in BlockVerify, and digital identity verification in Internet of Vehicles (IoV) blockchain [7].

Due to its decentralized, secure, and autonomous advantages, blockchain technology has been applied to distributed information and communication technology (ICT) systems, particularly, the IoT systems, since the era of the fifth generation (5G). For instance, IBM has built a blockchain platform offering blockchain solutions for different IoT applications such as smart healthcare, intelligent manufacturing, and the telecommunications industry [47].

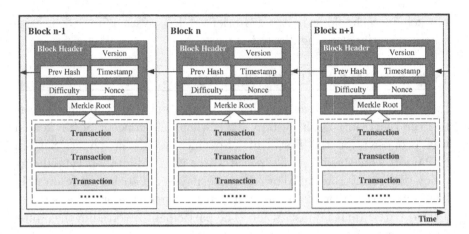

Figure 11.1 Structure of blockchain.

With a brief overview of the history of blockchain technology and related applications, one can see that the essence of blockchain is a decentralized database duplicated and shared among all distributed nodes in a network. To elaborate more on its principle, the general structure of blockchain is sketched in Figure 11.1. It can be seen that each block in the chain comprises a block header and a set of transactions. Both the block header and transaction are particularly predefined data structures. The block header is the descriptive information of the block. As depicted in Figure 11.1, the block header includes a block version indicating the validation rule, the hash value of its previous block, timestamp, the difficulty of generating the block, the number of transactions, and Merkle root concatenating the hash values of all the transactions in the block. Transactions are the executed records of the data exchanges that occurred in the network.

Figure 11.1 further unveils that the blocks are chained in chronological order by referring to the hash value of its previous block. Due to the chaining structure, when a block has been added to the chain, there will be a growing number of new blocks chained after it as time goes on. Under such a situation, the block would be impractical to be modified because the adversaries need to regenerate all the blocks chained after it with overwhelming effort. Thus, the transactions recorded on the blockchain can be protected from tampering.

More details of blockchain technology are further provided in Section 11.1. In Section 11.1.1, the structure of the blockchain is presented in detail with the overall block structure, block header, and transaction. Consensus mechanisms are further introduced in Subsection 11.1.2, where the approach of generating a new block and adding it to the blockchain is analyzed. Finally, the technical advantages of blockchain technology are comprehensively summarized in Section 11.1.3.

11.1.1 Structure of Blockchain

The structure of the blockchain is presented here from the perspective of a fundamental block, including the overview of block structure, block header, and transaction.

11.1.1.1 *Overview of Block Structure*

Block is the fundamental component of a blockchain, which is generally consisted of a block header and a set of transactions, as depicted in Figure 11.1. In practical applications, a block is normally defined as a container type of data structure, and the structure of the block is tailored according to the requirements of certain systems.

The specific structures of blocks tailored for the originated Bitcoin network and an IoT-enabled smart home system are depicted in Table 11.1 [2] and Figure 11.2 [13], respectively.

As shown in Table 11.1, the block structure in the Bitcoin network consists of four main fields. The first field is the block size assigned with 4 bytes. The second field is the block header with a fixed size of 80 bytes. The third and fourth fields are transaction counter and transactions and refer to the number and records of transactions comprised in the current block, respectively. The size of the transaction counter field varies from 1 to 9 bytes, while the size of a transaction is a variable with an average value of 400 bytes. The size of different fields infers that the records of transactions occupy most of the memory in a block.

In Figure 11.2, blockchain technology is applied to an IoT-enabled smart home system for secure access control of IoT devices and data. As shown in Figure 11.2, a block comprises a block header, a policy header, and a set of transactions. The block header contains the hash value of its

TABLE 11.1 Structure of Block in Bitcoin Network

Field	Block Size	Block Header	Transaction Counter	Transactions
Size (bytes)	4	80	1–9 (Variant)	Variable

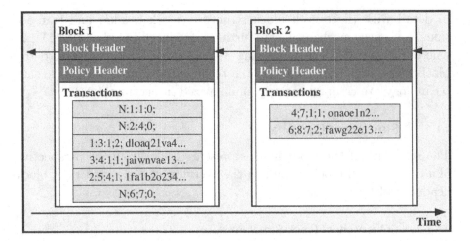

Figure 11.2   Structure of blockchain in a IoT-enabled smart home system.

previous block to maintain the chain. The policy header is designed for the control of IoT devices. In this example of a IoT-enabled smart home system, the transactions are the records of communications between IoT devices.

11.1.1.2 Block Header

A block header contains the descriptive information of the block, which is consisted of the hash value of its previous block and optional self-defined fields.

The hash value of the previous block is used to establish the chain of blocks. Specifically, each block is identified by its block height and hash value as depicted in Figure 11.3.

Every time when a new block is generated and added to the chain, the block height is incremented by one automatically. The blocks are chained in chronological order. Particularly, the first block is termed as genesis block and the height of the genesis block is 0. In terms of the hash value, it is generated by hashing either the block header or the whole block, which depends on the specific design. For example, the 256-bit hash value in Figure 11.3 is generated by hashing the block header with SHA256, namely, hash value = SHA256 (block header); SHA256 is the secrete hash algorithm.

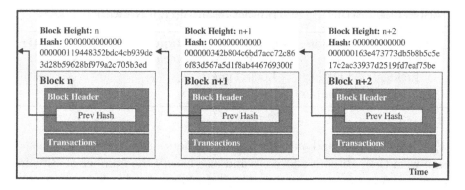

Figure 11.3 Identifications of a block: block height and hash value.

The self-defined fields are optional. If necessary, the fields are defined according to the specific requirements of applications. In the IoT-enabled smart home system, the block header merely contains the hash value of the previous block. While in the Bitcoin network, in addition to the hash value of the previous block, some other fields are defined in the block header as well, to identify the timestamp and contributions of generating such a block.

11.1.1.3 *Transaction*

The essence of a transaction in blockchain is transferring the ownership of a certain amount of property from one participant to another. The property can be Bitcoins in the Bitcoin network, sensing and communication data in IoT systems, and any other actual or virtual commodities. In blockchain technology, a transaction is defined as a data structure with a stereotype. Identities of the and the trading between the nodes are recorded in the transaction. A transaction can be created by any node, not necessarily the ones that exactly participate in the transaction. The created transaction is then openly broadcast throughout the network so that every node in the network can witness the transaction. Only the nodes that participated in the transaction can transfer the property by providing their private keys, due to that the ownership of property is locked by the digital signatures generated from the private keys.

The structure of transactions defined in the Bitcoin network is shown in Table 11.2 [2] and Figure 11.4. The field of version is used to track the upgrade of software. The input counter and output counter illustrate the number of inputs and outputs included in the transaction, respectively.

TABLE 11.2 Structure of Transaction in Bitcoin Network

Field	Version	Input Counter	Input	Output Counter
Size (bytes)	4	1–9 (Variant)	Variable	1–9 (Variant)
Field	Output	Lock Time		
Size (bytes)	Variable	4		

The fields of input and output identify the nodes involved in the transaction with their Bitcoin addresses and also clarify the number of Bitcoins transferred between the nodes. The lock time refers to the earliest time that a transaction can be added to the blockchain.

Another example is the structure of transactions defined in the IoT-enabled smart home system, which is presented in Table 11.3 [13]. Transactions are the records of communications between IoT devices. Four types of transactions are designed, *i.e.*, genesis, access, store, and monitor. A genesis transaction is generated when a new device joins the system. Store transactions are generated by IoT devices to store data. An access transaction is generated when the data storage is accessed. Monitor transactions are periodically generated to monitor the information of IoT devices. In addition, the transaction number is used to identify the current transaction, and the field of the previous transaction refers to the transaction number of the previous transaction generated by the same IoT device. Particularly, for a genesis transaction, the field of the

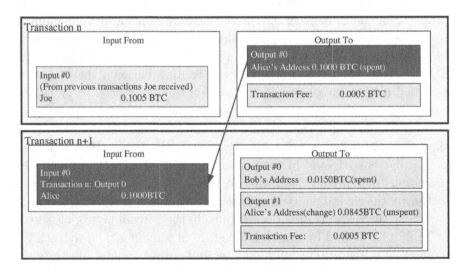

Figure 11.4 Diagram of transactions in the Bitcoin network.

TABLE 11.3   Structure of Transaction in Smart Home

Previous Transaction	Transaction Number	Device ID	Transaction Type	Corresponding Multisig Transaction if any (for keeping signature) of requester)
N = Genesis Transaction			Genesis: 0; Access: 1; Store: 2; Monitor: 3	

previous transaction is filled out with "N". Device ID is the system identification of the IoT device generating the current transaction. The last
field is used to collect the signature of the requester if the transaction is
generated out of the local blockchain network.

11.1.2   Decentralized Consensus Mechanism for Blockchain

The general structure of a block has been presented in Section 11.1.1.
This section introduces how to generate a new block and add it to the
chain through the decentralized consensus mechanisms.

A decentralized consensus mechanism defines the conditions needed
to be reached to make an agreement among all the networking nodes
on validating the blocks to be added to the chain. Thus, the decentralized consensus mechanism is the exact mechanism in blockchain technology that validates and secures the transactions and blocks in substitution of the traditional central authority and third-party verification. In
blockchain technology, the process of generating a new block and adding
it to the chain is termed as "mining", which means that mining is the
process of reaching the conditions predefined in the decentralized consensus mechanism to validate and add a new block to the chain. The
nodes participating in the block mining are termed as "miners".

In blockchain technology, the design of decentralized consensus mechanisms should meet the fundamental requirements listed as follows [15].

- **Termination:** For each node in the blockchain network, a block
  holding a set of new transactions in its transaction pool should be
  either discarded or accepted into the chain.

- **Agreement:** A block holding a set of new transactions should be
  either accepted or discarded by all the nodes in the blockchain
  network.

- **Validity:** If a block passes the verification of all the nodes in the blockchain network, it should be accepted into the chain.

- **Integrity:** In the blockchain network, all accepted transactions should be consistent, and all accepted blocks should be hashed and chained in chronological order.

Several decentralized consensus mechanisms have been proposed, such as proof-of-work (PoW), proof-of-stake (PoS), proof-of-activity (PoA), and practical Byzantine fault tolerance (PBFT) [52]. Due to the different predefined conditions of consensus, the related approaches of mining in these algorithms are different.

In the originated Bitcoin network, the PoW is used as the decentralized consensus mechanism, which is analyzed here as an instance [2]. In the PoW, a cryptographic puzzle is defined as finding a certain integer, nonce, that can make the hash value of the block header less than or equal to a predefined threshold, difficulty target, namely, Hash (nonce+other contents in the block header) ≤ difficulty target. During the process of mining, the miners have to spend their computational power on repeating the hashing calculations with different values of nonce until the result meets the difficulty target. The nonce and difficulty target are included in the block header of the newly mined block as proof, so that other nodes can verify the work of the miner.

As introduced in Section 11.1.1.3, transactions are broadcast over the Bitcoin network for tamper resistance. A miner adds transparent transactions to its transaction pool. By the time of the generation of the last block, a new round of mining competition among the miners starts. The miner encapsulates the transactions collected in its transaction pool into a new block and spares its computational power on solving the cryptographic puzzle. The miner who solves the puzzle in the shortest time wins the competition and subsequently broadcasts the newly mined block to the network. When a peer node in the network receives the new block, verification of the block is executed locally, namely, the work of the miner identified in the block is verified. If the block passes the verification, the node further propagates the block to other peers in the network. In the meantime, the node adds the verified block to its own copy of the blockchain, where the longest chain accumulated with the largest amount of computational power is selected. If the node is also a competing miner, it stops the mining efforts on the same block height and begins the next round of mining regarding the received block as the previous block.

Owing to the chaining structure and the decentralized consensus mechanism, when a block has been added to the chain for a while, there will be a number of blocks chained after it. Under such a situation, the block would be impractical to be modified because the adversaries need to regenerate all the blocks chained after it with overwhelming expenses. Therefore, the transactions recorded on the blockchain can be protected from being altered or removed by unauthorized parties.

11.1.3 Technical Advantages of Blockchain

The well-designed structure of blockchain and decentralized consensus mechanism have brought the following technical advantages [8]:

- **Decentralization:** As investigated in Section 11.1.2, the exploitation of decentralized consensus mechanisms allows the decentralized validation of transactions between any two participants in the blockchain network, without referring to any central authority or third-party authentication. The service cost, performance bottleneck, and the risk of single-point failure brought by the centralized validation can be thus mitigated.

- **Transparency:** For the public blockchain, users can access and participate in the blockchain network with equal rights. Within the blockchain network, the newly generated transactions are globally broadcast and validated, which implies that the transactions stored on the blockchain are transparent to all the authenticated users.

- **Anonymity:** Although the recorded transactions are transparent, the privacy of the participants in the blockchain network can be partially protected by anonymizing their blockchain addresses. It is emphasized as partial protection, due to that blockchain addresses are essentially traceable by inference.

- **Immutability:** Owing to the chaining structure and the decentralized consensus mechanism, when a block has been added to the chain for a while, there will be a number of blocks chained after it. Under such a situation, the block would be impractical to be modified because the adversaries need to regenerate all the blocks chained after it with overwhelming expenses.

- **Cryptographically Security:** Asymmetric cryptographic techniques can be applied to the practical applications of blockchain technology. Public and private keys generated by a certain asymmetric cryptographic algorithm can be used for transaction encryption. A transaction is encrypted by the private key. It can only be accessed and verified by the authenticated participants with the public key, which further enhances the security of the blockchain.

11.2 APPLICATIONS OF BLOCKCHAIN TECHNOLOGY IN 6G-ORIENTED IOT SYSTEMS

With the technical advantages investigated in Section 11.1, blockchain technology has already been developed and applied in many other fields beyond the originated cryptocurrency. The decentralized and secure characteristics of blockchain have made it a natural and promising solution to the technical issues in IoT systems as well. The leading technology companies and startups have spared no efforts on empowering future IoT with blockchain technology. Furthermore, a large number of scientific papers from academic institutions have been published on the investigation of applying blockchain technology to IoT systems.

Therefore, a comprehensive survey on the applications of blockchain technology in sixth-generation (6G)-oriented IoT systems is conducted in this section, where research efforts from both the industry and academia are investigated. According to the different usages of the blockchain in the IoT systems, applications are classified into the following categories, data privacy and security management, distributed data sharing and storage, smart contract and property trading, intelligent manufacturing and asset tracking, and intelligent and autonomous vehicular IoT. The statement of each category comprises both the theoretical principle and the application instances.

11.2.1 Data Privacy and Security Management

Privacy and security are critical issues in IoT systems all the time, due to that IoT devices not only register user's personal data but also monitor user's activities. In the current IoT system, privacy and security are normally managed by centralized authority or third-party verification, which unveils the whole system to the potential risk of central failure. The decentralized and secure characteristics of blockchain technology can facilitate privacy and security management in IoT systems from the following aspects.

- **Data Integrity:** In the context of security, data integrity means that the data collected from IoT devices should be protected from being altered or removed by unauthorized parties. By taking advantage of the tamper-proof feature of blockchain technology, IoT data integrity can be guaranteed.

- **Identity Management:** In traditional IoT systems, the identification of an IoT device is normally managed by the centralized authentication server. The potential risk is that if the authority is compromised, the legal IoT devices would possibly be blocked away while the adversaries would be able to obtain the access. By utilizing blockchain technology, a user's identity can be easily established by its unique cryptographic address, and the identity is managed by tailored transactions, identity verification, and identity revocation without the third-party authority. Blockchain-based identity management has already been applied in [51], where Pretty Good Privacy (PGP) is built on a Distributed Web of Trust, in which a user's trustworthiness is established by others who could vouch through a digital signature for that user's identity and no central certification or third-party authority is involved.

- **Dynamic Key Management:** Public and private key pairs are critical for an individual user in an IoT system, since personal identity, data, and all belongings are encrypted and managed by the key pairs. By utilizing the inherent cryptographic mechanism, blockchain technology can provide a more efficient solution to dynamic key management. For instance, an IoT device management scheme has been developed together with the Ethereum-based smart home system [19]. In the scheme, the key is managed by the Rivest–Shamir–Adleman (RSA) public key cryptosystems, where public keys are stored on the blockchain and private keys are saved on individual IoT devices. Data exchanges within the system are secured by certain keys.

Guardtime [46] is a company working on keyless signature infrastructure (KSI) using blockchain technology for customers with the necessity of data integrity and security, where all data requests are hashed and recorded into a time-stamped Merkle tree. Merkle tree is a binary hash tree for storing and searching data elements, as depicted in Figure 11.5. By the exploitation of timestamps, the Merkle tree is managed in chronological order and data can be further secured from being altered.

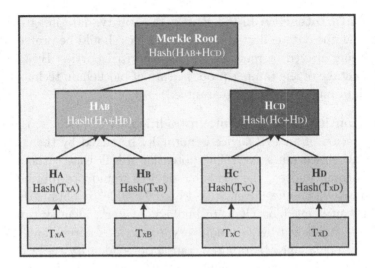

Figure 11.5 Diagram of Merkle tree.

KSI blockchain has already been developed to overcome the loss of data control and protect the integrity of digital data in IoT systems [54].

11.2.2 Distributed Data Sharing and Storage

In traditional centralized IoT systems, the cloud server is the centralized data storage and processing center. Due to that, most IoT users cannot afford the cost of building their own cloud server, so data collected from the IoT devices are normally uploaded to a third-party cloud server. Failure or compromise of the centralized cloud server can make the whole system exposed to various malicious cyber-attacks.

With the development of blockchain technology, the blockchain-enabled distributed data sharing and storage is becoming a substitution for the cloud server-based centralized data center, as depicted in Figure 11.6, so that the risks of single-point failure and privacy exposure can be mitigated [56].

The technical advantages of distributed data sharing and storage are summarized as follows [18].

- **Decentralization and Robustness:** In a cloud-orchestrated IoT system, the cloud platform is the centralized data center. If the data center is compromised, all the data of the IoT system are unveiled to different kinds of attacks. By using the blockchain-enabled distributed data storage, the data are stored on individual

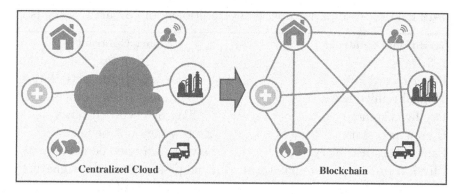

Figure 11.6 Migration from centralized cloud storage to blockchain-based decentralized data sharing and storage.

and distributed nodes, which makes the whole network more robust to central failure.

- **Privacy and Security:** All the data are encrypted and stored in distributed storage spaces. No third party, such as the owner or manager in the centralized cloud platform, can access the distributed stored data. Hence, the privacy and security of data are enhanced.

- **Price:** As compared to Amazon's centralized cloud server which costs $25 per TB per month, blockchain storage provided by Storj costs only $4 per TB per month.

Filecoin [45] is a decentralized data sharing and storage provider. The decentralized storage network of Filecoin is developed based on the combination of blockchain technology and IPFS (interplanetary file system, a distributed web-based file sharing platform [49]).

Storj [50] is also a company working on blockchain-based decentralized data storage. In the Storj network, each user shares his/her own spare disk space. All the distributed disk spaces are jointly utilized as the decentralized cloud platform. Data uploaded by users are sliced, encrypted, and stored in the distributed spaces. In this way, single-point failure occurring on the traditional centralized cloud platform can be prevented. The data are then secured and managed by the cryptographic keys that are privately owned by the users. Moreover, to ensure the data download speed, data are generally stored in the neighborhood spots.

TABLE 11.4   Comparison between Traditional and Smart Contracts

Traditional Contract	Smart Contract
1–3 Days	Minutes
Paper record	Code-based transaction
Manual remittance	Automatic remittance
Pay by currency	Pay by cryptocurrency
Escrow necessary	Escrow may not be necessary
Lawyers are necessary	Lawyers may not be necessary
Physical presence (wet signature)	Virtual presence (digital signature)
Expensive	Fraction of the cost
No safety guarantee	Transaction security guarantee

11.2.3   Smart Contract-Based Property Trading

With the advent of blockchain technology, a related technique termed smart contract has been further developed, which makes it a suitable solution to decentralized smart property trading in IoT systems. The concept of a smart contract was initially defined as a set of promises recorded in digital form, namely, converting papery contracts into code and embedding it into hardware so that the transactions could be secured without the necessity of a trusted third party [38]. With the exploitation of the blockchain as the public ledger, the concept of smart contracts has come to reality [59].

Specifically, a potential contract is jointly created and coded by the traders, and the code is encrypted and recorded on the public ledger. The public ledger is shared among all the participants in the network, which ensures that the contract is transparent and hard to tamper with. The contract is automatically executed when all the predefined conditions are met. Only the participants involved can decrypt the contract so that the trading property can be protected. The comparison between the traditional contract and smart contract is listed in Table 11.4 [5, 29].

A general smart contract-based property trading model for IoT systems is depicted in Figure 11.7. The seller and buyer as trading entities conduct the transaction through smart contracts without third-party authority. While for IoT systems, the trading commodities can be any kind of property, including virtual properties such as IoT data and services [21, 32]. In terms of payment, it can be equivalent properties or cryptocurrencies such as Bitcoin and IoTcoin.

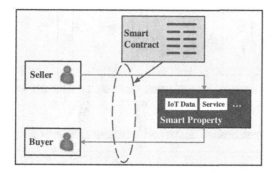

Figure 11.7 Smart contract-based property trading model for IoT systems.

Filament [44] has built a Blocklet Mobility Platform, where smart contracts enable automation on everything from trusted wallets for usage-based pricing to trusted vehicle history for the increased residual value.

11.2.4 Intelligent Manufacturing and Asset Tracking

IoT technology has already been applied to the manufacturing industry for logistics tracking and quality assurance. By integrating blockchain technology into IoT systems, not only the procedure of logistics but also the whole life cycle and information of a product can be tracked. Blockchain and IoT jointly enabled systems can potentially facilitate the manufacturing industry in the following aspects.

- **Asset Tracking:** By using blockchain-enabled smart contracts as public ledgers, all the information about a product in its life cycle is traceable. Specifically, materials, manufacturing date, and factory information of a product are recorded on the blockchain by manufacturers. Information on the logistics and retailers is recorded on the blockchain as well. Thus, the information of a product is accessible to customers, and certifications for products are more trustworthy, so that the customers can be protected from counterfeits.

- **On-demand Manufacturing:** In the blockchain and IoT-enabled intelligent manufacturing, manufacturers and customers are considered peer participants in the P2P blockchain network and keep updating their local copies of the blockchain. Products can be

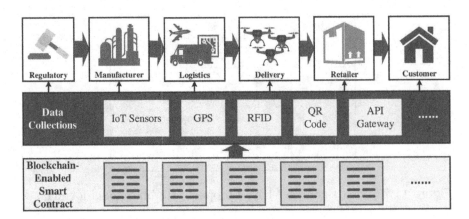

Figure 11.8 Blockchain and IoT-enabled supply chain and asset tracking.

manufactured according to the demands of customers recorded on the blockchain.

Figure 11.8 shows a diagram of blockchain and IoT-enabled manufacturing supply chain and asset tracking [34]. The manufacturing information on a product is recorded on the blockchain. The product is then delivered from the manufacturer to the retailer and later to the customer. The trading of this product is enabled by smart contracts, and the information on the whole procedure of logistics is recorded on the blockchain. Therefore, all the manufacturing and logistic information of the product is traceable.

Everledger [43] is a blockchain-enabled company that provides the certification of diamonds. Through the exploitation of blockchain, a permanent ledger was built for recording the production and transaction history of diamonds. The unique identifications of the diamond such as weight, size, and color are hashed and recorded on the permanent ledger as well.

11.2.5   Intelligent and Autonomous Vehicular IoT

Intelligent and autonomous vehicular IoT, also renowned as the IoV, is an emerging and promising branch of IoT in the coming 6G era. As reported at World Intelligent Vehicle Conference 2020, the number of intelligence-connected vehicles (ICVs) worldwide is expected to reach 74 million in 2025. IoV security is directly related to the privacy of drivers and passengers and road safety. Thus, IoV security is one of the most significant issues that must be resolved before the pervasive deployment of IoV. Blockchain technology with its intrinsic advantages in security and privacy may enhance the performance of IoV from the following aspects [35]:

- **Spectrum Sharing:** The number of mobile users meets explosive growth in the 6G era. Thus, the efficient utilization of the radio spectrum is a necessity. Spectrum sharing enables the access of the same frequency band by multiple users. In traditional spectrum sharing, there is an administrator who leases out the spectrum to users, where the tampering of lease records is a major security concern. The technical advantage of blockchain on immutability makes it promising to protect the records from being tampered with, which can secure the spectrum sharing in IoV systems with a large number of ICVs.

- **Network Function Virtualization (NFV):** With the development of NFV, the efficiency of a network can be improved by the virtualization of the network resources, due to that the resources of the cloud server, edge servers, and users can be virtualized and reallocated to meet the ultra-reliable and low-latency requirements of the tasks in IoV. The technical advantages of blockchain on immutability and anonymity can protect the security and privacy of virtual network slices.

- **Network Architecture:** The introduction of artificial intelligence (AI) technology to IoV brings the potential of enabling intelligent services for moving vehicles. However, traditional centralized AI model training suffers from high latency for cloud server access and data privacy leakage of vehicle clients. To address these issues, federated learning (FL) has been introduced to IoV for distributed privacy-preserving model training. The intrinsic decentralization characteristic of blockchain has made it a solution to the network architecture for IoV in support of the FL and distributed AI. Figure 11.9 depicts the hierarchical blockchain framework for IoV [4].

IoV blockchain [48] is a blockchain underlying infrastructure that provides the foundation for the IoV. The infrastructure chain provides

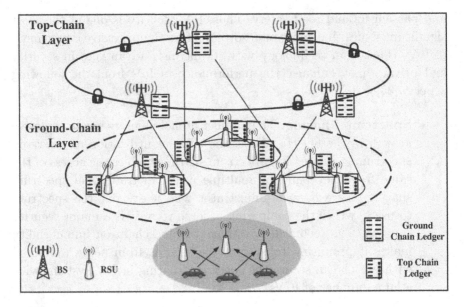

Figure 11.9   Hierarchical blockchain framework for IoV.

the automotive industry and customers with decentralized application services and extends to travel-related multi-scenario applications, such as the digital identity system, e-commerce system, data flow system, and software development kit (SDK) interface. Moreover, the IoV Token is released as the encrypted digital currency and also the certification of partnership. It provides such services to the user groups of 2 billion vehicles worldwide and the trillion-dollar scale automobile consumer market. For now, IoV blockchain has attracted many smart IoV ecological partners including automobile manufacturers BYD, Mercedes-Benz, BMW, and Tesla, mobility server provider Uber, oil industry company Mobil, and so on.

In addition to data privacy and security management, distributed data sharing and storage, smart contract-based property trading, intelligent manufacturing, and intelligent and autonomous vehicular IoT summarized in this section, there are many other applications of blockchain technology in IoT systems, such as smart agriculture [41, 53], smart healthcare [40], and crowdsourcing [57]. The applications of blockchain technology in IoT systems still meet explosive growth in the coming 6G era [23].

11.3 FUTURE RESEARCH DIRECTIONS

As surveyed in Section 11.2, blockchain technology has been gradually applied to IoT systems. However, for the integration of blockchain technology into 6G-oriented IoT systems, several technical concerns still need to be further addressed, so as to enhance the system's performance. In this section, future research directions are investigated.

11.3.1 Malicious Attacks

Although it has been applied to IoT systems for data privacy and security management, as analyzed in Section 11.2.1, blockchain technology has met some malicious attacks. These attacks subsequently make blockchain-enabled IoT systems vulnerable to the risks of privacy leakage and network breakdown. Thus, the following malicious attacks need to be addressed to enhance the innate privacy and security of the blockchain:

- **Anonymity Cracking:** In blockchain technology, transparent transactions are openly broadcast and can be visualized and recorded by all the participants. The technical advantage of transparency brings the risk of anonymity cracking. Based on the observation of transparent transactions, adversaries can infer the actual identities behind the anonymous participants by pattern recognition. In blockchain-enabled IoT systems, malicious attackers can easily grab the information of users, as IoT data are highly related to the personal identities and daily activities of users. Therefore, anonymity protection is a fundamental necessity for privacy preservation in the blockchain.

- **Authentication Risk:** In blockchain-enabled systems, a private key held by the users is used for authentication. However, the renowned attacking event occurring to Mt.Gox, a Bitcoin wallet company, where all the private keys of customers were stolen, reminds the fact that the elliptic curve cryptography used in blockchain technology is not strong enough to protect the private keys [25]. Therefore, the private key protection or even a novel authentication mechanism needs to be seriously investigated while applying blockchain technology to the 6G-oriented IoT systems.

- **Fifty-One Percent Attack:** The security of the blockchain is built on the decentralized consensus mechanism. The decentralized consensus mechanism brings the technical advantage of decentralization while potentially leading to a risk of a 51% attack in the meantime. When a malicious attacker controls over 51% of the computational power, it may be able to manipulate the blockchain and almost guarantees its successfulness [16]. Fifty-one percent attack is a critical challenge since the advent of blockchain technology. For Bitcoin, with the boosting difficulty of mining, it is almost impossible for a single or few malicious attackers to manipulate over 51% computational power of the Bitcoin network and launch the attack. However, for blockchain-enabled IoT systems, the small scale of the system cannot guarantee its security, and the 51% attack is still a critical consensus risk.

- **Malleability Attack:** The aim of data integrity protection in the blockchain is to guarantee that all the transactions recorded in the blocks are not altered or tampered with by unauthorized parties. However, a malleability attack severely challenges the data integrity protection in the blockchain [9]. In a malleability attack, malicious attackers cheat the transaction issuer by intercepting, modifying, and rebroadcasting a transaction, which can finally lead to the termination of all new transactions.

With the development of blockchain technology, malicious attacks have attracted research efforts from both industry and academic institutions. For anonymity protection, the following categories of approaches have been proposed: (i) *mixing services*, such as centralized mixing service providers [17] and decentralized multi-party computation [58], (ii) *ring signature*, such as CryptoNote [42], and (iii) *non-interactive zero-knowledge proof*, such as ZeroCoin [33]. To resolve the issue of authentication risk, research works have been done by proposing novel identity management schemes such as rotating asymmetric keys [22] and novel authentication mechanisms such as automatic authentication [12]. In terms of the 51% attack, there are only a few contributors working on tailoring the mechanisms for IoT systems against such an attack [28, 36]. To resolve the malleability attack in blockchain-enabled IoT systems, an improved ElGamal cryptosystem has been proposed [26]. Malicious attacks on blockchain-enabled IoT systems have attracted wide attention and achieved great progress. However, investigations on the 51% attack and malleability attack are not as many as the other attacks, which still need to be further studied.

11.3.2 Heterogeneous Devices and Scalability

The advent of 5G beyond (B5G) and 6G wireless communication technologies drastically empowers the intelligent connections of human beings, physical objects, and even virtual objects. Under such a situation,
the number and heterogeneity of IoT devices increase dramatically. The
capabilities of IoT devices are different in computation, communications,
storage, power supply, *etc.* For some IoT devices, the computational
power and memory space are limited, which cannot support a complex
mining mechanism or record a full version of the blockchain. Therefore,
the different capabilities of IoT devices need to be considered while integrating blockchain technology into 6G-oriented IoT systems.

Besides the inherently limited resources of IoT devices, the size of
the blockchain keeps growing all the time, and due to that, transactions are kept generated, packed into a block, and added to the chain.
The insistently growing blockchain size leads to the overwhelming cost
of time and memory on the initial download and synchronization of
blockchain, which makes many more IoT devices cannot afford the resources of communication bandwidth, memory space, and even power
supply. The explosive growth of blockchain size has brought a critical
technical challenge of scalability to IoT systems.

Considering the heterogeneity of IoT devices and the issue of scalability, the development of lightweight versions of blockchain for IoT systems
is a necessity. In recent years, the lightweight blockchain for IoT systems
has already attracted wide attention. Research efforts have been spared
on IoT applications and IoT system performance enhancement, such as
lightweight blockchain for IoT-enabled smart healthcare [20], lightweight
blockchain for industrial IoT, and lightweight blockchain-enabled performance enhancement for resource-constraint IoT [3], [24]. In the coming
6G era, the number of connected objects and the blockchain size expect
to meet explosive growth, which undoubtedly makes the issues of heterogeneity and scalability more critical and needs more investigations from
both industry and academic institutions.

11.3.3 Parallel Transactions and Side-Chains

With the pervasive deployment of IoT systems, the number of IoT devices and the corresponding data generated by these devices meet explosive growth. The traditional blockchain with a single main chain can
hardly support the highly distributed and heterogeneous IoT devices

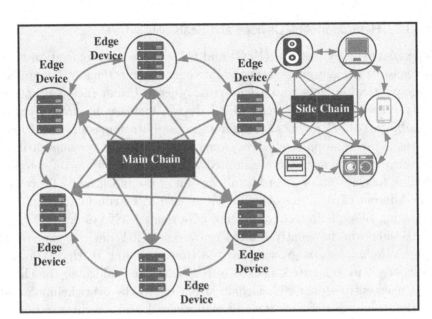

Figure 11.10   Hierarchical blockchain-enabled IoT system architecture.

and data. Under such a situation, the development of a hierarchical blockchain with parallel transactions and side-chains is a necessity.

In the hierarchical blockchain, there is a main chain for the collection of the essential infrequent global events across the whole system, and side-chains are developed for local events of the local sub-networks in the system. For instance, as shown in Figure 11.10, the hierarchical blockchain-enabled IoT system architecture consists of two layers, the P2P network of edge computing devices as the main chain layer and the hybrid network consisting of edge and IoT end devices as the side-chain layer. Edge devices function as the interface between main and side-chains.

- **Edge Computing Device:** The main chain in the hierarchical architecture is managed by the edge computing devices, where the full records of the blockchain are synchronized among the edge devices.

- **IoT End Device:** The edge device and the IoT end devices are formed up as a local sub-network. The side-chain is managed among these devices and interacts with the main chain through the edge device.

In terms of the performance of the hierarchical blockchain, to ensure data integrity, the main chain and side-chains must be interconnected, and the hash values of side-chains need to be synchronized to the main chain through tailored transactions. To reduce the capacity requirements and preserve consistency across the system, the main chain should be synchronized at a lower frequency than the side-chains.

11.3.4 System Performance and Overhead

As investigated in Section 11.2, blockchain technology can facilitate IoT systems from different aspects. However, the workload of mining and storage of blockchain have brought an extra burden to IoT systems and directly affect the system performance.

- **Latency:** The latency of task processing is critical for IoT applications with the requirement of real-time data analytics, such as smart healthcare, smart city, and industrial IoT [55]. However, the traditional decentralized consensus mechanism and the related mining method can lead to long latency for the generation and addition of a new block, which can hardly meet the real-time requirements of the time-critical IoT applications [1].

- **Energy Consumption:** The low-cost IoT devices are normally built with limited resources on computation and power supply. However, the maintenance of a blockchain is energy-consuming, particularly in mining and communications. In the mining process, all the volunteer participants compete to solve the cryptographic puzzle, which occupies a high proportion of the local computational power and consumes a large amount of energy. For communications, in P2P-based blockchain-enabled IoT systems, the IoT devices have to keep awake all the time for listing, transmitting, and receiving messages, where the energy leakage is fast [6]. Therefore, the high energy consumption brought by blockchain is a critical challenge for resource-constrained IoT systems.

- **Throughput:** With the explosive increment in the number of IoT devices, the number of transactions to be processed in IoT systems increases in an exponential trend. However, in the traditional decentralized consensus mechanism, PoW, only a maximum of seven transactions can be processed per second, which can hardly meet the throughput in the enlarging IoT systems [15].

A promising solution to improve the performance of blockchain-based IoT systems is to develop consensus algorithms with lower cost and higher efficiency. In recent three years, lightweight blockchain for IoT applications and system performance enhancement have been investigated [10, 11, 14, 27, 30]. While meeting the ultra-reliable and low-latency requirements of 6G-oriented IoT systems, lightweight consensus algorithms and lightweight versions of blockchain still need to be further studied in the future.

11.4 SUMMARY

In this chapter, blockchain technology for 6G-oriented IoT systems has been presented from the perspectives of principle, applications, and future research directions. First of all, the principle of blockchain technology is introduced with the structure of blockchain, related decentralized consensus mechanisms, and its technical advantages. Afterwards, applications of blockchain technology to 6G-oriented IoT systems are surveyed, where data privacy and security management, distributed data sharing and storage, smart contract-based property trading, intelligent manufacturing and asset tracking, and intelligent and autonomous vehicular IoT are summarized with both theoretical principle and application instances. Lastly, the future research directions of applying blockchain technology to 6G-oriented IoT systems have been well analyzed as follows, namely, heterogeneous devices, malicious attacks, blockchain size and scalability, parallel transactions and side-chains, and the network performance including throughput, latency, and energy efficiency.

BIBLIOGRAPHY

[1] Subhi M Alrubei, Edward A Ball, Jonathan M Rigelsford, and Callum A Willis. Latency and performance analyses of real-world wireless IoT-blockchain application. *IEEE Sensors Journal,* 20(13):7372–7383, 2020.

[2] Andreas M Antonopoulos. *Mastering Bitcoin: Unlocking digital cryptocurrencies.* O'Reilly Media, Inc., 2014.

[3] Eranga Bandara, Deepak Tosh, Peter Foytik, Sachin Shetty, Nalin Ranasinghe, and Kasun De Zoysa. Tikiri—Towards a lightweight blockchain for IoT. *Future Generation Computer Systems,* 119:154–165, 2021.

[4] Haoye Chai, Supeng Leng, Yijin Chen, and Ke Zhang. A hierarchical blockchain-enabled federated learning algorithm for knowledge sharing in internet of vehicles. *IEEE Transactions on Intelligent Transportation Systems*, 22(7):3975–3986, 2020.

[5] M. Cherednychenko. What is smart contract and how can your business benefit from it? (Online), 2021, Available: https://softensydevs. medium.com/what-is-smart-contract-and-how-can-your-business-benefit-from-it-2a97512c4948, [accessed on 10th August 2022].

[6] Ryan Cole and Liang Cheng. Modeling the energy consumption of blockchain consensus algorithms. In *2018 IEEE International Conference on Internet of Things (iThings) and IEEE Green Computing and Communications (GreenCom) and IEEE Cyber, Physical and Social Computing (CPSCom) and IEEE Smart Data (SmartData)*, pages 1691–1696. IEEE, 2018.

[7] Michael Crosby, Pradan Pattanayak, Sanjeev Verma, Vignesh Kalyanaraman, and Nachiappan. Blockchain technology: Beyond bitcoin. *Applied Innovation*, 2(6–10):71, 2016.

[8] Hong-Ning Dai, Zibin Zheng, and Yan Zhang. Blockchain for internet of things: A survey. *IEEE Internet of Things Journal*, 6(5):8076–8094, 2019.

[9] Christian Decker and Roger Wattenhofer. Bitcoin transaction malleability and mtgox. In *European Symposium on Research in Computer Security*, pages 313–326. Springer, 2014.

[10] BD Deebak, Fida Hussain Memon, Sunder Ali Khowaja, Kapal Dev, Weizheng Wang, Nawab Muhammad Faseeh Qureshi, and Chunhua Su. Lightweight blockchain based remote mutual authentication for AI-empowered IoT sustainable computing systems. *IEEE Internet of Things Journal*, 2022, doi: 10.1109/JIOT.2022.3152546.

[11] Ronald Doku, Danda B Rawat, Moses Garuba, and Laurent Njilla. Lightchain: On the lightweight blockchain for the internet-of-things. In *2019 IEEE International Conference on Smart Computing (SMARTCOMP)*, pages 444–448. IEEE, 2019.

[12] Ali Dorri, Salil S Kanhere, and Raja Jurdak. Towards an optimized blockchain for IoT. In *2017 IEEE/ACM Second International Conference on Internet-of-Things Design and Implementation (IoTDI)*, pages 173–178. IEEE, 2017.

[13] Ali Dorri, Salil S Kanhere, Raja Jurdak, and Praveen Gauravaram. Blockchain for IoT security and privacy: The case study of a smart home. In *2017 IEEE International Conference on Pervasive Computing and Communications Workshops (PerCom Workshops)*, pages 618–623. IEEE, 2017.

[14] Ali Dorri, Salil S Kanhere, Raja Jurdak, and Praveen Gauravaram. LSB: A lightweight scalable blockchain for IoT security and anonymity. *Journal of Parallel and Distributed Computing*, 134:180–197, 2019.

[15] Tiago M Fernández-Caramés and Paula Fraga-Lamas. A review on the use of blockchain for the internet of things. *IEEE Access*, 6:32979–33001, 2018.

[16] J. Frankenfield. 51% Attack. Investopedia. (Online), 2019, Available: https://www.investopedia.com/terms/1/51-attack.asp, [accessed on 10th August 2022].

[17] Ethan Heilman, Leen Alshenibr, Foteini Baldimtsi, Alessandra Scafuro, and Sharon Goldberg. Tumblebit: An untrusted bitcoin-compatible anonymous payment hub. *Cryptology ePrint Archive*, 2016.

[18] Z. Herbert. Why blockchains are the future of cloud storage. (Online), 2017, Available: https://blog.sia.tech/why-blockchains-are-the-future-ofcloud-storage-91f0b48cfce9, [accessed on 10th August 2022].

[19] Seyoung Huh, Sangrae Cho, and Soohyung Kim. Managing IoT devices using blockchain platform. In *2017 19th International Conference on Advanced Communication Technology (ICACT)*, pages 464–467. IEEE, 2017.

[20] Leila Ismail, Huned Materwala, and Sherali Zeadally. Lightweight blockchain for healthcare. *IEEE Access*, 7:149935–149951, 2019.

[21] Yuna Jiang, Yi Zhong, and Xiaohu Ge. Smart contract-based data commodity transactions for industrial internet of things. *IEEE Access*, 7:180856–180866, 2019.

[22] David W Kravitz and Jason Cooper. Securing user identity and transactions symbiotically: IoT meets blockchain. In *2017 Global Internet of Things Summit (GIoTS)*, pages 1–6. IEEE, 2017.

[23] Aparna Kumari, Rajesh Gupta, and Sudeep Tanwar. Amalgamation of blockchain and IoT for smart cities underlying 6G communication: A comprehensive review. *Computer Communications*, 172:102–118, 2021.

[24] Chunlin Li, Jing Zhang, Xianmin Yang, and Luo Youlong. Lightweight blockchain consensus mechanism and storage optimization for resource-constrained IoT devices. *Information Processing & Management*, 58(4):102602, 2021.

[25] R. McMillan. The inside story of mt. gox, bitcoin's $460 million disaster. *Wired*, 3, March 2014.

[26] Maya Mohan, MK Kavithadevi, V Jeevan Prakash. Improved ElGamal cryptosystem for secure data transfer in IoT networks. In *2020 Fourth International Conference on I-SMAC (IoT in Social, Mobile, Analytics and Cloud)(I-SMAC)*, pages 295–302. IEEE, 2020.

[27] Sachi Nandan Mohanty, KC Ramya, S Sheeba Rani, Deepak Gupta, K Shankar, SK Lakshmanaprabu, and Ashish Khanna. An efficient lightweight integrated blockchain (ELIB) model for IoT security and privacy. *Future Generation Computer Systems*, 102:1027–1037, 2020.

[28] Ho-se Moon, Jaegeun Song, Hyeonwoo Shin, and Juwook Jang. Home IoT device management blockchain platform using smart contracts and a countermeasure against 51% attacks. In *2022 4th Asia Pacific Information Technology Conference*, pages 191–195, 2022.

[29] A. Morrison. How smart contracts automate digital business. (Online), 2016, Available: http://usblogs.pwc.com/emerging-technology/howsmart-contracts-automate-digital-business/, [accessed on 10th August 2022].

[30] Dongjun Na and Sejin Park. Fusion chain: A decentralized lightweight blockchain for IoT security and privacy. *Electronics*, 10(4):391, 2021.

[31] Satoshi Nakamoto. Bitcoin: A peer-to-peer electronic cash system. (Online), 2008, Available: https://bitcoin.org/bitcoin.pdf, [accessed on 10th August 2022].

[32] Tahmid Hasan Pranto, Abdulla All Noman, Atik Mahmud, and AKM Bahalul Haque. Blockchain and smart contract for IoT enabled smart agriculture. *PeerJ Computer Science*, 7:e407, 2021.

[33] Zerocoin Project. https://zerocoin.org/, [accessed on 10th August 2022].

[34] Abirami Raja Santhi and Padmakumar Muthuswamy. Influence of blockchain technology in manufacturing supply chain and logistics. *Logistics*, 6(1):15, 2022.

[35] Krushali Shah, Swapnil Chadotra, Sudeep Tanwar, Rajesh Gupta, and Neeraj Kumar. Blockchain for IoV in 6G environment: Review solutions and challenges. *Cluster Computing*, 25:1927–1955, 2022.

[36] Rakesh Shrestha and Seung Yeob Nam. Regional blockchain for vehicular networks to prevent 51% attacks. *IEEE Access*, 7:95033–95045, 2019.

[37] Statista. Blockchain technology market size worldwide 2018-2025. [Online], 2022. Available: https://www.statista.com/statistics/647231/worldwide-blockchain-technology-market-size/, [accessed on 10th August 2022].

[38] Nick Szabo. Smart contracts: Building blocks for digital markets. *EXTROPY: The Journal of Transhumanist Thought*, 18(2):28, 1996.

[39] Faisal Tariq, Muhammad R. A. Khandaker, Kai-Kit Wong, Muhammad A. Imran, Mehdi Bennis, and Merouane Debbah. A speculative study on 6G. *IEEE Wireless Communications*, 27(4):118–125, 2020.

[40] Noshina Tariq, Ayesha Qamar, Muhammad Asim, and Farrukh Aslam Khan. Blockchain and smart healthcare security: A survey. *Procedia Computer Science*, 175:615–620, 2020.

[41] Quang Nhat Tran, Benjamin P Turnbull, Hao-Tian Wu, AJS De Silva, Katerina Kormusheva, and Jiankun Hu. A survey on privacy-preserving blockchain systems (PPBS) and a novel PPBS-based framework for smart agriculture. *IEEE Open Journal of the Computer Society*, 2:72–84, 2021.

[42] CryptoNote Official Website. https://cryptonote.org/, [accessed on 10th August 2022].

[43] Everledger Official Website. https://diamonds.everledger.io/, [accessed on 10th August 2022].

[44] Filament Official Website. https://filament.com/, [accessed on 10th August 2022].

[45] Filecoin Official Website. https://filecoin.io/, [accessed on 10th August 2022].

[46] Guardtime Official Website. https://guardtime.com/, [accessed on 10th August 2022].

[47] IBM Blockchain Platform Official Website. https://www.ibm.com/blockchain/platform, [accessed on 10th August 2022].

[48] IOV Blockchain Official Website. http://blockchain.iovscan.com/iov/en/, [accessed on 10th August 2022].

[49] IPFS Official Website. https://ipfs.io/, [accessed on 10th August 2022].

[50] Storj Official Website. https://storj.io/, [accessed on 10th August 2022].

[51] Duane Wilson and Giuseppe Ateniese. From pretty good to great: Enhancing PGP using bitcoin and the blockchain. In *International Conference on Network and System Security*, pages 368–375. Springer, 2015.

[52] Yang Xiao, Ning Zhang, Wenjing Lou, and Y Thomas Hou. A survey of distributed consensus protocols for blockchain networks. *IEEE Communications Surveys & Tutorials*, 22(2):1432–1465, 2020.

[53] Xing Yang, Lei Shu, Jianing Chen, Mohamed Amine Ferrag, Jun Wu, Edmond Nurellari, and Kai Huang. A survey on smart agriculture: Development modes, technologies, and security and privacy

challenges. *IEEE/CAA Journal of Automatica Sinica*, 8(2):273–302, 2021.

[54] Hao Yin, Dongchao Guo, Kai Wang, Zexun Jiang, Yongqiang Lyu, and Ju Xing. Hyperconnected network: A decentralized trusted computing and networking paradigm. *IEEE Network*, 32(1):112–117, 2018.

[55] Tianqi Yu and Xianbin Wang. Real-time data analytics in internet of things systems. *Handbook of Real-Time Computing*. Springer, 2020.

[56] Xiaochen Zheng, Jinzhi Lu, Shengjing Sun, and Dimitris Kiritsis. Decentralized industrial IoT data management based on blockchain and IPFS. In *IFIP International Conference on Advances in Production Management Systems*, pages 222–229. Springer, 2020.

[57] Saide Zhu, Zhipeng Cai, Huafu Hu, Yingshu Li, and Wei Li. zkCrowd: A hybrid blockchain-based crowdsourcing platform. *IEEE Transactions on Industrial Informatics*, 16(6):4196–4205, 2019.

[58] Jan Henrik Ziegeldorf, Fred Grossmann, Martin Henze, Nicolas Inden, and Klaus Wehrle. Coinparty: Secure multi-party mixing of bitcoins. In *Proceedings of the 5th ACM Conference on Data and Application Security and Privacy*, pages 75–86, 2015.

[59] Weiqin Zou, David Lo, Pavneet Singh Kochhar, Xuan-Bach Dinh Le, Xin Xia, Yang Feng, Zhenyu Chen, and Baowen Xu. Smart contract development: Challenges and opportunities. *IEEE Transactions on Software Engineering*, 47(10):2084–2106, 2019.

CHAPTER 12

6G and IOT Use Cases

Asif Ali

Advanced Communication Lab, Department of Electronics and Communication Engineering, National Institute of Technology Srinagar, Srinagar, Jammu and Kashmir, India

Syed Mujtiba Hussain

Advanced Communication Lab, Department of Electronics and Communication Engineering, National Institute of Technology Srinagar, Srinagar, Jammu and Kashmir, India.
Department of Computer Science and Engineering, Islamic University of Science and Technology Awantipora, India

G R Begh

Advanced Communication Lab, Department of Electronics and Communication Engineering, National Institute of Technology Srinagar, Srinagar, Jammu and Kashmir, India

CONTENTS

12.1	Introduction	316
12.2	Comparison of 5G and 6G	318
	12.2.1 On the Basis of Use Cases	318
	12.2.2 Technology Trends	319
12.3	6G Wireless Communication Technology for IOT	320
	12.3.1 Edge Intelligence	320
	12.3.2 Super Reliable and Ultra-low-Latency Network (SRULLN)	321
	12.3.3 Extreme Coverage Communication Environment	322
	12.3.4 Ambient and Backscattered Communication	323
	12.3.5 Highly Secure and Reliable Communication	323
	12.3.6 Terahertz Communication	324
12.4	IOT Use Cases and Likelihood of 6G Wireless Communication	324
	12.4.1 Extremely Low-Latency Healthcare System	324
	12.4.2 Smart City	325

DOI: 10.1201/9781003282211-12

	12.4.3	Enhanced Smart Industry	325
	12.4.4	Enhanced Vehicular Internet of Things (VIoT) and Autonomous Driving	326
	12.4.5	Satellite Internet of Things (SIoT)	327
	12.4.6	Haptic Communication	327
	12.4.7	Robotics	329
	12.4.8	Internet of Bio-Nano-Things	329
	12.4.9	Augmented Reality and Virtual Reality	330
	12.4.10	Holographic Telepresence (Teleportation)	330
	12.4.11	Smart Education/Training	330
12.5	Security and Privacy in 6G for IoT		331
12.6	Research Challenges and Future direction		332
	12.6.1	Energy Efficiency in 6G-IoT	332
	12.6.2	Hardware Constraints of IoT Devices	333
	12.6.3	Security and Privacy Challenges in 6G-IoT	334
	12.6.4	Standard Specifications for 6G-IoT	335
12.7	Conclusion		335
Bibliography			336

E VER SINCE THE inception of wireless communication, networks are undergoing substantial changes from time to time, determined by the unprecedented growth in data traffic and ever-increasing demand for fast, sustainable, and reliable exchange of information. Enormous amount of data traffic is generated from the Internet every day, particularly driven by the ever-increasing popularity of smart devices. Over the years, different generation networks have come into existence to meet the ever-increasing traffic demands. Lately, sixth generation (6G) wireless communication networks based on Internet of Things (IOT) are expected to transform consumer services and applications. The underlying technology for IOT to exist everywhere will be artificial intelligence (AI). A comprehensive study on the inception and co-existence of 6G and IOT is performed to investigate the rising potential of 6G technologies in IOT networks.

12.1 INTRODUCTION

Internet of Things (IoT) is a kind of network in which sensor devices sense the information and transmit it for processing through gateways. Advancement in wireless communications has aided the expansion of IoT capabilities to interconnect millions of devices [1, 2]. To connect a physical environment with Internet, IoT plays an important role by enabling innovative software solutions which have improved the quality of living for human beings. These applications necessitate high data rates, coverage, latency, and so on in terms of performance. They are computationally demanding and more data intensive, well exceeding the capabilities of super-reliable low-latency network (sRLLN) and massive machine-type communication (mMTC). Where it becomes difficult to manage a vast number of IoT devices efficiently. Also, serious security issues arise due to the generation of a huge amount of data. With the advancement in IoT, fifth-generation (5G) will gradually reach its limits and be unable to support the majority of technical criteria's of innovative applications such as highly dynamic, autonomous and fully intelligent services, as previous generations have [3, 4]. The rapid expansion of automated and intelligent IoT networks is likely to outpace the capabilities of 5G, so we need to extend the capabilities of 5G to 6G, to enable massive IoT.

The technology of wireless communication has advanced every decade, resulting in new-generation systems. Figure 12.1 shows how technology has changed from first generation to fifth generation and future

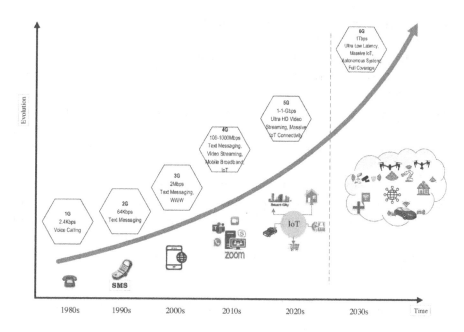

Figure 12.1   Innovation in technology and services toward 6G-IoT.

6G. It is focused to furnish an unbelievable service quality and will improve the experience of users in existing IoT networks due to its exceptional capabilities over previous network generations, like extremely high throughput, ultra-low-latency communication, satellite-based customer service, autonomous and massive networks [5, 6].

12.2 COMPARISON OF 5G AND 6G

We will discuss the need for 6G on the basis of use cases and technological trends

12.2.1 On the Basis of Use Cases

5G represented a tremendous step toward evolving a delay-sensitive tactile access network, by redesigning the core network through the usage and management of millimeter wave in NR (new radio). The exponentially increasing demand for data rates by data centric, autonomous applications, fully connected, intelligent digital world may surpass the limits of 5G network. So in order to meet these demands fundamentally new communication technology and network architecture will be needed. 6G might very significantly benefit from a still higher spectrum such as through Terahertz and optical communication [4, 7].

Even though 5G developments are toward the more improved wireless network, the diversity of future intelligent network applications and the need for 3D connectivity push for innovative cell-less architecture and design framework. By 2030, we anticipate 6G to transmit intelligence from centrally controlled computing facilities to end terminals. Improved services, integration of new and existing different use cases, acceleration of data analysis, and the evolution of integrated devices will be required [1, 8].

5G provides various solutions to many of the high-speed low-latency social problems that include distance education, telemedicine and autonomous operation of various devices. In the 2030s, more widespread adoption of solutions and more advanced interaction will necessitate comprehensive problem addressing and development. Everything will be required to access an ultra-real interaction from anywhere at any time. This will significantly reduce social and cultural differences between different habituations, rural or urban, thereby prevent urbanization and support local development [2, 9].

Wearable device advanced functionalities, such as extended reality (XR) (virtual reality [VR], mixed reality [MR], augmented reality [AR]) devices, pictures and holograms with extremely high-resolution, five-sense communications, will proliferate, and communications among people and between humans and different objects will become a reality [6].

Due to the requirements of numerous sensor networks, autonomous objects, and autonomous construction sites, a communication space in an environment void of humans is also required. Therefore, every location on the land, in the sky, and at undersea will be a communication zone.

In comparison to its 5G counterpart, sixth generation (6G) is anticipated to offer detrimental communication technology and advanced networking infrastructures to realize millions of innovative IoT applications by introducing stringent and strict network requirements in a holistic manner. 6G communication systems evolve toward a more powerful and stronger IoT environment that builds a fully connected and smart cyber world toward the foreseen social, economic, and environmental ecosystems in near future, 2030 [4].

In the future, the transition from VR/AR, wearable devices, and tactile networking to five-sense communications, holographic interactions, will result in a spike of large amounts of IoT data being generated in near real-time; more computing power will be required for data processing and analysis. The innovative IoT-enabled applications in 6G will also demand more rigorous latency requirements. As more data will be involved, there will be a greater requirement for data security and privacy protection [2].

By communicating and processing vast amounts of data without delay between digital and physical space, a tighter collaboration between the two spaces will be created, and eventually, fusion without a gap between the places would be realized [10]

12.2.2 Technology Trends

One of the most important technical metrics for the change of communication generation from 1G to 6G has been the peak data rate. Without a doubt, 6G will increase the peak data rate attained, which is estimated to exceed Tb/s. Since low latency is critical in many wireless applications, 6G is expected to provide ultra-low-latency requirement. 6G will provide unprecedented energy consumption difficulties due to its ultra-high throughput, ultra-wide bandwidth, and ultra-massive universal wireless access points (APs). Various sensors will become increasingly common in people's lives, necessitating a large amount of energy to run and maintain these sensor networks. As a result, green communication is especially important for 6G. Potential applications involving complicated operations, such as autonomous driving, autonomous factories, and remote surgery, necessitate a 6G localization requirement to ensure user safety and protect values and people. The reliability requirements of various 6G systems are expected to vary, with the most severe example being one billion bits sent with a only one-bit mistake and a 0.1 ms latency. Human activities are being stretched to more severe environments as knowledge and technology improve, such as higher altitudes, outer space, seas, and deep beneath the sea. Communication nodes, particularly IoT devices, will be dispersed across a larger region. So 6G is expected to have extreme coverage environment. 6G should provide an extra level of privacy and security compared to 5G as more and more personal information will be involved in future with new applications emerging [6].

12.3 6G WIRELESS COMMUNICATION TECHNOLOGY FOR IOT

The underlying technologies that will ensure future 6G-IoT systems and applications are shown in Figure 12.2.

12.3.1 Edge Intelligence

In intelligent systems, AI is expected to be extended to the edge nodes, which includes the co-existence of AI, communications, and edge computation, resulting in an entirely new technology known as edge intelligence. Edge intelligence leads to dynamic spectrum access, thereby allowing for rapid and reliable intelligence processing of data at edge nodes such as platoons, street vehicles, and roadside units (RUSs). Edge intelligence's benefits are also being considered for industrial robotics in industrial automation. Furthermore, AI functions may be executed on pervasive edge devices to provide large-scale intelligence services powered by big data analytics. Based on sophisticated communication technologies, the 6G-based big data analysis solution may considerably improve large-scale data transfer and data processing speeds. In the future 6G era, fog computing may be integrated with AI approaches to create fog intelligence, offering low-latency IoT services. Distributed IoT devices can further collaborate with a data collector to execute neural network training in the framework of intelligent IoT networks, where devices simply communicate parameters and raw data sharing is not required. One

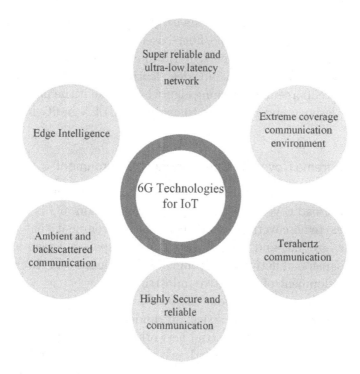

Figure 12.2 Understanding aspects of 6G for IoT.

of the potential technologies in edge intelligence is AI on chips, which can speed the growth of IoT systems in the upcoming 6G technology. The advancement of edge intelligence is heavily reliant on edge nodes, which play most important roles in the learning of tasks, which includes categorization and prediction. For a secure and trustworthy edge intelligence environment, a more secure solution is to be developed in 6G-IoT networks [11].

12.3.2   Super Reliable and Ultra-low-Latency Network (SRULLN)

Wireless connections that interconnect AI and gadgets in cyber-physical integration are analogous to the communication of data by the human nervous system. An always consistent end-to-end (E2E) low latency appears to be a basic condition for actualizing services in real time and is highly interactive. The desired E2E latency for 6G is about 1 ms or less. With SRULLN, autonomous vehicular systems will be able to communicate information reliably among cars, infrastructure, and people.

Integration of SRULLN into the smart grid is also a current application, with the goal of replacing cable/fiber-based systems for real-time protection and management of dispersed grid lines and stations. Furthermore, recent advances in AI that include deep learning and fast network control using deep reinforcement learning algorithms make it the perfect tool for analyzing latency and reliability to enable SRULLN in 6G-IoT [10, 12].

12.3.3 Extreme Coverage Communication Environment

In order to effectively enable future IoT applications, we need to have incredibly broad coverage and ubiquitous connection. 6G is expected to have an extreme coverage communication environment that can cover space, air, ground, and underwater. The widespan coverage area helps in the expansion of activity environments for humans and things, as well as the development of innovative industries.

As a result, a cell-free, four major connectivity for 6G-IoT may be constructed, with four tiers: space, air, underwater, and land. The areas that can't be served by terrestrial networks, space communication, provides wireless service via low earth orbit (LEO), geostationary earth orbit (GEO), and medium-earth orbit (MEO), or satellites, with an added advantage that it is used to enable large-bandwidth space networks for satellite–ground connectivity. Unmanned aerial vehicles (UAVs) serve as mobile base stations, providing coverage and connection for inaccessible areas and simultaneously assisting public safety networks and emergency circumstances where ultra-reliable and low-latency communication (URLLC) is necessary. UAVs can also establish direct air–ground linkages for cooperative sensing and data transfer in 6G-IoT by integrating with terrestrial base stations. Terrestrial communication seeks to provide wireless coverage and connection for human activities on the ground by interconnecting physical base stations, computer servers, and mobile devices. The THz band will be used in 6G-IoT to improve bandwidth efficiency and increase data rates, particularly in ultra-dense manets with millions of users. Underwater IoT devices such as submarines participating in broad-sea and deep-sea operations provide underwater communication. The communication between the control hubs and underwater IoT devices necessitates communication from both sides [9, 13].

12.3.4 Ambient and Backscattered Communication

Wireless networks powered by ambient waves are potent to change the IoT world, enabling unprecedented device downsizing and low-maintenance distributions even in harsh locations. Recent improvements in backscatter connectivity (BC) have enabled the use of electromagnetic (EM) waves in IoT devices, generated by the plethora of wireless networks around us. BC decouples the creation of carrier signals from the process of communication. As a result, the carrier may be produced by an external device, and BC-enabled devices may absorb a portion of it to become active, as well as regulate it on the fly, allowing data to be attached to it. This significantly simplifies the BC-device hardware by eliminating the requirement for a battery as well as the carrier-generating circuitry. Ambient BC (aBC) offers great deployment adaptability by using ambient EM waves generated by various devices. Despite this, aBC must contend with the intrinsically incoherent and chaotic nature of ambient waves, resulting in inconsistent and unpredictable functioning. Programmable wireless environments (PWEs), which were recently proposed, may operate as natural enablers for stable aBC systems. PWEs are E2E systems that turn the propagation of EM waves in space into a software-defined and even a predictable process. An aBC device does not generate radio frequency (RF) signals, according to the backscattering paradigm. Rather, aBC networks are based on modified reflection of RF waves arriving at the aBC device from external sources. The external RF signal itself serves as the aBC device's power source, which even the device simultaneously receives, harvests, and modifies for retransmission. The lack of batteries and RF transmitters reduces the hardware of aBC devices, allowing them to operate in extremely low-power conditions, such as sub-10 mW [14, 15].

12.3.5 Highly Secure and Reliable Communication

6G systems are expected to face increased risks of assaults and threats and achieving a high level of security in IoT networks is of great concern. Maintaining data privacy in open systems, with the primary example being vehicular data sharing, is a major concern. As an emerging technology, blockchain is capable of providing novel solutions to successfully address privacy and security issues. The blockchain, in its most basic form, is a distributed, immutable, and public database in which no central authority is required to handle the data. Blockchains are characterized as either public or private. A public blockchain enables anybody to conduct transactions and participate in decision-making. However, private blockchains, on the other hand, are the approved networks maintained by a central institution. The approval to join a blockchain network passes through a defined validation method. The blockchain has a variety of desirable properties enabled by its working idea, including decentralization, traceability, reliability, and integrity. These characteristics make blockchain an attractive contender for security and privacy provision in 6G-IoT networks [11].

12.3.6 Terahertz Communication

The FCC of the United States suggests that frequencies higher than 5G, such as 95 GHz to 3 THz, be examined for 6G. Even, in compared to 5G, a surprisingly large frequency bandwidth may be employed in such high-frequency bands ranging from the top portion of the millimeter wave band to the "terahertz wave" band, and extreme high data speeds surpassing 100 Gbps are being investigated. It should be noted that, in addition to cellular, WiFi, and Bluetooth, phones will incorporate other modes of operation from numerous frequencies, with a vast quantity of accessible bandwidth of up to many THz and a data rate of up to Tbps. The technological study is required, such as advancements in RF device technology and use based on the aforementioned new network topology. Furthermore, when additional frequency bands, such as millimeter wave and terahertz wave bands, are added to the present frequency band, very large frequency bands will be used, in contrast to the past. As a result, there appear to be many related study fields, such as optimizing the selected application of multiple bands based on the application, reexamining the frequency reuse method between cells, upgrading the duplexing method in the uplink and downlink, and reexamining the low-frequency band utilization method [3, 16].

12.4 IOT USE CASES AND LIKELIHOOD OF 6G WIRELESS COMMUNICATION

Applications of 6G will be more diverse; few of the applications of 6G for IoT are discussed below.

12.4.1 Extremely Low-Latency Healthcare System

6G will transform the healthcare industry by removing time and space barriers through remote surgery and ensuring healthcare workflow optimizations. Aside from the high cost, the main limitation at the moment is the lack of real-time tactile feedback. Furthermore, the proliferation of eHealth services will put the ability to meet their stringent quality of service (QoS) requirements, such as continuous connection availability (99.99999 percent reliability), ultra-low latency (sub-ms), and mobility support, to the test. These key performance indicators (KPIs) will be guaranteed by improved spectrum availability, paired with the advanced intelligence of 6G networks, as well as 5–10× gains in spectral efficiency [9, 17].

12.4.2 Smart City

The enhanced characteristics of 6G will result in a considerable improvement in life quality, intelligent monitoring, and automation, enabling the faster construction of super-smart cities and households. A city is considered smart when it is capable of operating intelligently and autonomously by gathering and examining large amounts of data from a variety of industries, ranging from urban planning to garbage collection, in order to make effective use of public resources, improve the quality of services provided to citizens, and reduce the operational costs of government agencies. Cities will emerge smarter in 6G due to the use of smart mobile devices, self-driving cars, and other technology. A smart home is not only a residential or commercial structure outfitted with Internet-connected smart devices that enable people to handle and track a variety of appliances and systems from their smartphones, but it is also an intellectual entity capable of making instantaneous and collective decisions. People can also use their brains to handle light, heat, or multimedia entertainment (brain–computer interfaces, BCI), or they can leave it all to AI, which can study their activities and make their life safer and easier. Because of 6G, the smart home will become a reality. Given a large number of sensors and intelligent terminals, building a smart city/home offers substantial obstacles to 6G connectivity and coverage [6].

12.4.3 Enhanced Smart Industry

The Industry 4.0 revolution, which began with 5G and comprises the digital transformation of manufacturing via cyber physical systems and IoT services, will be realized fully with 6G. By removing the boundaries that exist between the actual factory and the cyber computational environment, Internet-based diagnostics, maintenance, operation, and direct machine communications will become more cost-effective, adaptable, and efficient. Automation has unique communication requirements in terms of consistency and isochronous connectivity, which 6G is well suited to meet with its disruptive combination of technologies. For example, industrial control necessitates real-time operations with guaranteed s delay jitter and Gbps peak data rates for AR/VR industrial applications (e.g., for training, inspection) [18, 19].

12.4.4 Enhanced Vehicular Internet of Things (VIoT) and Autonomous Driving

6G technology advancements have greatly transformed VIoT networks, revolutionizing intelligent transportation systems (ITSs). In 6G-based VIoT networks, the research employs mMTCs to enable vehicle-to-everything (V2X) connectivity for the communication of short vehicular information payloads by a large number of vehicles without human interaction. To do so, the peculiarities of V2X are exploited to achieve a balance of reliability, latency, and scalability via a vehicle discovery approach in which a discovery entity at the sink collects information about the vehicles' proximity. As a result, signature properties such as time slots and hash functions are tuned to optimize the discovery scheme, with the goal of minimizing the false-positive probability when scheduling radio resources for V2X data communications within the limited spectrum budget. Data rate prediction in future 6G-based VIoT is a difficult task owing to the complex interdependencies between factors like mobility, channels, and networking. Machine learning (ML) can be an effective approach to mimicking the possible behavior of network-aided throughput prediction in future 6G vehicular networks. To fully realize the potential of vehicular intelligence in VIoT, edge intelligence functions based on ML are integrated into road RUSs, which are capable of estimating traffic volume and weather forecasts based on the aggregation of local observations from vehicles.

Autonomous driving (AV) will play a crucial role in enhancing automobile quality, fuel efficiency, and the safety of roads in future intelligent vehicular networks. 6G technologies are expected to open up new opportunities for addressing the stringent service requirements of AV applications for reliability and high-speed connectivity. Because each vehicle is viewed as an entity with full control and recognition in the interconnected transport system, it is critical to explore the communication performances of the vehicle-to-vehicle (V2V) networks in order to fully implement AV in the 6G. In this context, collaborative driving is enabled through information sharing and vehicle coordination, and deep learning (DL) can be a suitable solution for performing fast prediction of V2V communication performance constraints for intelligent regulation of intervehicle distances. Edge intelligence is also critical for delivering smart functions at the network edge, such as RSUs for AV system control. AV controllers, for example, can be hosted at edge servers coupled with DL processors to process vehicular data for self-driving decision-making and high-definition (HD) mapping for navigation. In addition, federated learning (FL) can be utilized to provide coordinated learning and centralized vehicular communications between automobiles and edge servers while maintaining user privacy and reducing network overheads incurred by sharing raw data [20, 21].

12.4.5 Satellite Internet of Things (SIoT)

In the 6G technology for compressive IoT coverage, it is necessary to integrate satellite communications into existing wireless networks, offering rise to a new domain known as SIoT. Satellites are partitioned into three network tiers to provide worldwide services to terrestrial IoT consumers: LEO, MEO and GEO. In comparison to LEO and MEO, the LEO system has attracted more attention in 5G network generation research because of its lower orbit height and ideal characteristics for IoT connections, such as shorter communication delay and lower path loss. However, in the 6G era, numerous satellites can be launched at hundreds of orbits above the earth, allowing LEO systems authentically realize wide coverage and greater efficiency through frequency reuse. Furthermore, it is planned to establish intersatellite links to assure intersatellite communications based on THz bands, which can accommodate more satellites and achieve higher link performances due to their much wider bandwidth, in comparison to current spectrum resources in mmWave communication and optical communication counterparts in the 5G era [22, 23]. Figure 12.3 shows how 6G-based satellite communication for IoT helps in full coverage.

12.4.6 Haptic Communication

Currently, rehabilitation exercises do not have the capability of observing the patient's motion and ensuring that they are completed correctly. The use of a device with IoT capabilities enables the analysis of the

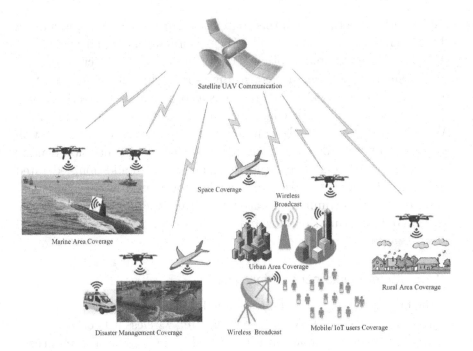

Figure 12.3 6G-based satellite communication.

reach of the motions, the force used, and the exercise completion. Using a range of procedures and assistive equipment, the major goal of rehabilitation and therapy is to improve the extremities and prevent additional disability. The inclusion of touch in the visual perception of an object has a significant impact on how that object is perceived. Haptics is an area of technology that interacts with computer applications through touch and control. In a virtual environment, a person can sense and move three-dimensional objects by utilizing a haptic controller. This technology can detect texture, shape, and weight among other things. Robotic-assisted therapy provides benefits such as a combo of visual and tactile cues, the opportunity to track the patient's development, and better engagement and repetition of certain actions. Haptic devices can provide tactile and kinetic sensations. Sensor and motor advances have enabled the development of tiny, low-cost, high-performance devices for human–machine interactions. For remote operations, we can employ haptic devices to generate kinetic sense and faux experience in a virtual world. Because they broaden the opportunity and target for remote palpation or surgery in the medical area, as well as remote inspection for infrastructure maintenance, these services are effective in addressing a number of societal challenges, such as rural depopulation [24, 25].

12.4.7 Robotics

Mobile robots can be included as entities in the new IoT paradigm, and they can be beneficial for expanding sensing and manipulating capabilities to remote regions where sensor networks are impractical. Much research has been conducted in the field of indoor localization using infrastructure that is not internal to the navigating robot, such as wireless networks, visible light communication (VLC), ultra-wideband (UWB) technology, V2V communications, and infrastructure-to-vehicle (I2V) communications. Outdoor localization is straightforward and can be performed with high precision utilizing Global Positioning Systems (GPSs) or independent cellular systems. Creating a navigation system that can move smoothly between outside and interior locales, on the other hand, is far more complex. It needs the deployment of a sophisticated perception system made up of sensors of several modalities that work well in specific situations, as well as the capability to discern which sensors provide the most important information in the current circumstance [26, 27].

12.4.8 Internet of Bio-Nano-Things

Bio-Nano Things index: Bio-Nano Things are described as uniquely identifiable basic structural and functional units that operate and interact in the biological environment within the purview of the Internet of Bio-NanoThings (IoBNT). Bio-Nano Things are designed to perform behaviors and functions typical of embedded computer devices in the IoT, such as actuation, processing, sensing, and interacting with one another. Because they are formed from biological cells and enabled by synthetic biology and nano technology. Intra-body sensing and actuation, in which Bio-Nano Things within the human body work together to collect health-related data, transfer it to an external healthcare provider over the Internet, and execute commands from the same source, such as pharmaceutical production and release. In Intra-body connection control Bio-Nano Things would heal or prevent connectivity breakdowns between our internal organs, such as those based on the endocrine and neurological systems, which are at the basis of many disorders. Environmental control and cleaning, in which Bio-Nano Things put in the environment, such as a natural ecosystem, detect hazardous and polluting agents and work together to change these agents through bioremediation, such as bacteria employed to clean up oil spills [28]

12.4.9 Augmented Reality and Virtual Reality

The ability of video-over-wireless, one of the highest data-hungry applications at the time, was unleashed by 4G networks. The growing popularity of multimedia services and streaming now justifies the deployment of the additional spectrum (mmWaves) to ensure greater capacity in 5G. This multi-Gbps opportunity, on the other hand, is bringing in new data-heavy applications rather than bi-dimensional audiovisual content: AR/VR development will be accelerated by 5G. The proliferation of AR/VR applications, similar to video-over-wireless absorbed 4G networks, will deplete the 5G spectrum and necessitate a system capacity greater than 1 Tbps, rather than the 20 Gbps target defined for 5G. Furthermore, because AR/VR cannot be compressed to fulfill the latency demands that enable real-time user engagement in the immersive environment (coding and decoding take time), the per-user data throughput must surpass Gbps, rather than the more relaxed 100 Mbps 5G target [7].

12.4.10 Holographic Telepresence (Teleportation)

The human urge to connect globally with increasing fidelity will face considerable communication issues in 6G networks. The data rate requirements for a 3D holographic display are as follows: 4.32 Tbps would be required for a raw hologram with no compression, colors, full parallax, and 30 frames per second. The required latency will be in the sub-millisecond range, necessitating thousands of synchronized view angles rather than the few required for VR/AR. Furthermore, to completely realize a realistic remote experience, all five human senses will be digitized and transported through future networks, raising the entire desired data rate [29].

12.4.11 Smart Education/Training

6G wireless networks will enhance smart education/training because breakthroughs like holographic communications, five-sense communications, high-quality VR/AR, mobile-edge computing, and AI will assist in the development of smart education/training systems. Using the aforementioned methodologies allows students to study buildings and designs in 3-D form and even be lectured by a famous teacher from a remote location, resulting in immersive and interactive online education. Traditional training methods enable students to recall more information, decrease costs, and avoid being in risky locations by demonstrating processes (live) through holography and engaging with objects or other teachers. 6G can also aid in the development of smart classes, in which data is gathered by sensors and delivered to clouds or edge clouds for analysis. The findings can then be used to improve educational quality and interact with students more effectively [6].

12.5 SECURITY AND PRIVACY IN 6G FOR IOT

Enterprises and communities will benefit from 6G technology in the 2030s by providing high-secure and reliable network connectivity. Mobile data is expected to expand to 5016 EBpm by 2030, as per a research done by the International Telecommunication Union (ITU). Also the number of worldwide mobile users is expected to reach 13.8 billion by 2025 and 17.1 billion by 2030, according to estimates. Big data is created by enormous IoT devices, which may include sensitive or private information. It is vital to use data from IoT devices in an adequate way to avoid the disclosure of user personal data without the users' permission. Current techniques for guaranteeing privacy, security, and information management rely significantly on the supervision of third-party entities, several of which end users can entirely believe. 6G wireless communication is thought to enhance the future of a highly linked data-dependent society through the full automation of all forms of wireless networks extended over space, ground, air, and undersea. The key advantage of 6G technology is its ability to handle enormous amounts of data effectively and give exceptional bandwidth per device. Blockchain is ideal for enormous IoT because it improves interoperability, dependability, security, privacy, and scalability with its inherent superior qualities, such as cryptographic security, decentralized consensus, trust-less system, distributed nature, and nonrepudiation.

There are various benefits of combining blockchain with IoT. All network activities may be recorded as blockchain transactions, resulting in distributed system management. All activities will be verified without the need for a third party, which is more cost effective. A chronological timestamp in each data block preserved in a blockchain will ensure data traceability. The security of IoT data is ensured by blockchains' inbuilt asymmetric encryption techniques, hash functions, and digital signatures. The privacy and security of enormous IoT will be maintained in this way. Interoperability across different IoT platforms can be achieved by having a blockchain-composite layer on the top of an overlay point-to-point network with uniform access to different systems by providing a unified authorization mechanism, authentication, and billing system. However, blockchain requires a lot of resources in terms of processing, transmission, and caching for block formation, verification, ledger storage, and node consensus. It is necessary to examine how to assure the benefits of implementing blockchain into IoT [30].

12.6   RESEARCH CHALLENGES AND FUTURE DIRECTION

Main research challenges for the implementation of 6G are discussed below.

12.6.1   Energy Efficiency in 6G-IoT

How to attain high energy efficiency in future 6G-based IoT systems will be a major challenge. To ensure network operations, data communications and service delivery services, such as vehicular data sharing in autonomous vehicles and packet delivery in space with UAV communications, necessitate large energy resources. Furthermore, each base station in wireless cellular networks typically uses 2.5 kW to 4 kW, implying that the establishment of vast 6G-IoT networks with hundreds of stations leads in massive energy consumption, which increases carbon emissions. To achieve green 6G-IoT networks, it is necessary to design energy-efficient communication protocols through optimization.

IoT systems may harvest power from atmospheric environments, such as thermal power, vibration, wind, and sun, to provide their transmissions and processing services, and energy harvesting approaches to harness renewable energy resources would be highly valuable to construct green 6G-IoT systems. A solar energy harvesting solution, for example, is being considered for harvesting Internet of Things (HIoT) networks, in which implantable sensors can harvest solar power from sunlight to serve sensory data transmissions via a Bluetooth low-energy module in a translucent silicon housing for HIoT-based healthcare monitoring. Experiments are carried out using a wireless-embedded sensor prototype with a solar panel and AP that runs for 10 minutes, demonstrating steady energy harvesting while extending the lifetime of the HIoT-based healthcare system. Future researchers should look at energy efficiency issues in 6G-IoT networks at higher elevations, such as satellite networks with base stations and flying devices, where energy harvesting is reliant on device trajectories and ambient surroundings [31, 32].

12.6.2 Hardware Constraints of IoT Devices

Another potential problem in communications and computations in 6G-based IoT networks is the hardware constraint of IoT devices. Embedded sensors and mobile devices, for example, should be able to simultaneously run smart AI functions to achieve edge intelligence and implement data transfer with URLLC in intelligent 6G-based healthcare. Certain IoT sensors are unable to match the required computational needs due to hardware, memory, and power limits. Communication overhead is also incurred when data is exchanged between IoT sensors and the network server, which increases as task sizes grow. As a result, new hardware design for future smart and powerful IoT devices is required. A software-based DL accelerator, for example, might be used to enable data training on mobile sensor devices. The basic concept is to leverage a collection of diverse processors (e.g., graphic processing units (GPUs)), with each computing unit utilizing various computational resources for handling different evaluation phases of DL models. Using different approaches, deep architectural decomposition and layer compression tried to optimize the hardware utilization for data training without losing accuracy. Further research is required to deliver hardware-based AI training solutions on nano IoT devices and integrated wearables in future intelligent 6G IoT networks, such as intelligently enhanced life assistance services.

As a result, risk mitigation must be taken into account in particular to provide high levels of security and privacy for 6G-IoT. By creating composition theorems with advanced mathematical answers, perturbation techniques such as differential privacy or dummy can be used to protect training datasets against data breaches in edge intelligence-based 6G-IoT networks. Differential privacy, for example, is achieved by introducing artificial noise (e.g., Gaussian noise) to the gradients of neural network (NN) layers in order to shield training data and buried personal information from external attacks while ensuring convergence. An innovative privacy-preserving data aggregation mechanism is also implemented into the fog computing architecture to meet differential privacy in the sense that the aggregation results are near to the actual findings, but attackers cannot extract the ground truth in the shared gradients. Under various privacy budget settings, differential privacy aids in achieving a six percent higher information security degree than standard

Laplace differential privacy approaches, according to implementation results. However, the use of differential privacy comes at the cost of a reduction in training quality. It is suggested that in the future, accuracy-aware differential privacy designs will be developed to achieve a balance between privacy protection and training quality [33, 34].

12.6.3 Security and Privacy Challenges in 6G-IoT

The 6G technology implementation will transform IoT networks and services with various network advantages such as huge wireless coverage, ultra-low latency and extreme reliability. Furthermore, vulnerabilities associated with wireless interface attacks, such as integrity at computing units/servers, threats to access network architectural integrity, and denial of service (DoS) to software and data centers, may make 6G incorporation into IoT networks troublesome. In massive IoT access networks, such as the heterogeneity of IoT devices and access mechanisms, as well as massive device connectivity, provide additional security problems, as handover from one technology to another increase the risk of attacks. As the number of connections between devices and computational nodes in the network periphery expands, eavesdropping attacks, spoofing attacks, and DoS attacks become viable in data transmission and management systems. Additionally, AI capabilities can be implemented at dispersed edge nodes to actualize smart 6G-IoT networks, where data training in a spectrum access system can be controlled by adding fake signals or adjusting parameters. As a result, a hostile attack can use a large portion of the spectrum illegally by denying it to other users. Attackers can also take advantage of AI functions' dispersed information training nature and dependency on edge computing to launch various assaults, such as malicious data injection, spoofing, and data poisoning, which have a detrimental influence on AI function training outputs in intelligent 6G-IoT systems. Edge intelligence may also be vulnerable to security threats since AI functions are distributed at the network edge, where assaults can deploy data breaches or alterations, while remote 6G core network controller management is constrained. Furthermore, data privacy leakage induced by other parties and adversaries during information communication between satellite base stations, UAVs, and terrestrial IoT users can obstruct the implementation of satellite-UAV-IoT communications over untrustworthy settings in air [35, 36].

12.6.4   Standard Specifications for 6G-IoT

With sophisticated wireless networking features, 6G technologies have the ability to change the picture of IoT markets and reinvent IoT ecosystems. The creation of 6G-IoT systems, on the other hand, necessitates the engagement of all interested parties, including customers, service providers, and network operators. The shortage of quality system may impede the implementation of 6G features and technologies in client IoT systems. Furthermore, the advent of vertical 6G-IoT use cases in future intelligent networks necessitates considerable design changes to existing mobile networks in order to handle a varied range of rigorous needs at the same time (e.g., e-healthcare, self-driving, etc.). Due to the dependency on other critical services such as computing and 6G server-IoT unit communication protocols in such a scenario, network standards play a vital role in deploying 6G-IoT ecosystems on a broad scale. MODBUS is a popular protocol that is used to connect computer servers, industrial electrical devices, and sensor devices in IoT contexts. MODBUS is built on a number of enabling protocols, including TCP/IP, UDP, and remote terminal unit (RTU). It is based on mesh networking topologies and can provide corporate communications as well as supervisory control over industrial radio bands The European Telecommunications Standards Institute's Industry Specification Group recently released the campaign called ETSI Multi-access Edge Computing, which aims to enhance seamlessly edge computing and communication frameworks for incorporating different edge-based IoT applications originating from external and service providers in future WiFi networks. This would make it easier to provide IoT services like video analytics, AR, data caching, and content distribution. In the nearish future, interested parties should focus their efforts on developing new standard specifications for new space-air-ground-underwater communications, such as new IoT satellite IoT communications, which will be critical for the future deployment of new commercial IoT applications such as space travel and deep-sea marine services [37, 38].

12.7   CONCLUSION

Due to the appealing features of 6G compared to previous generations of wireless networks, it has recently ignited a considerable interest in both academics and industry. We begin by introducing some of the most basic 6G technologies that will power future IOT networks. Further, we will discuss the possibility and scope of 6G for IOT use cases for several key domains that include healthcare, smart city, smart industry, smart transport that includes unmanned and autonomous driving and SIoT. In healthcare IoT, it will enable remote monitoring, evaluation, and treatment of patients. In industrial IoT, manufacturing and industrial processes are enhanced using smart sensors and actuators. AV enables users to perceive the surroundings using on-board sensors. SIoT gives remote access to all those objects and things which are virtually inaccessible. The security and privacy concerns in 6G will be viewed vis-à-vis the 5G security protocols. Finally, we highlight research challenges and the potential directions for future research in this promising area. 6G-IoT systems and services are still in their early stages of development. As a result, 6G is expected to change present IoT network systems and deliver a new quality of service and user experience in future developments.

BIBLIOGRAPHY

[1] Xiaohu You, Cheng-Xiang Wang, Jie Huang, Xiqi Gao, Zaichen Zhang, Mao Wang, Yongming Huang, Chuan Zhang, Yanxiang Jiang, Jiaheng Wang, et al. Towards 6g wireless communication networks: Vision, enabling technologies, and new paradigm shifts. *Science China Information Sciences*, 64(1):1–74, 2021.

[2] W Saad, M Bennis, and M Chen. A vision of 6G wireless systems: Applications, trends, technologies, and open research problems (2019). *arXiv preprint arXiv:1902.10265*, 1902.

[3] Federica Rinaldi, Helka-Liina Maattanen, Johan Torsner, Sara Pizzi, Sergey Andreev, Antonio Iera, Yevgeni Koucheryavy, and Giuseppe Araniti. Non-terrestrial networks in 5G & beyond: A survey. *IEEE Access*, 8:165178–165200, 2020.

[4] Takehiro Nakamura. 5G evolution and 6G. In *2020 IEEE Symposium on VLSI Technology*, pages 1–5. IEEE, 2020.

[5] Dinh C Nguyen, Ming Ding, Pubudu N Pathirana, Aruna Seneviratne, Jun Li, Dusit Niyato, Octavia Dobre, and H Vincent Poor. 6G internet of things: A comprehensive survey. *IEEE Internet of Things Journal*, 9(1):359–383, 2022.

[6] Fengxian Guo, F Richard Yu, Heli Zhang, Xi Li, Hong Ji, and Victor CM Leung. Enabling massive IoT toward 6G: A comprehensive survey. *IEEE Internet of Things Journal*, 8(15):11891–11915, 2021.

[7] Lin Zhang, Ying-Chang Liang, and Dusit Niyato. 6G visions: Mobile ultra-broadband, super internet-of-things, and artificial intelligence. *China Communications*, 16(8):1–14, 2019.

[8] Qiang Liu, Songlin Sun, Heng Wang, and Shaowei Zhang. 6G green IoT network: Joint design of intelligent reflective surface and ambient backscatter communication. *Wireless Communications and Mobile Computing*, 2021:1–10, 2021.

[9] Zhengquan Zhang, Yue Xiao, Zheng Ma, Ming Xiao, Zhiguo Ding, Xianfu Lei, George K Karagiannidis, and Pingzhi Fan. 6G wireless networks: Vision, requirements, architecture, and key technologies. *IEEE Vehicular Technology Magazine*, 14(3):28–41, 2019.

[10] Ioannis Tomkos, Dimitrios Klonidis, Evangelos Pikasis, and Sergios Theodoridis. Toward the 6G network era: Opportunities and challenges. *IT Professional*, 22(1):34–38, 2020.

[11] Ahmed Al-Ansi, Abdullah M Al-Ansi, Ammar Muthanna, Ibrahim A Elgendy, and Andrey Koucheryavy. Survey on intelligence edge computing in 6G: characteristics, challenges, potential use cases, and market drivers. *Future Internet*, 13(5):118, 2021.

[12] Zakria Qadir, Hafiz Suliman Munawar, Nasir Saeed, and Khoa Le. Towards 6G internet of things: Recent advances, use cases, and open challenges. *arXiv preprint arXiv:2111.06596*, 2021.

[13] L Chettri and R Bera. A comprehensive survey on internet of things (IoT) toward 5G wireless systems. *IEEE Internet of Things Journal*, 7(1):16–32, 2020.

[14] Christos Liaskos, Ageliki Tsioliaridou, Sotiris Ioannidis, Andreas Pitsillides, and Ian F Akyildiz. Realizing ambient backscatter communications with intelligent surfaces in 6G wireless systems. *IEEE Wireless Communications*, 29(1):178–185, 2022.

[15] Sarah Basharat, Syed Ali Hassan, Aamir Mahmood, Zhiguo Ding, and Mikael Gidlund. Reconfigurable intelligent surface-assisted backscatter communication: A new frontier for enabling 6G IoT networks. *arXiv preprint arXiv:2107.07813*, 2021.

[16] Basel Barakat, Ahmad Taha, Ryan Samson, Aiste Steponenaite, Shuja Ansari, Patrick M Langdon, Ian J Wassell, Qammer H Abbasi, Muhammad Ali Imran, and Simeon Keates. 6G opportunities arising from internet of things use cases: A review paper. *Future Internet*, 13(6):159, 2021.

[17] Qi Zhang, Jianhui Liu, and Guodong Zhao. Towards 5G enabled tactile robotic telesurgery. *arXiv preprint arXiv:1803.03586*, 2018.

[18] Jay Lee, Behrad Bagheri, and Hung-An Kao. A cyber-physical systems architecture for Industry 4.0-based manufacturing systems. *Manufacturing Letters*, 3:18–23, 2015.

[19] Martin Wollschlaeger, Thilo Sauter, and Juergen Jasperneite. The future of industrial communication: Automation networks in the era of the internet of things and Industry 4.0. *IEEE Industrial Electronics Magazine*, 11(1):17–27, 2017.

[20] Weijie Yuan, Shuangyang Li, Lin Xiang, and Derrick Wing Kwan Ng. Distributed estimation framework for beyond 5G intelligent vehicular networks. *IEEE Open Journal of Vehicular Technology*, 1:190–214, 2020.

[21] Chen Li, Weisi Guo, Schyler Chengyao Sun, Saba Al-Rubaye, and Antonios Tsourdos. Trustworthy deep learning in 6G-enabled mass autonomy: From concept to quality-of-trust key performance indicators. *IEEE Vehicular Technology Magazine*, 15(4):112–121, 2020.

[22] Jianhang Chu, Xiaoming Chen, Caijun Zhong, and Zhaoyang Zhang. Robust design for noma-based multibeam leo satellite internet of things. *IEEE Internet of Things Journal*, 8(3):1959–1970, 2020.

[23] Ziye Jia, Min Sheng, Jiandong Li, Dusit Niyato, and Zhu Han. Leo-satellite-assisted uav: Joint trajectory and data collection for internet of remote things in 6g aerial access networks. *IEEE Internet of Things Journal*, 8(12):9814–9826, 2020.

[24] Nneka L Ifejika-Jones and Anna M Barrett. Rehabilitation— Emerging technologies, innovative therapies, and future objectives. *Neurotherapeutics*, 8(3):452–462, 2011.

[25] Lennart Thurfjell, John McLaughlin, Johan Mattsson, and Piet Lammertse. Haptic interaction with virtual objects: The technology and some applications. *Industrial Robot: An International Journal*, 29(3)210–215, 2002.

[26] Miteshkumar Patel, Maani Ghaffari Jadidi, Jacob Biehl, and Andreas Girgensohn. System and method for automating beacon location map generation using sensor fusion and simultaneous localization and mapping, June 9 2020. US Patent 10,677,883.

[27] Hongwei Du, Chen Zhang, Qiang Ye, Wen Xu, Patricia Lilian Kibenge, and Kang Yao. A hybrid outdoor localization scheme with high-position accuracy and low-power consumption. *EURASIP Journal on Wireless Communications and Networking*, 2018(1):1–13, 2018.

[28] Ian F Akyildiz, Max Pierobon, Sasi Balasubramaniam, and Y Koucheryavy. The internet of bio-nano things. *IEEE Communications Magazine*, 53(3):32–40, 2015.

[29] Xuewu Xu, Yuechao Pan, Phyu Phyu Mar Yi Lwin, and Xinan Liang. 3d holographic display and its data transmission requirement. In *2011 International Conference on Information Photonics and Optical Communications*, pages 1–4. IEEE, 2011.

[30] Shubhani Aggarwal, Neeraj Kumar, and Sudeep Tanwar. Blockchain-envisioned uav communication using 6G networks: Open issues, use cases, and future directions. *IEEE Internet of Things Journal*, 8(7):5416–5441, 2020.

[31] Fadi M Al-Turjman, Muhammad Imran, and Sheikh Tahir Bakhsh. Energy efficiency perspectives of femtocells in internet of things: Recent advances and challenges. *IEEE Access*, 5:26808–26818, 2017.

[32] Nan Zhao, Shun Zhang, F Richard Yu, Yunfei Chen, Arumugam Nallanathan, and Victor CM Leung. Exploiting interference for energy harvesting: A survey, research issues, and challenges. *IEEE Access*, 5:10403–10421, 2017.

[33] Xi Yang, Michail Matthaiou, Jie Yang, Chao-Kai Wen, Feifei Gao, and Shi Jin. Hardware-constrained millimeter-wave systems for 5G: challenges, opportunities, and solutions. *IEEE Communications Magazine*, 57(1):44–50, 2019.

[34] Han Cai, Chuang Gan, Ligeng Zhu, and Song Han. Tinytl: Reduce memory, not parameters for efficient on-device learning. *Advances in Neural Information Processing Systems*, 33:11285–11297, 2020.

[35] Pawani Porambage, Gürkan Gür, Diana Pamela Moya Osorio, Madhusanka Liyanage, Andrei Gurtov, and Mika Ylianttila. The roadmap to 6G security and privacy. *IEEE Open Journal of the Communications Society*, 2:1094–1122, 2021.

[36] Sina Shaham, Ming Ding, Bo Liu, Shuping Dang, Zihuai Lin, and Jun Li. Privacy preservation in location-based services: A novel metric and attack model. *IEEE Transactions on Mobile Computing*, 20(10):3006–3019, 2020.

[37] Jeongho Yeo, Taehyoung Kim, Jinyoung Oh, Sungjin Park, Younsun Kim, and Juho Lee. Advanced data transmission framework for 5G wireless communications in the 3GPP new radio standard. *IEEE Communications Standards Magazine*, 3(3):38–43, 2019.

[38] Marsa Rayani, Roch H Glitho, and Halima Elbiaze. Etsi multi-access edge computing for dynamic adaptive streaming in information centric networks. In *Globecom 2020-2020 IEEE Global Communications Conference*, pages 1–6. IEEE, 2020.

Index

1G, 1, 181, 208, 319
2G, 1, 181
3D MIMO, 156, 163–165
3G, 1, 181
3GPP, 76, 139, 140, 182
4G, 1, 162, 178, 182, 330
5G, 1, 3–9, 49, 78, 81, 133, 134,
      135, 138, 151, 159, 161,
      162, 178, 179, 181, 183,
      205, 206, 207, 210, 212,
      228, 232, 234, 236, 284,
      317–320, 324, 325, 327,
      330, 336
6G, 2, 3–12, 40–42, 46, 48, 49, 75,
      119–121, 123, 127,
      133–136, 139, 141, 144,
      157–159, 168, 178, 184,
      205–214, 218–220, 223,
      224, 227–230, 233, 234,
      236–240, 260, 284, 294,
      300, 305, 308, 316–327,
      331–336

A

Access control attack, 206
Adaptive modulation, 1
Aerial access network, 120
Aerial computing, 119, 120, 126,
      127, 130, 134–138
Alternating optimization, 20, 264
Angle of arrival, 31, 62, 75, 77
Angle of departure, 62, 75
Anonymity cracking, 303

Artificial noise, 216, 218, 221, 222,
      226, 233, 333
Asset tracking, 294, 299, 300, 308
Augmented reality, 169, 319
Average secrecy capacity, 48

B

B5G, 134, 141, 305
Base stations, 8, 20, 47, 159, 161,
      162, 267, 322, 333, 334
Block coordinate descent, 20, 45
Blockchain, 7, 12, 122, 143, 205,
      224, 229, 230, 238,
      285–303

C

Cell-free massive MIMO, 209, 217,
      240
Central limit theorem, 66
Channel state information, 18,
      104, 218, 264
Cloud radio access networks, 184,
      205
Code division multiple access, 181
Cognitive radio, 6, 46–48
Collective AI, 4–6
Complementary metal-oxide-
      semiconductor, 9,
      157–158
Complex circle manifold, 20
Compressed sensing, 76, 81, 91–93
Cooperative beamforming, 226,
      233

Cramer-Rao bound, 84
CSI, 18, 31, 32, 104, 218–221, 225, 264

D
D2D, 46, 154, 161, 162
Delta OMA, 184
Discrete Fourier transform, 67
Distributed denial of service, 205, 334
Doppler frequency, 62, 63
Doppler spectrum, 67
Dynamic subarray m-MIMO, 261, 269, 276, 278

E
Edge computing, 2, 8, 11, 119, 120, 126, 130, 134–138
Electromagnetic field, 63
Electromagnetic spectrum, 152, 179
Energy consumption, 170, 173, 174, 209, 222, 307
Energy efficiency, 19, 23, 26, 33, 40, 41, 42, 46, 119, 121, 141, 159, 165, 184, 267, 269, 308, 332
Energy harvesting, 8, 41, 42, 80, 332
Enhanced mobile broadband, 76
Exhaustive search, 25, 108
Extended reality, 75, 135, 209, 319
Extremely low-power communications, 210

F
Far-field, 60, 74, 81–86, 89–93, 110
Far-field transmissions, 60

Federal Communications Commission, 9, 324
Field programmable gate array, 81
Fixed subarray m-MIMO, 261, 266, 267, 269
Flying ad-hoc networks, 166
Fog-RAN, 7, 8
Fourier transform, 64, 67
Free space optical communication, 48, 49
Frequency division multiple access, 181–182
Fresnel approximation, 89
Fully connected m-MIMO, 261, 262, 278
Fully digital m-MIMO, 276

G
Geometric channel model, 22, 86
Global positioning systems, 329
GPS, 226, 234, 329

H
Hardware impairment, 11, 16, 18, 30
High-altitude computing, 122
High-altitude platforms, 220
Holographic communication, 135, 223
Holographic radio, 216, 223
Hybrid automatic repeat request, 48
Hybrid beamforming, 22, 23, 156, 216, 271
Hybrid m-MIMO, 261, 262, 271, 272, 274, 275, 278

I
Imperfect CSI, 29, 31, 33

Intelligent reflective surface, 10, 17, 60, 172
Internet of Bio-Nano-Things, 329
Internet of Things, 9, 26, 27, 41, 75, 119, 161, 166, 205, 284
Intrusion detection system, 227
IoBNT, 329
IRS, 10, 11, 17–33, 40–49, 60–69, 172, 216, 219, 222

K

Key performance indicators, 119, 325

L

Lagrange dual method, 27
Large intelligent surfaces, 5, 223
Licensed spectrum, 44
Li-Fi, 6, 8
Line-of-sight, 11, 17, 43, 48, 74, 120, 160, 187
LIS, 5, 223
Localization, 11, 32–33
Localization error bounds, 81, 84
Localization parameters, 76, 91, 93, 95
Low altitude computing, 122, 124
Low altitude platform, 120
Low earth orbit, 120

M

Machine type communications, 178
Machine-to-machine, 44, 134, 237
Majorization-minimization, 20
Malicious code injection, 205, 209, 213
Massive machine type communications, 76
Massive MIMO, 2, 209, 260, 263

Max-min fairness, 185
MEC, 17, 27, 123
Metasurfaces, 4, 5, 7, 40
Millimeter wave, 41, 60, 75, 150, 161, 163, 165, 212, 260, 318, 324
MIMO, 1, 2, 49, 59, 60, 75, 82–87, 156, 183, 208, 223, 260
MIMO-NOMA, 43
MIMO-OFDM, 270, 271
Mitola radio, 5, 6, 8
Mixed reality, 319
m-MIMO, 261–263
mmWave MIMO, 270
Mobile edge computing, 16, 17
Multiple access techniques, 180
Multiple input single output, 19, 84, 184, 185
Multi-user m-MIMO, 265

N

Nano antenna, 9, 153
Narrowband channel modeling, 61, 64
Near-field, 33, 60, 74, 83–86, 91, 93, 96, 102, 104, 106, 107, 110
Network function virtualization, 4, 205, 301
Next generation multiple access, 41
NFV, 4, 7, 205, 224–225, 228–229, 240, 301
NOMA, 11, 12, 17, 21, 23–26, 33, 41–44, 178, 180, 182, 184, 185, 195–197, 212, 215, 218–220, 240
Non-line-of-sight, 77, 151, 198
Non-payload communication link, 167

O

OFDMA, 182
Optical beamforming, 208
Orbital angular momentum, 6, 8, 205
Orthogonal multiple access, 23
Outage probability, 31, 40, 185, 190–196, 220

P

Packet error rate, 48
Packet switched networks, 1
Partial NOMA, 184
Particle swarm optimization, 101–102
Passive tunable elements, 81
Pervasive AI, 4
Physical layer security, 4, 5, 7, 16, 143, 212, 215
Power spectral density, 67
Primary user, 46
Programmable wireless environment, 7, 59

Q

Quality of experience, 6
Quality-of-service, 138

R

Radio access networks, 4, 122, 139
Radio frequency, 130, 151, 197, 205
Rate splitting multiple access, 12, 178, 183
Reconfigurable intelligent surface, 10, 74, 77, 78, 185, 219, 284
RIS, 10, 74, 77, 78, 185, 219, 284

S

Satellite computing, 122
Satellite internet of things, 326
Secondary user, 46
Secrecy outage probability, 28, 49
Self-organizing network, 4
Semidefinite relaxation, 20, 45
SIC, 25, 42, 218, 267
Signal space diversity, 48
Signal to interference and noise ratio, 188
Signal to noise ratio, 11, 31, 60, 74, 84
Single input single output, 19, 20, 83–85
Single-user m-MIMO, 265
SINR, 188
Smart city, 307, 325, 336
Smart contract, 230, 235, 294, 298, 299, 300
Smart factories, 120, 121, 126, 128, 130, 131
Smart grids, 119–121, 126, 128, 131
SNR, 11, 60, 61, 65, 66, 74, 84, 102, 106–110, 277
SON, 4
Space division multiple access, 183
Spatial division multiple access, 42, 183
Spectral efficiency, 11, 23, 31, 33, 41, 46, 48, 184, 263–267, 269–271, 325
Spectrum, 1, 6–11, 151, 178, 179, 181, 184, 196, 223, 301
Spectrum sensing, 46, 48
Spectrum sharing, 46, 224, 301, 318, 321, 331, 334
Spherical wave model, 85
Spread spectrum, 1

Successive convex approximation, 20, 184

Successive interference cancellation, 25, 42, 218, 267

SWIPT, 41

Sybil attack, 206, 208, 213, 225, 230

T

Terrestrial computing, 121–126

THz spectrum, 157, 158, 197, 198

Time division multiple access, 181

Transmit beamforming, 28

U

Ultra massive MIMO, 9, 10, 19, 20, 212

Ultrawide band, 86

Underwater optical wireless communication, 48

Underwater sensor networks, 233

Unlicensed spectrum, 44

Unmanned aerial vehicles, 11, 32, 120, 209, 322

URLLC, 10, 76, 216, 223, 224, 229, 322, 333

User localization, 85, 101

V

V2X, 7, 205, 326

Vehicular internet of things, 326

Virtual reality, 10, 25, 166, 220, 319, 330

Visible light communication, 206, 255, 329

VLC, 8, 206–208, 219, 233, 255, 329

W

Weighted minimum mean square error, 185

Weighted sum rate, 20, 46

Wide sense stationary, 66

Wireless power transfer, 6, 8, 26

Wireless propagation channel, 11, 27, 61, 69

Z

Zero forcing, 184, 221

Zero touch service, 234, 235, 237, 240

Index ▪ 354

Printed in the United States
by Baker & Taylor Publisher Services